Lecture Notes in Applied and Computational Mechanics

Volume 102

This series aims to report new developments in applied and computational mechanics - quickly, informally and at a high level. This includes the fields of fluid, solid and structural mechanics, dynamics and control, and related disciplines. The applied methods can be of analytical, numerical and computational nature. The series scope includes monographs, professional books, selected contributions from specialized conferences or workshops, edited volumes, as well as outstanding advanced textbooks.

Indexed by EI-Compendex, SCOPUS, Zentralblatt Math, Ulrich's, Current Mathematical Publications, Mathematical Reviews and MetaPress.

Peter Eberhard
Editor

Calm, Smooth and Smart

Novel Approaches for Influencing Vibrations by Means of Deliberately Introduced Dissipation

Springer

Editor
Peter Eberhard
Institute of Engineering and Computational Mechanics
University of Stuttgart
Stuttgart, Germany

ISSN 1613-7736 ISSN 1860-0816 (electronic)
Lecture Notes in Applied and Computational Mechanics
ISBN 978-3-031-36142-5 ISBN 978-3-031-36143-2 (eBook)
https://doi.org/10.1007/978-3-031-36143-2

This Springer imprint is published by the registered company Springer Nature Switzerland AG
The registered company address is: Gewerbestrasse 11, 6330 Cham, Switzerland

Preface

Priority Programme *Calm, Smooth and Smart—Novel Approaches for Influencing Vibrations by Means of Deliberately Introduced Dissipation* (SPP 1897)

The Senate of the Deutsche Forschungsgemeinschaft (DFG, German Research Foundation) established in 2015 a Priority Programme (in German: 'Schwerpunktprogramm') entitled *Calm, Smooth and Smart—Novel Approaches for Influencing Vibrations by Means of Deliberately Introduced Dissipation* (SPP 1897). The programme was scheduled to run for six years split into two calls for proposals for two three-year funding periods.

Over the last decades, the concept of lightweight design has become more and more important in engineering. Herein, it is the aim to reduce the mass of any kind of technical structure to a minimum in order to save resources, costs, and energy during both manufacturing and operation. Following the rules of this design principle also often means making technical components and the whole system more sensitive to unwanted vibrations. These vibrations can cause severe environmental and health issues, and hence should be minimized. This yet inevitable dependence shall now be eliminated by developing novel approaches for influencing vibrations that result from the research of the established Priority Programme and lead to a "calm, smooth and smart" behaviour of technical units.

"Calm" represents the demand to avoid or at least to severely reduce unwanted noise generated by technical installations. "Smooth" ensures a comfortable and jerk-free operation. Finally, "smart" means that the introduced damping devices not only help to achieve the desired vibrational behaviour of the overall technical systems, but also take over additional functional tasks.

The key to achieving a "calm, smooth and smart" characteristic of technical equipment is the time- and position-dependent application of dissipation in order to transform vibrational energy into heat or other forms of energy. But since there is, as yet, only limited knowledge on most of its physical phenomena and mechanisms, dissipation can hardly be introduced in a deliberate fashion in daily engineering.

For this reason, it was the main objective of this Priority Programme to form the basis for a deliberate utilization of dissipative processes which is not exclusively based upon trial and error and allows for the future development of vehicles, machines and facilities that are energy efficient and light, but at the same time, show a "calm, smooth and smart" behaviour as required.

It was the aim to pool the expertise of mechanics, mathematics, control engineering, tribology, fluid mechanics and material science in Germany, and to create new and strengthen already available networks in order to achieve the set goals. In the framework of this cooperation, the existing experiences were exchanged between the different working groups to generate synergies, to save time and costs and to raise the working efficiency. Moreover, it was intended to bring this new-born research spirit to international excellence in the field of innovative damping techniques.

In the first funding period, the Priority Programme tried to drive research towards the following directions concerning dissipation mechanisms and damping strategies:

- Systematic investigation of dissipative mechanisms and subsequent development of mathematical models used to describe them.
- Definition, analysis and validation of novel damping techniques as well as their effect on vibrations.
- Development of numerical methods that allow modelling of dissipation and damping devices in an engineering-compliant environment.
- Integration of submodels describing new kinds of damping-based vibration absorbers and of proper model order reduction techniques into the overall models of vehicles, machines and facilities.
- Experimental investigation of the influence of damping mechanisms and devices on the characteristics of overall system dynamics.

In the second funding period, a clear focus of the projects was on specific novel damping devices and/or novel applications to technical problems on the basis of the insights gained during the first phase.

Project proposals addressed most of these points, i.e. allow the description of the influence of new damping mechanisms/devices or refined modelling/description of damping on the overall dynamics of a mechanical or mechatronic system having influence towards the system properties "calm, smooth and smart". Furthermore, it was expected that the proposals provide clear visionary aims.

It is the goal of this book to summarize the findings of the Priority Programme and to make them available to the scientific community. Of course, during the two phases, many separate or joint scientific publications evolved from the different projects and many details were published, but here, an overview of the various projects and their influence on "calm, smooth and smart" was intended. Thus, we are grateful that all projects from the first and second phases provided their contributions to this book irrespective of being funded in one or both phases.

Besides the research in the separate projects, there were many cooperations and joint research works. This was also supported by many common events like annual status seminars, doctoral student meetings, working group meetings, sessions at meetings and many others.

Besides this book and the separate publications, for everybody interested and not just for the experts, also a video about the common topic was produced and made available at `https://www.itm.uni-stuttgart.de/spp_1897/videos` and every project produced additionally a fascinating video about its specific insights and approaches.

Of course, the corona crisis hit the second phase of the Priority Programme hard. For more than two years, we had to switch to video meetings and were not able to meet in person. Fortunately, most cooperations were already established, the project coworkers already knew each other and the scientific work and output hardly suffered.

We want to thank the programme committee who accompanied the Priority Programme from the preparation phase (P) through both phases (1, 2). The members Prof. von Wagner (Berlin, P, 1, 2), Prof. Fidlin (Karlsruhe, P, 1), Prof. Stykel (Augsburg, P) and Prof. Willner (Erlangen, P, 1) spent a lot of effort and time to make this research possible.

Further thanks go to Nadine Walker and Elizaveta Shishova, who not only scientifically contributed to the Priority Programme, but also organized, as coordinator assistant, all the meetings and events, and also took care of the web pages, collected and organized the videos and made this book possible.

Finally, we want to thank the German Research Foundation (DFG), who not only funded our research for over six years, but always helped in many aspects such as accompanying the discussions which led to the preparation of the initial proposal, organizing the review processes and providing us extreme flexibility, especially during the corona crisis. Colleagues from other countries always envy us for this wonderful research funding organization and they are completely right.

Personally, I want to give my special thanks to Dr. Simon Jörres, from the DFG, who was always available during any smaller and larger challenges. He skilfully guided the Priority Programme with a light hand and highly contributed to its success. It was my true pleasure to cooperate with him and his team.

Stuttgart, Germany
September 2022

Peter Eberhard
Coordinator of the SPP 1897

Contents

Particle Dampers—Vibration Reduction Through Distributed Dissipation Over Complex Particle Shapes

Andreas Schönle, Chandramouli Gnanasambandham, and Peter Eberhard

1 Introduction

The growing emphasis on lightweight construction has not only reduced the weight of technical systems drastically but also made them more vulnerable to unwanted vibrations. This fact combined with the growing complexity of these systems has lead to a rethinking of the purpose of damping devices and has paved way for the development of new methods to dissipate the unwanted vibrational energy. Their relatively simple design and their flexible ability to dissipate energy in a wide frequency range [1] have made solid particle filled dampers a popular alternative to conventional damping devices. Moreover, unlike viscous dampers, PDs do not rely on a fixed anchor point as an impulse source, which makes it even easier to mount them on technical systems. One of the earliest applications of PDs in the context of machine tools was presented in [2]. Some of the more recent applications include damping the structural vibrations of an oscillatory saw [3], noise reduction in transmission systems [4], reducing the horizontal vibrations in wind turbine towers [5], and to reduce vibrations on circuit boards of a spacecraft [6]. Furthermore, PDs have also been used to control vibrations in combustion discharge nozzles in industrial gas turbines [7].

The process of energy dissipation in PDs is very complex, as the damping performance depends on various parameters such as the strength of the forcing function (i.e. amplitude and frequency), size and geometry of the particles, inter-particle friction, material properties of the particle themself amongst others. PDs can be more easily put to practical use in technical applications if there is a deep understanding of the

A. Schönle · C. Gnanasambandham · P. Eberhard (✉)
Institute of Engineering and Computational Mechanics, University of Stuttgart, Pfaffenwaldring 9, 70569 Stuttgart, Germany
e-mail: peter.eberhard@itm.uni-stuttgart.de

A. Schönle
e-mail: andreas.schoenle@itm.uni-stuttgart.de

P. Eberhard (ed.), *Calm, Smooth and Smart*, Lecture Notes in Applied and Computational Mechanics 102, https://doi.org/10.1007/978-3-031-36143-2_1

underlying dissipation mechanisms. A systematic theoretical/experimental analysis of PDs in the context of free response behavior of a cantilever beam with a PD was presented in [8]. Here, an elementary analytical model to predict the macroscopic dissipative properties of PDs is reported, which was able to predict the experiments with reasonable accuracy. A parametric model of the nonlinear damping of PDs as an equivalent viscous damper was proposed in [9]. This research could be used in order to make predictions during the early design stage. Even though these models are reasonably good in predicting the macroscopic behavior of PDs, their predictions become fallible as soon as the particle-level parameters change. This is the case when PD contents has two distinctively different materials like solid and liquid, or when the particles have complicated shapes. This motivates further research to improve the understanding and applicability of PDs. Such research is to be conducted in this project using numerical simulations and laboratory experiments. The simulations are carried out using the particle simulation programm Pasimodo [10].

1.1 Dependencies of the PD Behaviour

The dependency of PD behaviour on particle movement can be illustrated by an experiment of rather simple design. In Fig. 2, one possible experimental setup is shown. Compressing the particles through the movable container walls ensures their immobility (left). Contrary to this, particles are free to move within the container volume in the second case (right). In Fig. 1 the experimental setup is illustrated schematically. Figure 3 shows the resulting measurement curves. Both of these measurements were conducted with an initial displacement of $A_0 = 40\,\text{mm}$. It can be seen in Fig. 3 that significant amounts of kinetic energy can be dissipated when

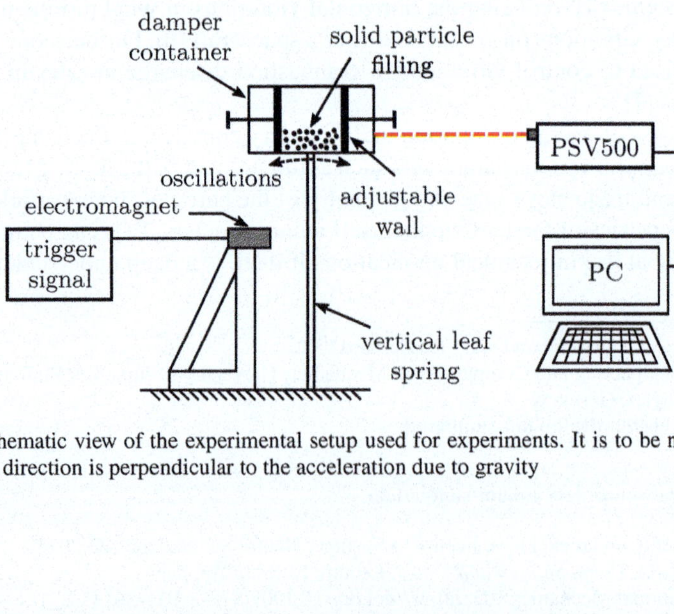

Fig. 1 Schematic view of the experimental setup used for experiments. It is to be noted that the oscillation direction is perpendicular to the acceleration due to gravity

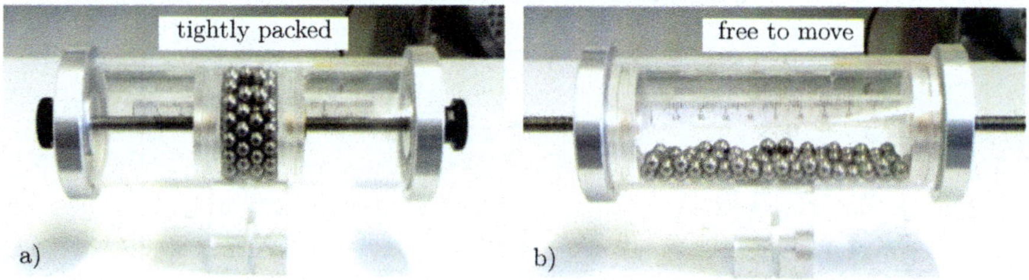

Fig. 2 The PD configurations with **a** the particles tightly packed and **b** whith the particles free to move and a container length of 100 mm

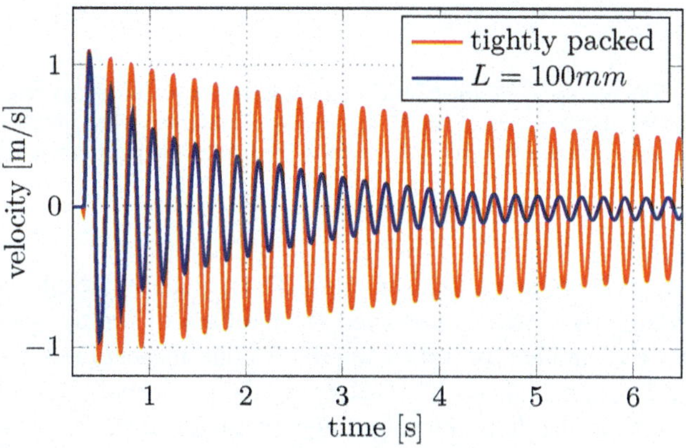

Fig. 3 Container velocities measured using an LDV for two configurations. First (red) the particles are tightly packed. Second (blue) particles are allowed to move freely. A considerable amount of kinetic energy is dissipated when particles are allowed to move freely due to inter particle collisions

the particles are allowed to move relative to each other. To quantify the dissipation present in the system, an effective decay rate parameter Λ is introduced. The decay rate during the ith cycle Λ_i is defined as the natural logarithm of two successive peak amplitudes. It is given by

$$\Lambda_i = \ln \frac{y(t_i)}{y(t_{i+1})}, \tag{1}$$

where y_i and y_{i+1} are the peak displacement of the damper container during the ith and i+1th cycle, respectively. In Fig. 4, the logarithmic decay rate for each cycle Λ_i is plotted over time for both cases. Since the particles were restricted to move relative to each other in the tightly packed configuration, the dissipation in this case, seen in Fig. 4 (red line), is very small and is essentially due to the intrinsic material damping present in the leaf spring.

However, it can also be observed from Figs. 3 and 4 that a residual amplitude exists in the case of the freely moving particles. Here the amplitude of the container

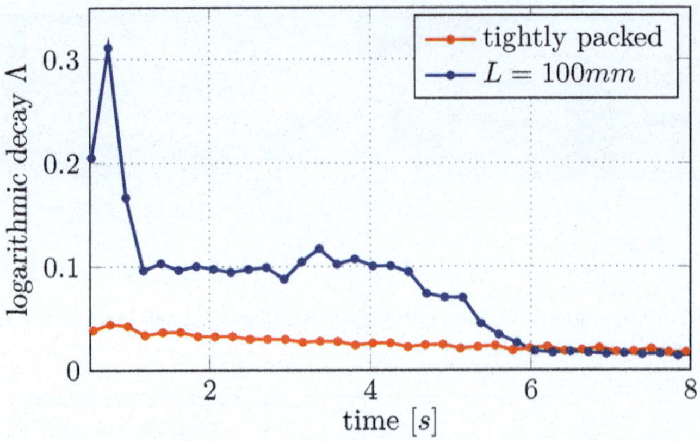

Fig. 4 The logarithmic decay Λ for both configurations. A constant decay rate, mostly due to the intrinsic material damping present in the leaf spring, is observed when the particles are tightly packed. Two distinctly different decay rates are observed when the particles are allowed to move freely

is not sufficient to keep the particles moving. Thus, hindering dissipation extending the material damping which is observed for the case of tightly packed particles. These insights give an idea for improving the overall dissipation behaviour of PDs, for a more detailed analysis see [11]. Keeping the particles in motion and finding measures to improve the dissipation for low amplitudes of excitation are, therefore, overall research objectives in this project.

Tools to grant detailed insights on how the taken measures improve the damping behaviour of PDs accelerate this process. This leads to a combined approach of numerical simulation and laboratory experiments. While the simulations offer access to the mechanisms within the PD, experiments ensure the agreement of the simulations with the actual system. A snapshot from a PD simulation with spherical particles can be seen in Fig. 5. The good agreement in Figs. 6 and 7 illustrates the feasibility of this approach for spherical particles using DEM.

2 Influence of Fillings

One possible way to increase the damping, even under low driving accelerations, is, to combine a liquid with solid fillings. Another influence factor is the shape of the particles as it alters the contact situation among particles. These influences are therefore of great interest for the overall performance of PDs.

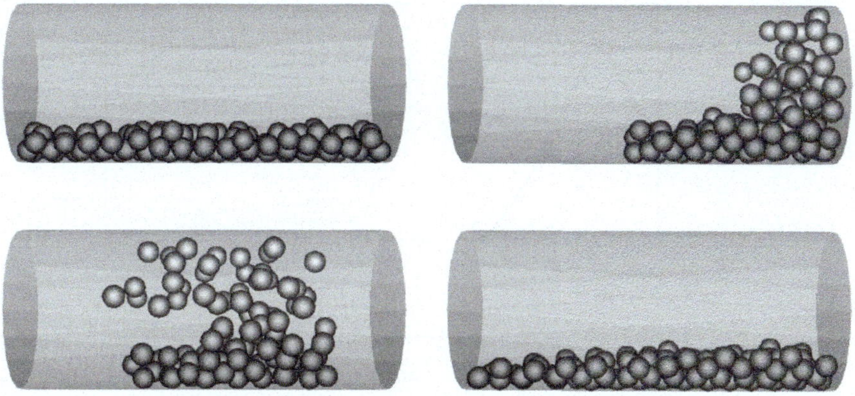

Fig. 5 DEM simulation snapshots showing the motion of the solid particles

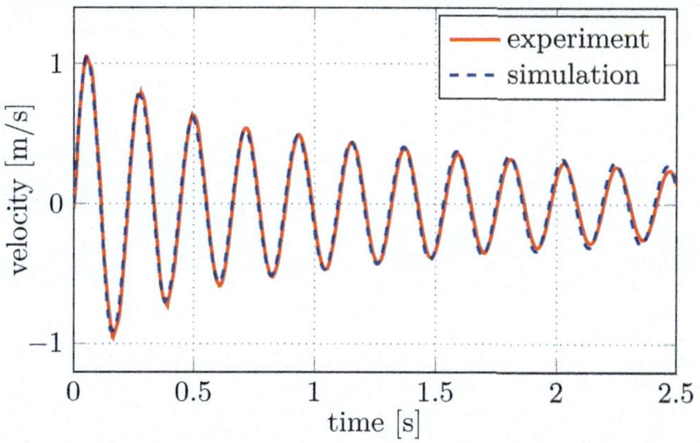

Fig. 6 Comparing DEM simulation results and experimental data for a damper filled with 100 aluminum spherical particles. A good agreement between experiment and simulation is observed

2.1 Influence of Liquid

Before investigating the effects of an added liquid, it is necessary to investigate the case where the damper is filled only with liquid contents. To simulate liquids in the PD the SPH method is used, as it is shown in [12]. The liquid motion resulting from these simulations is illustrated in Fig. 8. Figures 9 and 10 also show good agreement for this approach.

Combining solid particles with liquid in PDs has great potential in improving the damping behaviour. To study the effects of this approach the previously used simulation methods are combined. The damping behavior of a damper filled with a combination of solid particles and a liquid, as shown in Fig. 11, is investigated. Again this is done in both, numerical simulations and experiments. While the simulation

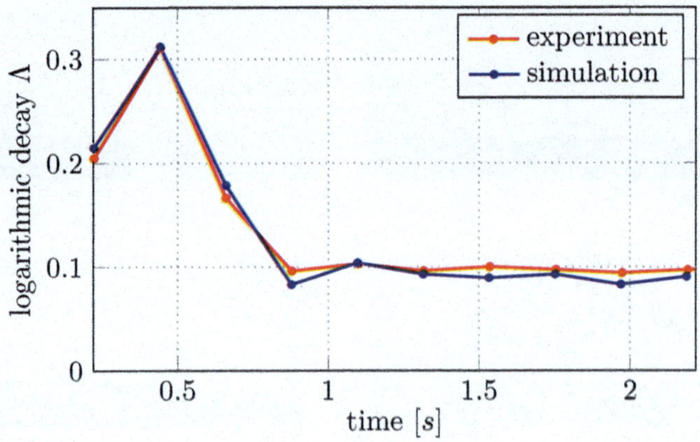

Fig. 7 Logarithmic decay rates predicted by DEM simulations and measured using experimental data for a damper filled with solid particles. Initially, due to the large relative motion, higher decay rates are observed

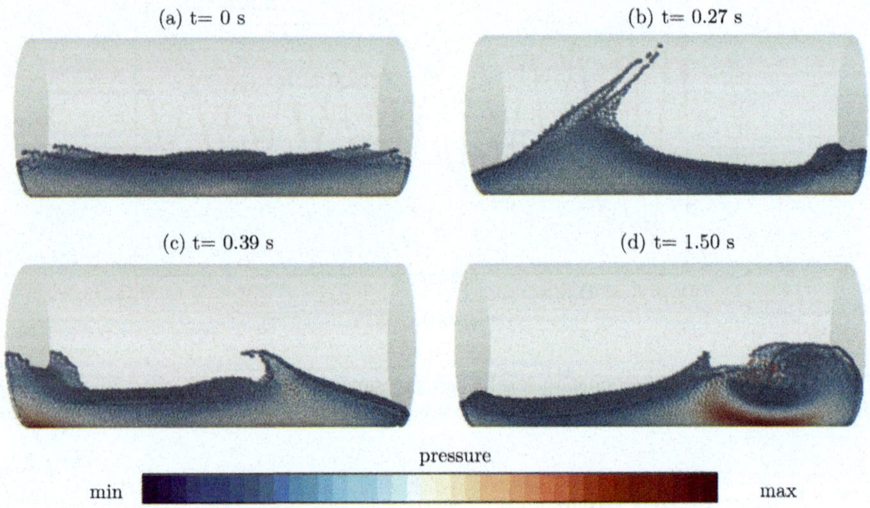

Fig. 8 SPH simulation snapshots showing the motion of the liquid modeled

grands important insights to understand the mechanisms leading to the improved damping behaviour, the experimental data is used for confirmation of the results. In the simulations the contact model is extended to cover the fluid-solid interactions. For this purpose, the damper container is filled with 100 aluminum particles and 30 ml of distilled water. In order to ensure a fair comparison between all the damper configurations, compensation masses were added during the laboratory experiments, so that all the configurations have the same static mass, measured using a weighing scale. The simulation results, showing the velocity amplitude decay for all three damper

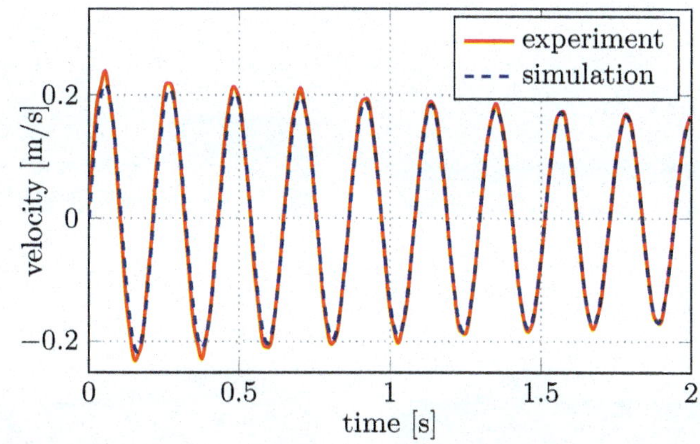

Fig. 9 Comparing SPH simulation results and experimental data for a damper filled with 30 ml distilled water. A good agreement between experiment and simulation is observed

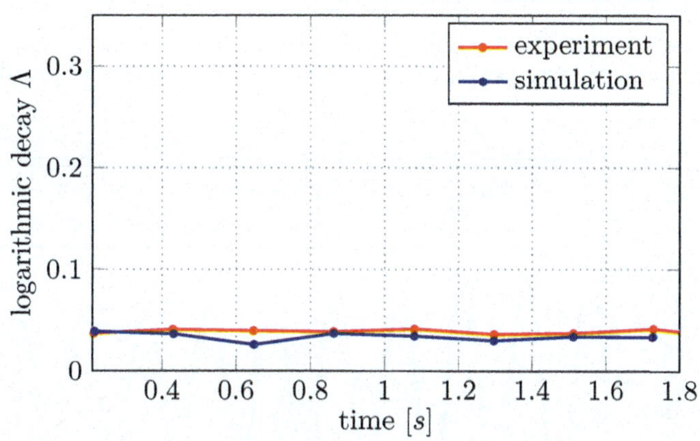

Fig. 10 Liquid filled dampers exhibit decay rates which are much smoother than for only solid particle filled dampers. Liquid filled dampers continue to dissipate energy even under low vibration amplitudes

configurations is seen in Fig. 12. As seen in an earlier investigation, the only-solid particle-filled damper seems to be not so effective under low acceleration conditions. Under low forcing condition, the solid particles seem to arrange themselves in an orderly packing. This makes it increasingly difficult for the particles to move relative to each other. As a consequence, the decay rate of solid-filled dampers in Fig. 13 (green) is observed to be lower than that of solid-liquid-filled dampers (blue). In fact, the worst performing of all three damper configurations is the one with only liquid filling. Even though there is a large relative motion observed between liquid layers, the resistance forces acting against the motion of the damper container are small, due to the lower momentum carried by the liquid.

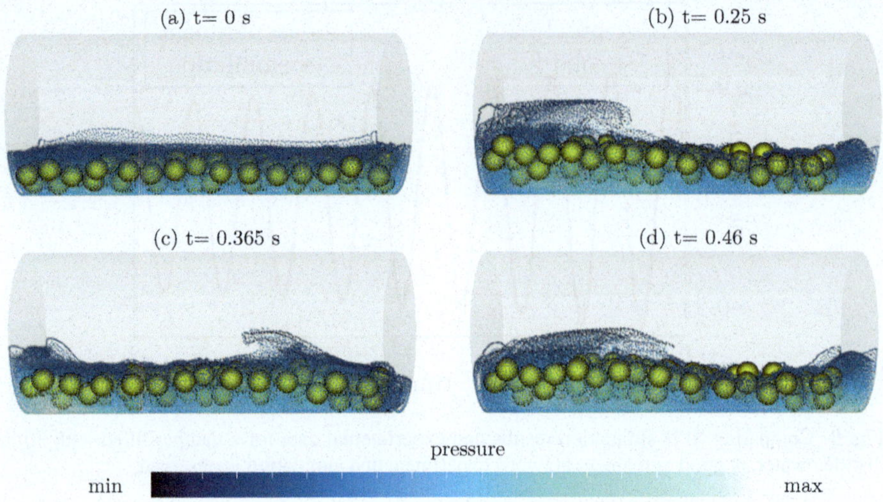

Fig. 11 Fluid-solid motion, predicted by coupled DEM-SPH simulations, at various time steps. The liquid is visualized using smaller coloured balls, whereas the solids are visualized as large yellow balls

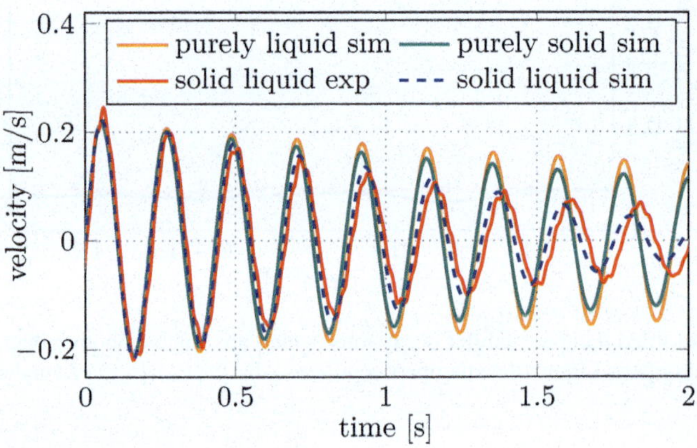

Fig. 12 The velocity of the damper container is compared for all the damper configurations. The blue dotted curve represents the container velocity predicted by coupled SPH-DEM simulations of a damper filled with solids and liquid particle

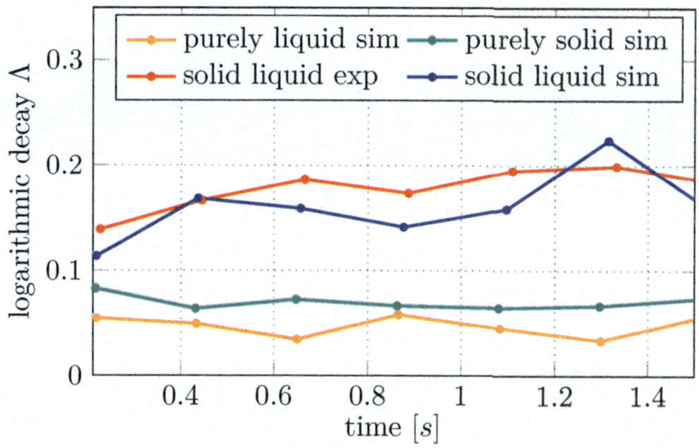

Fig. 13 Decay rates of all damper configurations are compared. Decay rates of solid filled dampers and liquid filled dampers are relatively similar. A substantial improvement in the damping performance is observed in the solid and liquid filled dampers

On the other hand, it can be seen that the damper configuration with a combination of both, solid-liquid contents indeed performs better than the other two configurations. Since the solid particles are surrounded by a liquid, it is much more difficult for the solids to remain in an ordered structure. This disorder makes them more susceptible to move relative to each other, which ultimately leads to stronger collisions and intern leads to more energy dissipation. In general, good agreement for the combined solid liquid simulation can be observed in Fig. 12. Therefore, the coupled SPH-DEM simulation is a useful tool to investigate particle dampers and the behaviour of their fillings. From Figs. 12 and 13 a significant increase in the decay of the vibration amplitude is seen.

This motivates further simulations and experiments to gain additional insights in the underlying effects resulting in such improvement. Possible influences considered are the solid-liquid fill ratio and the shape of the solid particles.

2.2 *Influence of Particle Shape*

The influence of the particle shapes is mainly attributed to the motion of the liquid through the particle filling. An experimental comparison of spheres and tetrapods as solid particles in a solid-liquid filled PD can be seen in Fig. 14. The tetrapods provide an evident advantage over simple spheres in the damping behaviour, which shows potential for further investigation. To study the effects leading to the observed advantages, detailed insights into the PD filling are advantageous. Therefore, the established SPH-DEM simulations are adapted for non-convex polyhedra, as it is described in [13]. In Fig. 15 (left), the velocity of the partially liquid-filled PD with

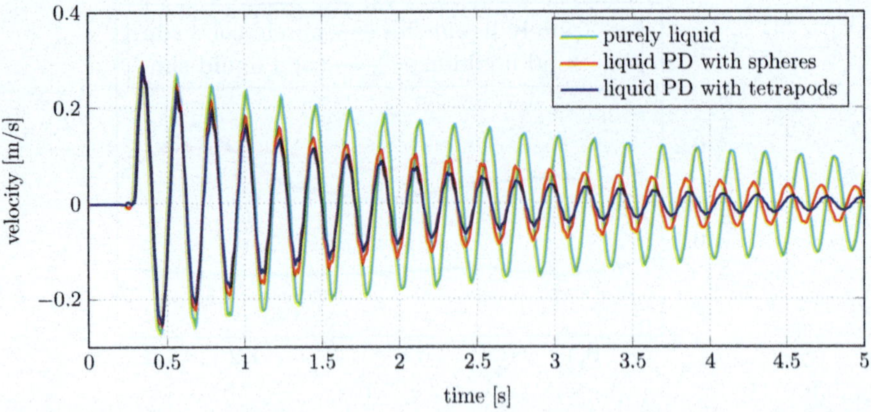

Fig. 14 Comparing solid particles with different shapes and additional liquid. The tetrapod shaped particles perform better compared to spherical particles or pure liquid. The overall mass is kept constant at all cases

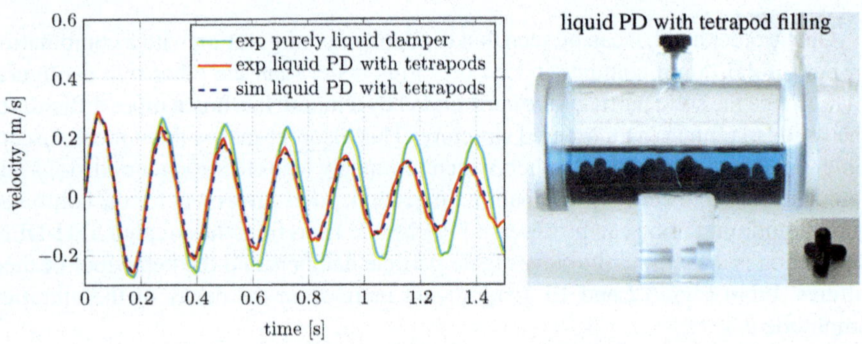

Fig. 15 The red line represents the container velocity measured during laboratory experiments for the liquid-filled PD case with tetrapods. The blue dotted line represents the container velocity predicted by coupled SPH-DEM simulation. An excellent agreement between experiments and simulations are observed

tetrapod shapes, measured during experiments and predicted by coupled SPH-DEM simulation is compared. In order to better understand the effect of solid particles, the velocity for a damper filled purely with a liquid measured during experiments is also plotted in Fig. 15 (left). Macroscopically seen, there is a good agreement between simulations and experiments, showing that coupled SPH-DEM simulations can adequately model the dynamics involved in a partially liquid-filled particle damper, also for the case of non-convex polyhedra as solid particles. It can be clearly seen, that the velocity decay is faster for the case with both liquid and solid filling, than for the purely liquid filled case, confirming the previous experimental findings. The gained insights through simulations allow for a more detailed investigation of the underlying damping effects.

The observed improvements can, therefore, be explained as follows. First, the solid particles due to the hydraulic forces applied by the liquid, remain agile even under lower vibration amplitude, thereby leading to more effective collisions and in turn more energy dissipation. Secondly, the liquid is squeezed between two approaching solid particles leading to shearing of liquid layers. This ultimately results in more energy dissipation. In this case, the non-convex tetrapod particles behave effectively as obstacles to waves created by liquid motion. These general findings concerning the filling of PDs also raise an additional research question. What should the solid-liquid fill ratio be in order to maximize the dissipation rate? In order to gain further insights regarding this question a numerical investigation is performed. In this numerical study the number of tetrapod solids are varied in three stages (0, 40, 60 solids) while the amount of liquid is kept at a constant volume of 30 ml. By this way, the solid-liquid ratio is implicitly varied. For this study, the density of each solid tetrapod particle is chosen to be 7850 kg m^{-3}. While setting up the simulations, compensation masses were added to the system mass so that all the configurations have the same static mass.

In Fig. 17 (left) the simulated velocity decay for different solid-liquid fill ratios is compared and the corresponding average logarithmic decay rate is visualized in Fig. 17 (right). Moreover, increasing the number of solids particles seem to substantially increase the decay rate. As the number of solid particles, also in the presence of a liquid, governs the number of solid particle collisions. Additionally, the liquid flow is observed to be more fierce with increase in the number of solid particles, leading to even more kinetic energy dissipation. This effect is highlighted by calculating the logarithmic decay. The most significant step is from a purely liquid filling to the use of tetrapods and additional liquid. But also the increase of particle numbers clearly shows to improve the damping effect. In Fig. 16 the simulation of the tetrapod and liquid filled particle damper is illustrated. It can be seen, that the tetrapods lead to a higher pressure in the fluid when the it sloshes through them, resulting in an increased damping effect. For more details of the simulation, see [14].

3 Obstacle Grids

An additional approach to increase damping in PDs are obstacle grids. These grids improve the interaction between the PD container and the particles by preventing the clustering of particles. This increases the energy transferred from the container to the particle filling and allows for greater particle movement. Obstacle grids, therefore, are especially advantageous for systems under forced excitation. For experimental investigations of this approach, a different testbench is introduced. The main difference lies in the change to a forced excitation through an electromagnetic shaker. The laboratory apparatus consists of a horizontally mounted steel beam of dimensions 540 mm × 20 mm × 2 mm, which is rigidly clamped on one side.

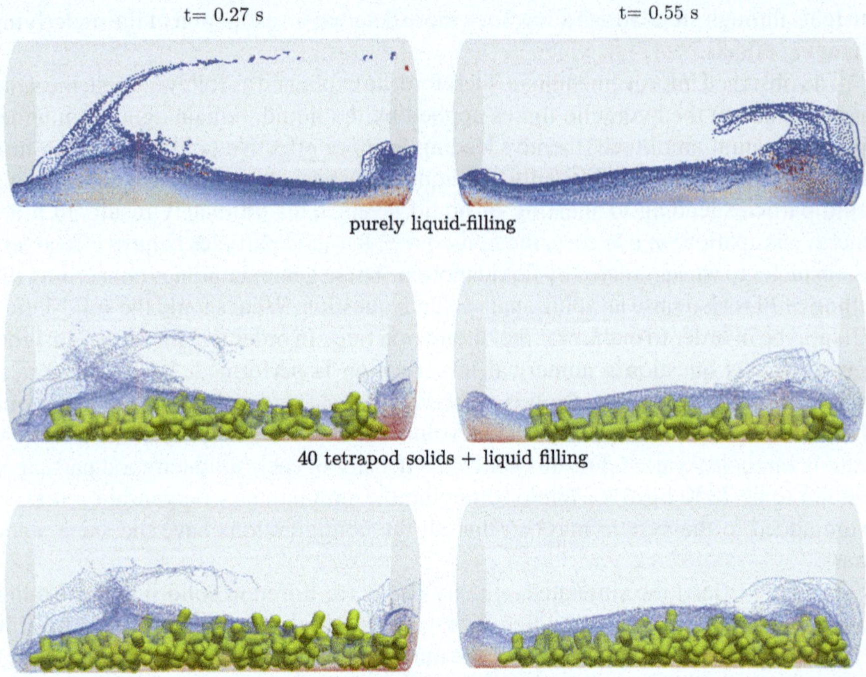

Fig. 16 The motion of damper contents visualized for (top row) purely liquid-filled damper, (middle row) 40 tetrapod solid + liquid and (bottom row) 60 tetrapod solids + liquid, at two different time instances. In all cases, the fluid is visualized as coloured balls, where the colour coding visualizes pressure from low (red) to high (blue)

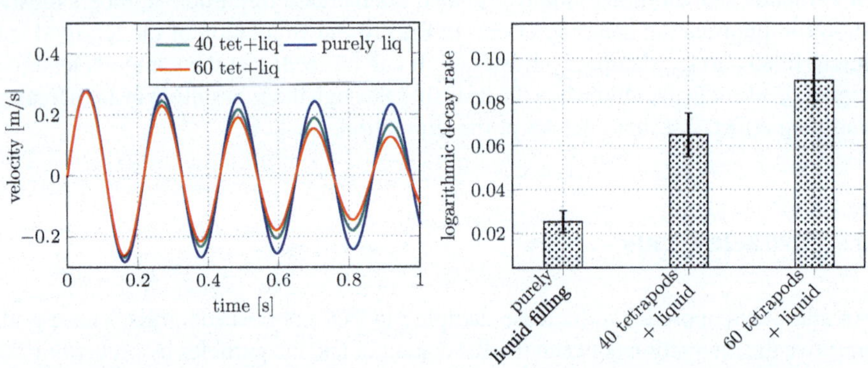

Fig. 17 (left) The velocity of the damper container for various solid-liquid fill ratios is compared. (right) The average logarithmic decay rate, computed at the end of every cycle, is visualized with respect to different solid-liquid fill ratios

The PD enclosure, weighing 58.8 g, is a transparent acrylic box of inner dimensions 35 mm × 35 mm × 20 mm, which is fixed to the side of the beam. This PD enclosure is filled with steel spheres of 2 mm diameter. The total mass of the spherical particles was kept constant at 0.03 kg throughout this investigation, which corresponds to 1880 spheres. The uniform obstacle-grid used for this investigation has several cells, each having a gap volume corresponding to a cube of size 8 mm. The size of each of these cubes is a characteristic dimension of the obstacle-grid and will be referred to as the cell-size of the obstacle-grid. The uniform obstacle-grid used in the experiments was manufactured using a Formlabs Form 2.0 Stereolithography (SLA) 3D printer. Additional information about the experimental setup can be found in [15]. In order to characterize dissipation in the system, a procedure similar to the one described in [16] is followed. In this procedure a dissipation parameter η is computed for a structure at resonance as the ratio of the measured average power dissipated over time period T and half the square of the absolute input velocity amplitude. The dissipation parameter η has a unit of Ns m^{-1}. Similar dissipation parameters have been used in [17, 18] for characterizing PDs. The dissipation parameter for a sinusoidal motion can be written as

$$\eta = \frac{\frac{1}{T}\displaystyle\int_{t}^{t+T} \dot{y}\, f_{\mathrm{e}} \mathrm{d}t}{\frac{A^2\omega^2}{2}}. \tag{2}$$

Here, A is the amplitude, ω is the frequency of the sinusoidal motion function. Moreover, f_{e} is the force applied on the PD enclosure due to particle interactions in the direction of oscillation. It is to be noted, that due to the discrete nature of the particle interactions the dissipated energy measurement varies over time. In order to take this into account, the dissipation measure for each PD configuration is calculated by averaging over several oscillation cycles. Throughout this investigation, the structure was excited at varying amplitudes at its second resonance frequency, in this case at 27 Hz and correspondingly a time period T of 0.037s. The second mode of the beam allows positioning of the PD at a precise location where there is minimum rotation and nevertheless maintain large driving amplitudes.

In Fig. 18, the driving forces and resulting forces on the container are shown for such a system containing of a leaf spring with mounted PD. The damping performance of three different configurations is experimentally compared. The first configuration consists of an empty PD enclosure without particles or an obstacle-grid. The empty PD configuration allows the quantification of the dissipation present due to intrinsic material behavior and other external effects. By this way, the additional energy dissipation contribution purely due to particle interactions can be better understood. The second configuration consists of a conventional PD solely filled with solid particles. In the third configuration, the damper contains in addition to solid particles a deliberately introduced 3D printed obstacle-grid. In order to ensure a fair comparison between all the damper configurations, compensation masses were added during the laboratory experiments, so that all the configurations have the same static mass, measured using a weighing scale. A comparison of typical experimental lines,

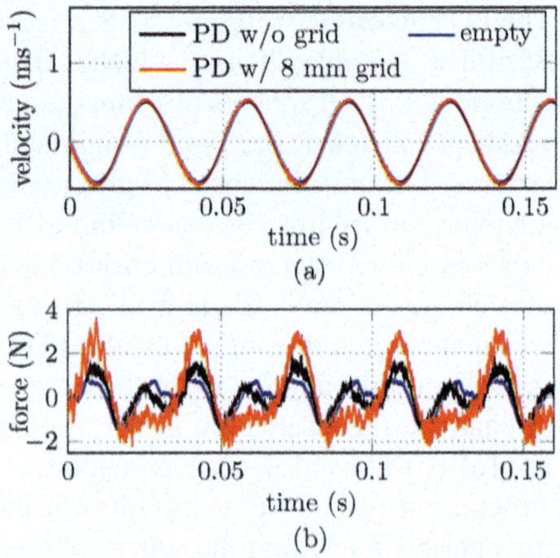

Fig. 18 Typical measurement obtained from laboratory experiments. **a** Driving velocity of the PD enclosure for the three configurations. **b** Container forces. It can be seen that the container forces are higher in the case where the PD is equipped with an obstacle-grid, indicating more dissipation

showcasing the velocity and force measurements, of the three damper configurations is shown in Fig. 18. In general it can be said that all the waveforms, irrespective of the configuration, are periodic indicating that a steady-state is being reached. Even though the velocity signal seems to be nearly sinusoidal, the force signal is complex, due to the presence of higher modes of the clamped beam. It can be seen in Fig. 18b, that for the same input velocity, the impedance forces (red curve) measured for a PD with an obstacle-grid seems to be drastically higher, indicating higher energy dissipation in this case compared to the other configurations. Similarly, the dissipation parameter η computed using Eq. 2 for all the PD configurations is presented in Fig. 19 as a function of the dimensionless acceleration amplitude $\Gamma = A\omega^2/g$, where g is the acceleration due to gravity. The translucent bands around the curves represent the variance present in several experiments for the measured dissipation parameter. It can be seen, that due to the absence of particles, the variance present in the empty PD case is negligible compared to other configurations. As a consequence, the dissipation in the empty PD case (blue line with triangle markers), which occurs due to intrinsic structural damping, is far lower compared to other configurations. A higher dissipation rate is observed for the case with particles than for the empty case. It is also seen in Fig. 19 (black line with diamond markers), that increasing the acceleration amplitudes seems to steadily increase dissipation, at least starting from $\Gamma = 6$, for the conventional PD case. For lower acceleration levels, between $\Gamma = 3$ and $\Gamma = 6$, the measured dissipation for the conventional PD case seems to have high variance. This is probably due to the very noisy force signals measured during these experiments. Perhaps the most interesting aspect of Fig. 19 is the dissipation behavior of a PD with a deliberately introduced obstacle-grid. The dissipation rate for a PD with obstacle-grids irrespective of the acceleration amplitude clearly outperforms the conventional PD without an obstacle-grid. This is arguably due to the increased particle activity and relative motion of the particles leading to more effective collisions and thereby increasing energy dissipation. Moreover, at an acceleration amplitude of $\Gamma = 10$ the dissipation rate for the PD with obstacle-grid is

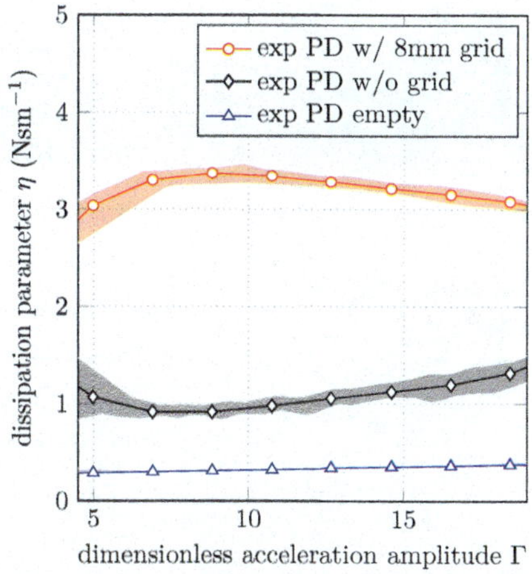

Fig. 19 Dissipation parameter for all the PD configurations as a function of the dimensionless acceleration amplitude $\Gamma = A\omega^2/g$. The translucent bands around the curves represent the variance observed during the experimental trials. The deliberate introduction of an obstacle-grid drastically increases dissipation

observed to reach an optimum. Additional information about the simulation model and the contact detection algorithm used can be found in [15].

The comparison of the resulting simulations with the previously conducted experiments in Fig. 20 (b) shows good agreement. This is the case for both configurations with and without the use of an obstacle grid. The previously mentioned increase in particle motion through the grid can clearly be observed in the simulation visualizations namely subfigures (d) and (e) in Fig. 20. This also highlights again the useful additional insights into the PD provided through numerical simulations. A more detailed discussion of these results and additional visualisations of the simulation data can also be found in [15].

Next, a numerical study is set up to investigate the effect of cell-size on the damping performance of a PD with an obstacle-grid. For this purpose, obstacle-grid geometries with two different cell-sizes, namely 5 mm and 8 mm, are investigated, as depicted in Fig. 21. In order to ensure a fair comparison, the PD is filled with the same amount (0.03 kg) of spherical steel particles for both simulation scenarios. In order to predict the effect of obstacle-grid cell size several simulations are performed. During each simulation, the PD equipped with an obstacle-grid of a particular cell-size is driven with a sinusoidal velocity at a constant frequency of 27 Hz for 5 cycles. The resulting container forces accumulated due to particle interactions are recorded at every numerical time step and are utilized to compute an average dissipation rate for each cycle according to Eq. 2. This procedure is performed for both obstacle-grid sizes and for several acceleration amplitudes Γ. The results are reported in Fig. 21c. As seen in Fig. 21c, the dissipation rate predictions for the two configurations are significantly different, even though all the parameters except the obstacle-grid cell-size are identical. Therefore, the cause of the observable difference in dissipation rate must be entirely due to the geometry of the obstacle-grid. It can be observed, that for both cases a clear peak in dissipation rate is observed at a specific acceleration amplitude. The peak dissipation rate predicted for a PD with 8 mm grid is higher at

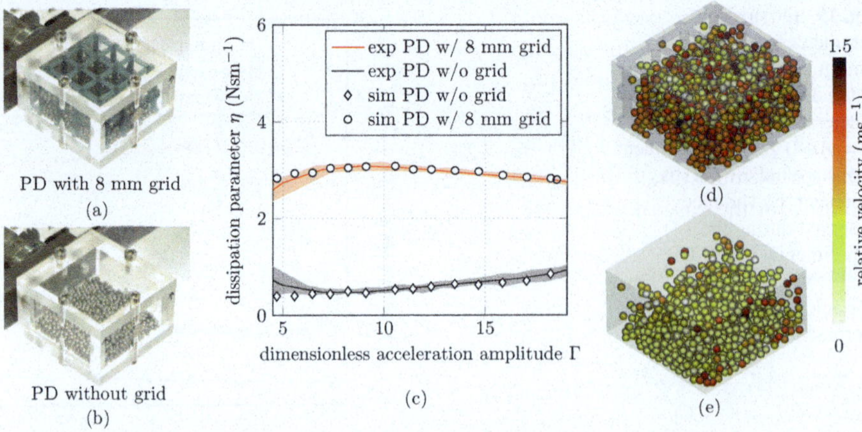

Fig. 20 **a** and **b** show the PD configurations tested during the laboratory experiments. **c** Comparison of the dissipation parameter predicted using DEM simulations and measured during lab experiments. A good agreement between experiments and simulations is observed. Visualizations of the particle motion predicted by DEM simulations **d** with and **e** without an obstacle-grid for $\Gamma = 10$. Here, the obstacle-grid **d** is made translucent for visualization purposes. Particles attain far higher relative velocities for the case with a obstacle-grid than without it

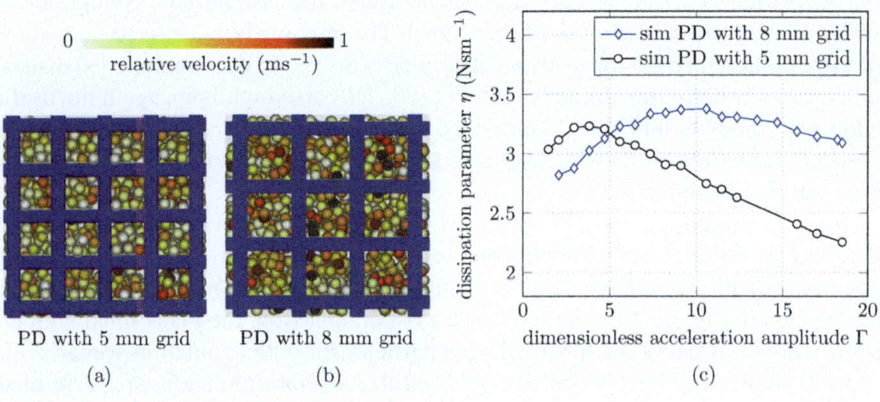

Fig. 21 A numerical study compares the damping performance of PDs with obstacle-grids of two different cell-sizes namely **a** 5 mm and **b** 8 mm. In this simulation, both configurations were subjected to $\Gamma = 10$ and the color gradient represents the relative velocity of the solid particles with respect to the PD container. **c** Results show the noticeably distinct dissipation rates predicted for the two cell-sizes at various acceleration levels. The dissipation rate for the PD with 5 mm grid peaks at a different acceleration level compared to the PD with 8 mm grid. This shows, that the obstacle-grid cell size plays a crucial role in the design of PDs with obstacle-grids

3.38 $\mathrm{Ns\,m^{-1}}$ than the peak dissipation of a PD with 5 mm grid which is found to be 3.23 $\mathrm{Ns\,m^{-1}}$. Moreover, the acceleration level at which the dissipation rate peaks is higher at $\Gamma = 10$ for PD with 8 mm grid than for PD with 5 mm grid at $\Gamma = 4$. This indicates that the cell size could be used as an additional tuning parameter to control the damping performance of PDs equipped with an obstacle-grid. More interestingly, after the maximum dissipation is reached, the rate at which the dissipation parameter decreases for the case of a PD with 5 mm grid is much steeper than for the PD with 8 mm grid. This shows that the PD with 8 mm grid seems to be less sensitive, in other words more robust, to changes in the driving acceleration level.

4 Broadband Properties

To study the broadband damping properties of a PD, a weakly damped frame structure as shown in Fig. 22 is designed and build. This frame structure exhibits multiple vibration modes even in the lower frequency range. When excited with a shaker, the

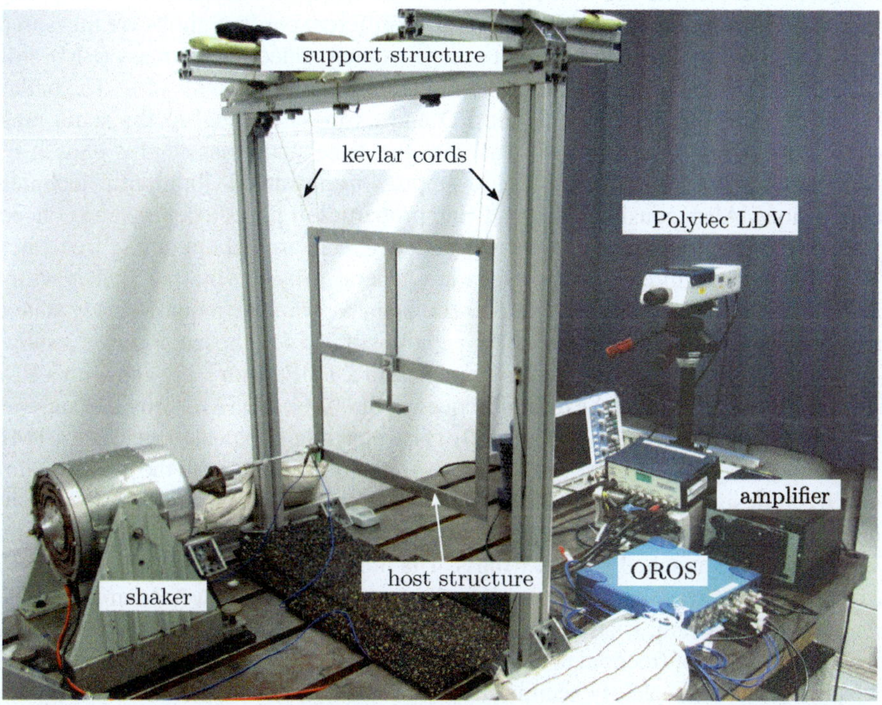

Fig. 22 Lab apparatus consisting of the host structure hung using kevlar cords. The host structure is excited using an electrodynamic shaker. A Laser Doppler Vibrometer is used to measure the structural response

frequency response of the host structure with a PD allows a systematic investigation of the broadband damping effect of PDs. In Fig. 22, the surrounding lab setup used for the experiments is also shown. The host structure is suspended by kevlar cords to replicate a free-free boundary condition. This gives the opportunity to study the dynamics of the structure independently from clamping to a surrounding suspension. The host structure is excited using an electromagnetic shaker Elektro-Mechanik Schmid SW100, which is driven by a power amplifier Brüel & Kjær Type 2706. The driving signal for the shaker is controlled using a Tectronix AFG2022B signal generator. A PCB 288D01 impedance sensor situated between the shaker and the host structure is used to measure the force and acceleration at the forcing point (input). The forces and accelerations at the input vary, but are in a range up to 55 N and 40 ms^{-2} peak values, respectively. For more information about the structure used, as well as, detailed pictures of the experimental setup refer to [21].

4.1 Comparison with Tuned Mass Damper

In order to investigate the influence of a PD on the host structure, the steady state frequency response functions (FRFs) of three different configurations are measured. The three configurations are: host structure with an added ballast mass (BM), host structure with a tuned mass damper (TMD) and host structure with a PD. The ballast block is in the same location as the PD and its mass is equal to the static mass of the particle damper. The TMD also has the same static mass and is chosen for comparison as it is a common device for suppressing unwanted vibrations in technical applications. Also, considering broadband properties, it poses a contrasting concept as it only works well within a narrow frequency band around the tuning frequency. The FRFs for the three configurations are generated by driving the shaker with a frequency sweep signal from 25 to 100 Hz. To make sure that a quasi-steady state is reached, a relatively long frequency sweep time of 375 s is chosen. Figure 23 shows the frequency response of the host structure with a TMD, with a PD and with a BM. It is clearly seen that near the design frequency of 60 Hz, the TMD provides superior vibration suppression compared to a PD. This behaviour is expected, because a TMD works by introducing a vibration node at the point of attachment to the host structure exactly at design or operating frequency. Therefore, a conventional TMD actually does not directly dissipate the vibrational energy but rather transfers the energy to the vibration of the attached auxiliary mass. The movement of this mass functions as a kinetic energy reservoir and also leads to dissipation through material damping of the deflected TMD beam, friction in the joints, etc.. Furthermore, a TMD introduces an additional degree of freedom to the host structure and thus, adds additional resonance frequencies. This is clearly seen in the additional resonance peak at 70 Hz for the case where the host structure is fitted with a TMD, see the blue dashed curve in Fig. 23.

Additionally, the introduction of a TMD lowers the natural frequencies of the reference host structure that are below the design frequencies. For instance, the nat-

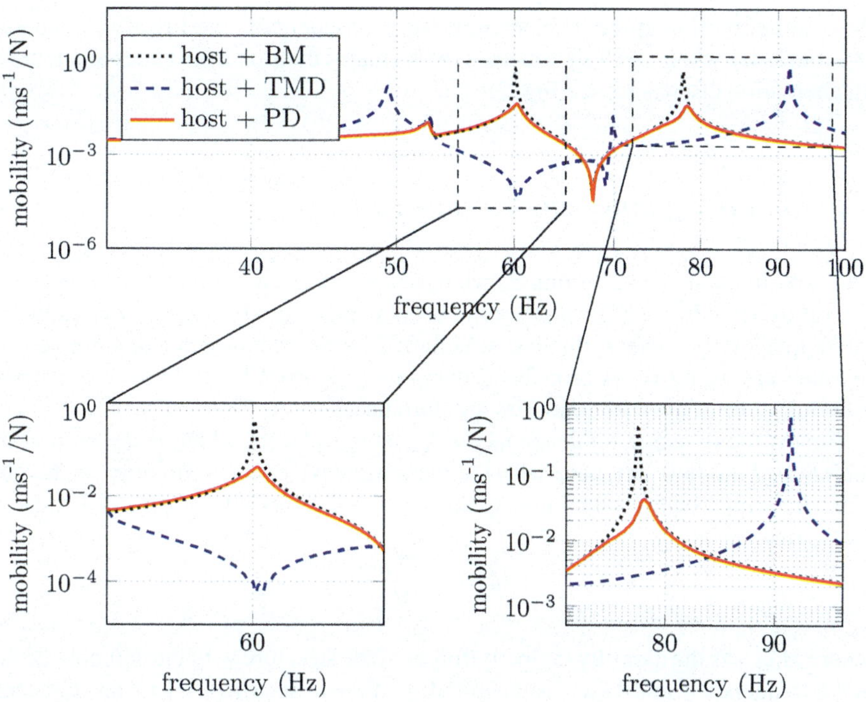

Fig. 23 Experimental frequency response of the host structure with a BM, TMD and PD. Near the design frequency of 60 Hz, the TMD provides superior vibration attenuation compared to a PD. However, the PD provides considerable vibration damping at multiple resonance frequencies at the same time

ural frequency at 60 Hz is now lowered to 45 Hz. On the other hand, the resonance frequencies of the reference host structure that are higher than the design frequency, for instance the frequency at 80 Hz, are raised with the introduction of a TMD, see Fig. 23 for high frequencies. On the whole, the TMD, even though it does a very good job in reducing vibration near the design frequency of 60 Hz, drastically alters the frequency response of the host structure and creates trouble at other frequencies. Another aspect of the TMD is, that its vibration attenuation property is highly sensitive to changes in stiffness and mass of the auxiliary system. In other words, changes in the TMD configuration, for instance due to fatigue, lead to a detuning of the TMD which could result in a sudden unwanted increase in vibrations. Therefore, care has to be taken when designing a TMD and it should only be used for a system that is subjected to a constant frequency excitation.

On the other hand, PDs provide considerable vibration damping not only at 60 Hz but also at other frequencies. Unlike a TMD, PDs due to inter particle collisions and friction, actually dissipate the vibrational energy of the host structure and convert it to other energy forms (for instance heat). As seen in Fig. 23, the energy dissipation in the PD is relatively insensitive to the excitation frequency and highly sensitive to the external motion at the attachment point. Consequently, the PD affects the host structure only where the host structure shows high vibration amplitudes and induces no alternation elsewhere, see Fig. 23.

4.2 Towards Quantifying Broadband Dissipation

The experimental investigations shown have opened up two crucial questions which are yet to be addressed. To what extent does a particular damping device influence the frequency response of the host structure? Moreover, how can a damping device be rated according to its broadband damping property? In order to answer these questions, two additional quantities are introduced.

Firstly, the dynamic influence factor S_{dev} is introduced, which is the ratio of the mobility of the host structure with the damping device to the mobility of the host structure with a ballast block having the same static mass of the device. The factor S_{dev} is defined as

$$S_{\text{dev},f} = \frac{M_{\text{dev},f}}{M_{\text{ref},f}}, \tag{3}$$

where $M_{\text{dev},f}$ is the mobility (velocity/force) of the structure with the damping device at the frequency f and $M_{\text{ref},f}$ is the mobility of the host structure with an equivalent mass block. The factor S_{dev} helps to quantify the effect of the particular damping device on the host structure. For instance, a high dynamic influence (greater than 1) indicates vibration amplification and a low value (smaller than 1), indicates vibration reduction. The dynamic influence factor S_{dev} applied to the investigated host structure with a PD and a TMD is shown in Fig. 24. It can be seen that the dynamic influence factor for the TMD has a very low value near the design frequency of 60 Hz, meaning vibration attenuation is only observed around the design frequency. Apart from the design frequency, especially at 49 Hz, 68 Hz, 91 Hz frequencies, the dynamic influence factor for the TMD case has high positive values, indicating even a vibration amplification at these frequencies, i.e. a worsening of the dynamic behaviour. From a practical point of view these vibration amplifications observed only in the TMD case are disadvantageous. This is because the TMD, apart from providing the vibration attenuation at the design frequency, fundamentally alters the frequency response of the host structure elsewhere and might even make it worse than that of the undamped structure.

Interestingly, the dynamic influence for the PD case is relatively smooth compared to the TMD case. The S_{PD} for the PD case attains values smaller than one (meaning vibration reduction) at frequencies where the host structure exhibits high vibration amplitudes. Usually, the value of S_{PD} is close to one. This means that the PD is a passive damping device which smoothens the resonance peaks without fundamentally altering or shifting the natural frequencies of the host structure.

Secondly, to analyse the broadband damping property of a device, another quantity namely the mean influence deviation σ_{dev} is introduced. The deviation σ_{dev} is defined

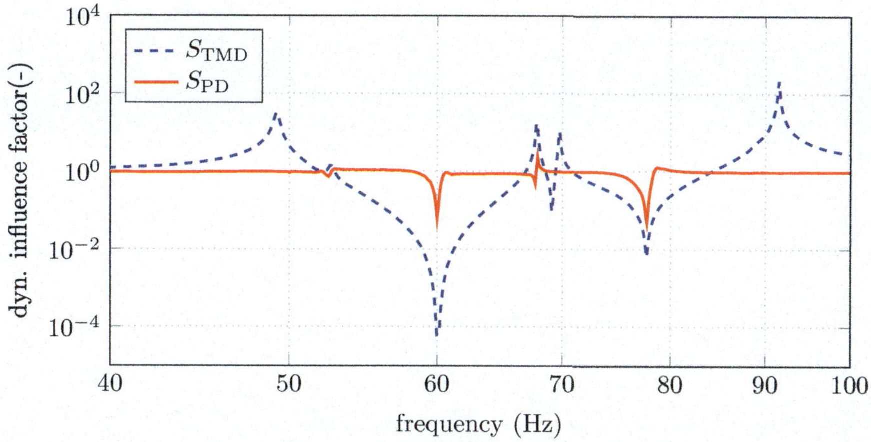

Fig. 24 Dynamic influence factor computed for the host structure equipped with a PD and a TMD. The S_{PD} shows that the PD influences the host structure only for high vibration amplitudes, whereas S_{TMD} shows that the TMD fundamentally alters the frequency response of the host structure

as the squared deviation of the dynamic influence factor from its mean behaviour of a particular damping device. This can be defined as

$$\sigma_{\mathrm{dev},f} = (S_{\mathrm{dev},f} - \bar{S}_{\mathrm{dev}})^2, \tag{4}$$

where $\bar{S}_{\mathrm{dev}}$ is the average dynamic influence factor of the host structure equipped with a particular damping device. In other words, σ_{dev} indicates the extent to which a device deviates from its mean response over frequency. A high value of σ_{dev} means, that the device changes its behaviour to a large extent. A perfect broadband damping device would have a value of zero, even though such a device would be impractical. Figure 25 shows the curves for σ_{TMD} and σ_{PD} computed for the host structure equipped with a TMD and PD, respectively. It is clearly seen that the deviation σ_{PD} for the PD case has a much lower numerical value over the entire frequency range when compared to a TMD. This means that the behaviour of a PD does not drastically deviate from its mean performance compared to a TMD. For the TMD case it can be seen that the vibration amplification around 49 Hz and 91 Hz are prominent in the influence deviation as well. However the attenuation around 60 Hz leads to no observable peak. This is due to numerically small numbers of S_{TMD} from which the constant $\bar{S}_{\mathrm{dev}}$ is subtracted. The formulation of $\sigma_{\mathrm{dev},f}$ in Eq. 4 especially highlights positive deviation from $\bar{S}_{\mathrm{dev}}$ which leads to undesired vibration amplification caused by the damping device.

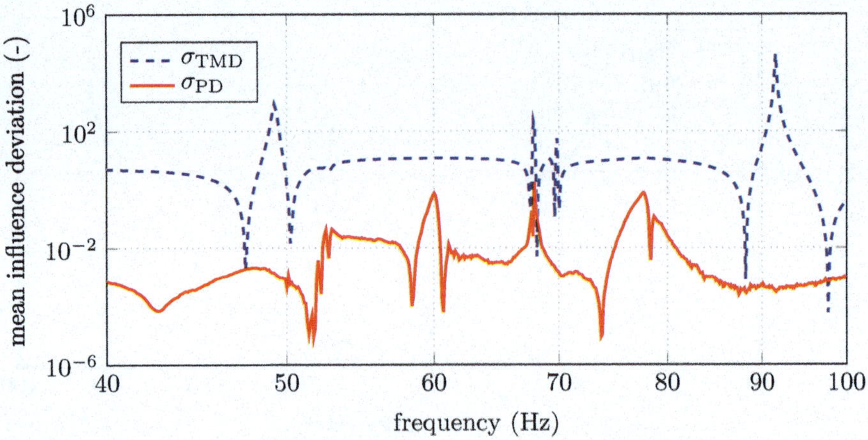

Fig. 25 Mean influence deviation computed for the host structure equipped with a TMD and a PD. The σ_{PD} has smaller values compared to σ_{TMD} indicating that the PD does not drastically deviate from its mean behaviour compared to a TMD

Therefore, the value S_{dev} provides quantitative insights on the extent to which a device influences the host structure and σ_{dev} provides insights regarding the broadband damping property of a damping device. So, S_{dev} and σ_{dev} together provide the right tools to quantitatively investigate damping devices or parameter changes systematically.

5 Conclusion

The emphasis of this research project was the improvement of PDs and their applicability. Therefore, several PD concepts were analysed through experiment and simulation. To improve the damping properties of PDs, the conceps investigated included, e.g., additional liquid, complex particle shapes, and obstacle-grids. These variations on PD fillings provided significant enhancement of damping under various loading conditions. Especially the combined approach of simulations supported by meaningful laboratory experiments allowed for a targeted investigation of the different concepts. The simulations allow detailed insights into the behaviour of the PD fillings and the experiments ensure the approximations made for the simulation model are realistic and yield good agreement with measured data. The ability to examine the particles during simulations in detail for every timestep is of great benefit for the understanding of the damping processes within the PD. This is especially important for a targeted design improving the applicability of PDs for a special use case. The general applicability was improved in this project by investigating damping properties over a wide frequency range with the help of test structures with dynamical behaviour relevant for industrial applications. In this context, the PD was also directly compared to a conventional TMD which is already widely used in technical applications. In this research project, the usage of PDs for targeted damping in structures and technical applications showed to be feasible and beneficial. Further research in this field seems promising to further improve the applicability of PDs and make them usable in various application fields over wide ranges of frequency and loading amplitudes. Enabling the industry to use PDs as standard damping devices seems of great interest in many fields.

Acknowledgements This research has received funding from the German Research Foundation (DFG) within the priority program SPP 1897 "Calm, Smooth and Smart: Neuartige Schwingungsbeeinflussung durch gezielt eingesetzte Dissipation" project EB195/25-1/2 "Partikeldämpfer—Schwingungsbeeinflussung durch verteilte Dissipation über komplexe Partikelformen" (project number 315008544). This support is highly appreciated.

References

1. Panossian, H.: Structural damping enhancement via non-obstructive particle damping technique. J. Vib. Acoust. **114**, 101–105 (1992)
2. Harris, C.M., Crede, C.E.: Shock and Vibration Handbook. McGraw-Hill, New York (1976)
3. Heckel, M., Sack, A., Kollmer, J.E., Pöschel, T.: Granular dampers for the reduction of vibrations of an oscillatory saw. Phys. A **391**(19), 4442–4447 (2012)
4. Xiao, W., Huang, Y., Jiang, H., Lin, H., Li, J.: Energy dissipation mechanism and experiment of particle dampers for gear transmission under centrifugal loads. Particuology **27**, 40–50 (2016)
5. Ma, C., Lu, Z., Wang, D., Wang, Z.: Study on the damping mechanisms of a suspended particle damper attached to a wind turbine tower. Wind Struct. **33**(1), 103–114 (2021)
6. Veeramuthuvel, P., Shankar, K., Sairajan, K.K.: Application of particle damper on electronic packages for spacecraft. Acta Astronaut. **127**, 260–270 (2016)
7. Tomlinson, G., Pritchard, D., Wareing, R.: Damping characteristics of particle dampers - some preliminary results. Proc. Inst. Mech. Eng. C J. Mech. Eng. Sci. **215**(3), 253–257 (2001)
8. Marhadi, K.S., Kinra, V.K.: Particle impact damping: effect of mass ratio, material, and shape. J. Sound Vib. **283**, 433–448 (2005)
9. Liu, W., Tomlinson, G.R., Rongong, J.A.: The dynamic characterisation of disk geometry particle dampers. J. Sound Vib. **280**, 849–861 (2005)
10. Pasimodo - Particle Simulations Software. www.itm.uni-stuttgart.de/software/pasimodo (last accessed on 07-06-2020)
11. Gnanasambandham, C., Schönle, A., Eberhard, P.: Investigating the dissipative effects of liquid-filled particle dampers using coupled DEM-SPH methods. Comput. Particle Mech. **6**, 257–269 (2019)
12. Gnanasambandham, C., Eberhard, P.: Predicting the influence of an added liquid in a particle damper using coupled SPH and discrete element method. PAMM Proc. Appl. Math. Mech. **17**, 31–32 (2018)
13. Gnanasambandham, C., Eberhard, P.: Investigating the effect of complex particle shapes in partially liquid-filled particle dampers using coupled DEM-SPH Methods. PAMM Proc. Appl. Math. Mech. **19** (2019)
14. Gnanasambandham, C., Eberhard, P.: Modeling a partially liquid-filled particle damper using coupled Lagrangian methods. In: International Centre for Numerical Methods in Engineering (CIMNE), Barcelona (2019)
15. Gnanasambandham, C., Fleissner, F., Eberhard, P.: Enhancing the dissipative properties of particle dampers using rigid obstacle-grids. J. Sound Vibrat. **484**, 115522 (2020)

16. Yang, M.; Lesieutre, G.; Hambric, S.; Koopmann, G.: Development of a design curve for particle impact dampers. In: Proceedings of the 11th Annual International Symposium on Smart Structures and Materials, pp. 450–465 (2004)
17. Salueña, C., Pöschel, T., Esipov, S.E.: Dissipative properties of vibrated granular materials. Phys. Rev. E **59**, 4422–4427 (1999)
18. Wong, C., Daniel, M., Rongong, J.: Energy dissipation prediction of particle dampers. J. Sound Vib. **319**, 91–118 (2009)
19. Ericson, C.: Real-Time Collision Detection. Elsevier, Amsterdam (2005)
20. Klosowski, J., Held, M., Mitchell, J., Sowizral, J.: Efficient collision detection using bounding volume hierarchies of K-Dops. IEEE Trans. Visual Comput. Graphics **1**, 21–37 (1998)
21. Schönle, A.; Gnanasambandham, C., Eberhard, P.: Broadband damping properties of particle dampers mounted to dynamic structures. Exper. Mech. (2022)

A Study on Friction Dampers and Their Contact Geometry Design

Jimmy Aramendiz and Alexander Fidlin

1 Introduction

Effective, targeted vibration damping with low efficiency reduction is an important goal for a modern design, especially for lightweight constructions and energy-efficient applications. An ideal damper dissipates energy only near the resonance frequency, when the undesired vibration amplitudes jeopardize the service life or function of the system. Such dampers increase the overall energy efficiency of a system.

To suppress undesired vibrations, viscous damping is introduced into the system in most applications, e.g. in the automotive industry in the form of a hydraulic shock absorber [2]. These damping mechanisms are always active and constantly dissipate energy even when this is not necessary. Thus, reducing the energy efficiency of the system. However, viscous damping is not the only way to reduce vibrations. The use of absorbers for vibration reduction is also common in engineering [1, 11]. Vibration absorbers achieve at their tuned operating frequency optimal vibration reduction. Nevertheless, at least one structural resonance must be overcome to reach the operating frequency. Furthermore, these devices are sensitive to the inevitable system parameter changes due to wear, time, and environmental conditions.

An additional alternative in vibration reduction are friction-based dampers, which are widely used in engineering. These dampers are used mainly in three different fields: in turbomachinery as so-called platform dampers, in civil engineering structures such as buildings and cables, and in railroad freight trucks. Platform dampers are in most cases metal elements which are pressed between two blade platforms by the centrifugal force in a gas turbine. These dampers are investigated mainly in

J. Aramendiz (✉) · A. Fidlin
Karlsruhe Institute of Technology, Kaiserstraße 10, 76131 Karlsruhe, Germany
e-mail: jimmy.aramendiz@kit.edu

A. Fidlin
e-mail: alexander.fidlin@kit.edu

P. Eberhard (ed.), *Calm, Smooth and Smart*, Lecture Notes in Applied and Computational Mechanics 102, https://doi.org/10.1007/978-3-031-36143-2_2

two variations: as curved friction dampers and as wedge (or cottage-roof) friction dampers [10]. In civil engineering friction dampers are mainly found in two forms. In some applications, planar contact surfaces are combined, whereas in other applications varying contact geometries are used. Applications with planar contact surfaces are found in [13], whereas investigations with geometric variations are presented in [9]. A special variant of friction dampers, the so called wedge dampers, are used in railroad trucks. These dampers are composed of a wedge placed between the bolster, which carries the wagon, and side frame, which is connected to the wheels. The geometric variations in the dampers' sliding contact partners "simulate" a viscous damping avoiding technological difficulties connected with the handling of fluids. Due to their robustness, low cost, and low maintenance they are quite popular and still investigated today [14].

A possible passive implementation of an energy-efficient vibration suppression can be realized by dry friction. The special character of dry friction with stick-slip transitions allows the design of elements that change their behavior. Frictional contacts, which can stick and slide, change the structure of a system and thus its behavior. The targeted design of the sticking and sliding dynamics allows systems to passively adapt to current operating conditions. This paper focuses on three such passive dry friction dampers: the dry friction lock-up damper, a prestressed sliding wedge damper, and the friction damper with polynomial contact geometry. The underlying operating principle of the dampers is based on the stick-slip properties of dry friction. This ensures sticking at low excitation amplitudes, which prevents any relative movement between the masses and thus no energy is dissipated. As soon as the breakaway force of the damper is overcome, the system moves in the stick-slip range. The breakaway condition ensures a selective energy dissipation, which only occurs at high vibration amplitudes. This work uses the simplest friction model namely the Coulomb model, in which Stribeck effects and a difference between static and dynamic friction coefficients are neglected.

The present paper is structured as follows. In Sect. 2 the dry friction lock-up damper is investigated. The equations of motion of the dampers are derived and numerical parameter studies are performed, which gives a first impression of the system dynamics. In addition, an analytical solution for the systems is derived and analyzed using the averaging method. Analogous investigations are carried out in Sect. 3. Analytical solutions for different polynomial degrees are derived in Sect. 4 for the dry friction damper with polynomial contact surface. Experimental results are presented in Sect. 5. The main findings are summarized in Sect. 6.

2 The Dry Friction Lock-Up Damper

The goal of the dry friction lock-up damper (Fig. 1) is to improve the dynamics of an existing main system. In this paper the main system is represented by the primary spring c_1 and the primary mass m_1. The lock-up damper, consisting of a dry friction element with friction force R, a secondary spring c_2, and a secondary mass m_2 is

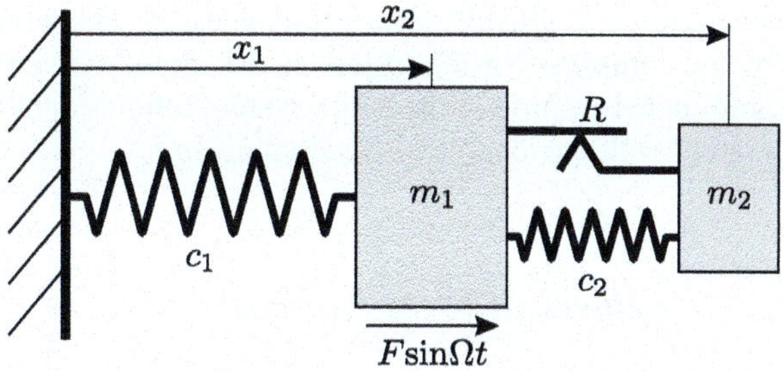

Fig. 1 Lock-up with a harmonic excitation on the primary mass

attached to the main system. Due to the neglecting of both Stribeck effects and a difference between static and sliding friction coefficients, the friction force R also corresponds to the breakaway force of the friction element. The friction element determines the stick-slip behavior of the system and thus also the conditions under which the secondary spring influences the system dynamics. While sticking both masses move together in the same way and the secondary spring is inactive. In the inactive state, the secondary spring can be in a deflected state, but this does not affect the motion of the system. The inactive secondary spring only affects the stiction force. As soon as the friction element allows relative movement, the structure of the system changes from a one degree to a two degrees of freedom oscillator. This leads to a change in the number and value of of the system's natural frequencies. The stick-slip properties divide the system behavior into linear and nonlinear ranges, each of which is determined by the closed and open state of the friction element respectively. Therefore the system's dynamics are characterized by these two ranges: the sticking range and the sliding range. The system is in the sticking range as long as the absolute value of the stiction force H is smaller than the breakaway force R ($|H| \leq R$) and the kinematic condition $\dot{x}_1 = \dot{x}_2$ is fulfilled. If one of these conditions is violated, the system switches to the sliding range. The equations of motion for both ranges are given by

while sticking

$$(m_1 + m_2)\ddot{x}_1 + c_1 x_1 = F \sin \Omega t, \tag{1}$$

$$H = m_2 \ddot{x}_2 + c_2(x_2 - x_1), \tag{2}$$

while sliding

$$m_1 \ddot{x}_1 + c_1 x_1 - c_2(x_2 - x_1) - R \,\mathrm{sgn}(\dot{x}_2 - \dot{x}_1) = F \sin \Omega t, \tag{3}$$

$$m_2 \ddot{x}_2 + c_2(x_2 - x_1) + R \,\mathrm{sgn}(\dot{x}_2 - \dot{x}_1) = 0. \tag{4}$$

The sticking range is described by one equation of motion and one algebraic equation for the stiction force, whereas the sliding range is characterized by two equations of motion. As described in [5], it is intuitive that the amplitude response of the whole system follows the amplitude response of the linear system up to a certain breakaway amplitude. After the breakaway point, a nonlinear region follows, where stick-slip and full sliding movements occur. The nonlinear range is only concluded by the complete closing of the friction point over an entire period. Afterwards, the amplitude response of the entire system follows the amplitude response of the linear system with one degree of freedom again.

2.1 Numerical Investigations

To gain a first insight into the dynamics of the lock-up damper, numerical parameter studies are carried out. The parameters of the main system are usually determined a priori and therefore cannot be modified. In these studies, both the mass and the spring stiffness are set to one without any restriction of the generality. Since the total mass of the system should not be significantly changed, the value of the secondary mass is chosen to be much smaller than the primary mass, e.g. one tenth of the primary mass. In order to reduce amplitudes over the entire frequency spectrum, the secondary stiffness is chosen to cancel out the resonance of the main system. Only the breakaway force of the friction element remains as a freely selectable design parameter. A parameter study on the influence of the friction force [12] is shown in Fig. 2 for the parameters

$$m_1 = 1\ \text{kg},\ m_2 = 0.1\ \text{kg},\ c_1 = 1\ \text{N/m},\ c_2 = 0.1\ \text{N/m},\ F = F_0 = 0.1\ \text{N}.$$

The friction force R determines the breakaway frequency and thus determines whether the system behaves mainly like a one or two degrees of freedom system. For a system with one peak it is necessary that the breakaway frequency is higher than the first resonance frequency of the two degrees of freedom system. On the basis of this

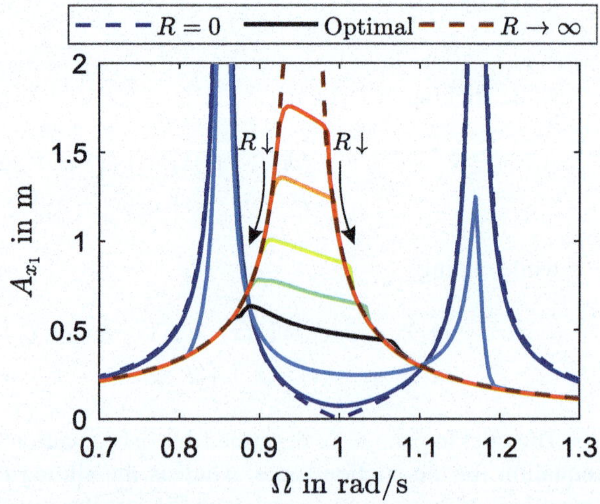

Fig. 2 Parameter study of the lock-up damper for $R \in [0\text{N(blau)}, 1.5\text{N(rot)}]$, $R_{opt} \approx 0.0369$ N

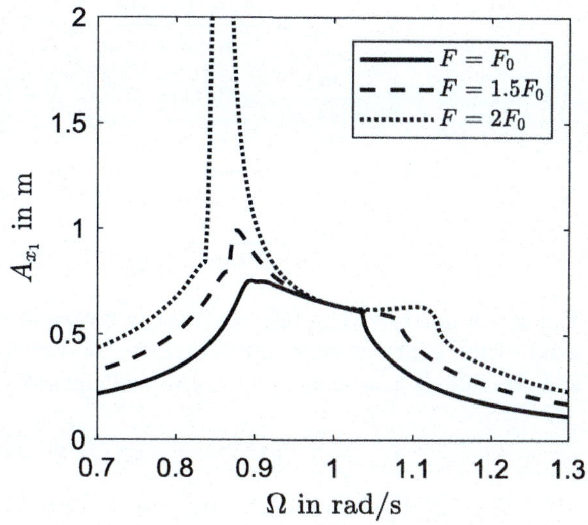

Fig. 3 Parameter study of the lock-up damper for different excitation forces with $R = 0.05$ N

parameter study the existence of an optimal friction force is identified. The optimal friction force leads to a minimization of the maximum amplitude in the frequency spectrum and must be tuned to the excitation force. This is clearly shown in Fig. 3. An increase of the excitation force with the same friction force leads to considerably larger amplitudes in a detuned lock-up damper. The reason for this is the relationship between the energy dissipated by the damper and the relative amplitude. The friction force is constant over the displacement. This leads to an energy dissipation proportional to the relative amplitude. In the detuned case the dissipated energy is not sufficient to limit the amplitudes. With the selected parameters, this effect can be seen when the excitation is doubled.

2.2 *Analytical Investigations*

After the insight into the dynamics of the lock-up damper, an analytical solution is advantageous for a deeper understanding of the system. A detailed description of this solution can be found in [6]. To derive an analytical solution for this system, the nonlinear equations of the system are considered. It is additionally assumed that permanent sliding occurs. The first step is to derive the nondimensional equations of motion of the system. The necessary transformations are

$$z_1 = \frac{m_1 x_1 + m_2 x_2}{m_1 + m_2},\ z_2 = x_2 - x_1, \tag{5}$$

$$\varepsilon = \frac{m_2}{m_1 + m_2} \ll 1,\ k = \sqrt{\frac{c_1}{m_1}},\ \lambda = \sqrt{\frac{c_2}{m_2}},\ p = \frac{\lambda}{k}, \tag{6}$$

$$\tau = kt,\ (\cdot)' = \frac{\mathrm{d}}{\mathrm{d}\tau}(\cdot),\ \eta = \frac{\Omega}{k},\ \mu = \frac{R}{m_2 k^2 (1-\varepsilon)}, \tag{7}$$

$$f_0 = \frac{F}{k^2 (m_1 + m_2)} = \varepsilon f,\ f = \mathcal{O}(1). \tag{8}$$

The newly introduced variables z_1 and z_2 represent the movement of the center of mass of the entire system and the relative movement between the masses. These transformations lead to the nondimensional equations

$$z_1'' + z_1 = \varepsilon(z_1 + z_2 + f \sin \eta\tau) - \varepsilon^2 z_2, \tag{9}$$

$$z_2'' + \frac{p^2}{1-\varepsilon} z_2 + \mu \operatorname{sgn}(z_2') = z_1 - \varepsilon \left(z_2 + \frac{f}{1-\varepsilon} \sin \eta\tau \right). \tag{10}$$

In order to obtain suitable equations for the averaging procedure, a Van der Pol transformation is additionally applied and the slowly changing amplitudes and phase differences of the variables are considered. This yields

$$z_1 = A \sin \varphi,\ z_1' = A \cos \varphi, \tag{11}$$

$$z_2 = B \sin \psi,\ z_2' = Bp \cos \psi, \tag{12}$$

$$\gamma = \varphi - \eta\tau,\ \varepsilon\delta_1 = 1 - \eta, \tag{13}$$

$$\theta = \psi - \varphi,\ \varepsilon\delta_2 = p - \eta, \tag{14}$$

$$A' = \varepsilon(f \sin \eta\tau + A \sin \varphi + B \sin \psi) \cos \varphi - \varepsilon^2 B \sin \psi \cos \varphi, \tag{15}$$

$$\gamma' = \varepsilon\delta_1 - \varepsilon(f \sin \eta\tau + A \sin \varphi + B \sin \psi) \sin \varphi + \varepsilon^2 B \sin \psi \sin \varphi, \tag{16}$$

$$\begin{aligned} B' = {} & \frac{\varepsilon}{p} \left((\tilde{A} \sin \varphi - \tilde{\mu} \operatorname{sgn}(\cos \psi) - (1 + p^2) B \sin \psi \right) \cos \psi \\ & - \frac{\varepsilon^2}{p} \left(p^2 B \sin \psi + \tilde{f} \sin \eta\tau \right) \cos \psi, \end{aligned} \tag{17}$$

$$\begin{aligned} \theta' = {} & \varepsilon\delta_2 + \frac{\varepsilon}{p} (\tilde{\mu} \operatorname{sgn}(\cos \psi) + (1 + p^2) B \sin \psi) \sin \psi \\ & - \frac{\varepsilon}{p} \tilde{A} \sin \varphi \sin \psi + \frac{\varepsilon^2}{p} \left(p^2 B \sin \psi + \tilde{f} \sin \eta\tau \right) \sin \psi. \end{aligned} \tag{18}$$

It should be noted that in the Eqs. 17 and 18 the scaling $A = \varepsilon \tilde{A}$, $\mu = \varepsilon \tilde{\mu}$ and $f = \varepsilon \tilde{f}$ was applied. The reason for this, is that Eq. 10 depends only on the ratios of these quantities and not on the actual quantities themselves. If these selected parameters are scaled with the same factor, the equation will qualitatively yield the same solution scaled only by the selected factor. For a more detailed explanation, the reader is referred to [6]. If a first and second order averaging procedure is applied to

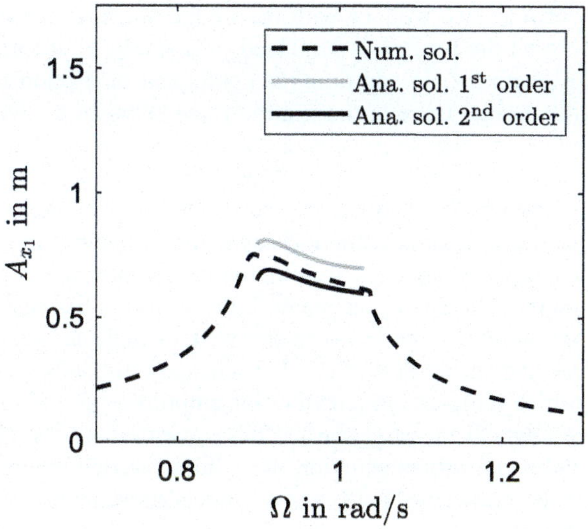

Fig. 4 Analytical solution of the lock-up damper

the Eqs. 15–18, the analytical solutions in Fig. 4 is obtained. The first-order solution represents the qualitative behavior of the amplitude response. The resonance peak of the system is limited and a declining characteristic curve in the resonance range can be observed. The second order solution provides a quantitative improvement of the result. It is also demonstrated, that the damper is very sensitive to changes of the excitation amplitude (or the changes of the friction coefficient). Therefore, its applicability is limited.

3 The Prestressed Sliding Wedge Damper

Similar to the lock-up damper, the aim of the prestressed sliding wedge damper is to improve the dynamics of a main system with spring stiffness c_1 and mass m_1. The damper is attached to the main system and consists of a prestressed mass m_2

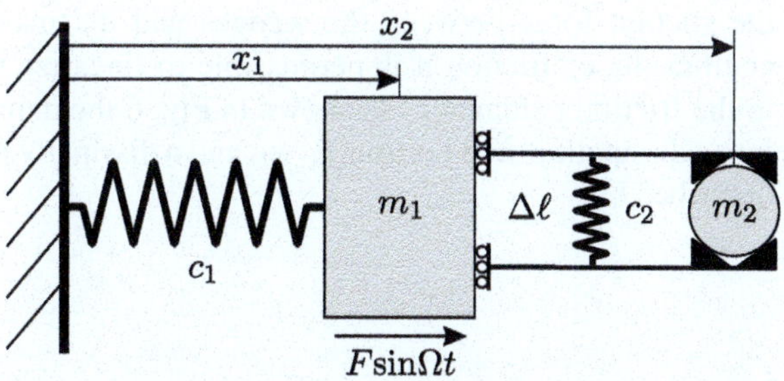

Fig. 5 The prestressed sliding wedge damper

between two wedges with the wedge angle α, see Fig. 5. The prestress force acting on the wedges is produced by a spring of stiffness c_2, which is prestressed by a length $\Delta\ell$. In addition, the coefficient of friction μ describes the relationship of the normal force to the friction force between the wedges and the mass. A relative movement of the masses presses the wedges apart symmetrically. Furthermore, the wedges are mounted on the main mass in such a way that they only transmit a force in the vibration direction. Because the wedges are pressed apart, this damper more robust compared to the lock-up damper. The friction force of the lock-up damper is constant, whereas the friction force in the prestressed sliding wedge damper depends on the relative displacement. If there is a relative displacement between the masses, the normal force between the wedges and the mass increases due to the geometry and spring deflection. This change ultimately leads to a variable friction force, which achieves limited vibration amplitudes at different excitation force amplitudes. Analytical investigations of this system without preload can be found in [4]. The industrial implementation of a sliding wedge damper without prestress in a drive train is described in [8] and is called a anti-clutch-judder-damper (germ.: Anti-Rupf-Tilger).

From a practical point of view, it makes sense to choose small coefficients of friction, because these lead to a longer service life of the damper [7]. However, a reduction of the coefficient of friction also reduces the energy dissipation. If this effect is not desired, compensation can be made by adjusting the angle α. The equations of motion of the system are given by

while sticking

$$(m_1 + m_2)\ddot{x}_1 + c_1 x_1 = F \sin \Omega t, \tag{19}$$

$$H = \frac{m_2}{m_1 + m_2}(F \sin \Omega t - c_1 x_1), \tag{20}$$

while sliding

$$m_1 \ddot{x}_1 + c_1 x_1 - F_{\mathrm{WD}} = F \sin \Omega t, \tag{21}$$

$$m_2 \ddot{x}_2 + F_{\mathrm{WD}} = 0, \tag{22}$$

$$F_{\mathrm{WD}} = 2c_2\ (2 \tan \alpha |x_2 - x_1| + \Delta\ell)\frac{\tan \alpha \operatorname{sgn}(x_2 - x_1) + \mu \operatorname{sgn}(\dot{x}_2 - \dot{x}_1)}{1 - \mu \operatorname{sgn}(\dot{x}_2 - \dot{x}_1) \tan \alpha \operatorname{sgn}(x_2 - x_1)}\ . \tag{23}$$

The stiction force H in Eq. 20 represents the necessary constraining force to prevent relative movement between the primary and secondary mass. Accordingly, it is not the stiction force between the wedges and the mass m_2. Furthermore, for small coefficients of friction it is permissible to linearize the damper force with respect to the friction parameter. As shown in Fig. 6 the damper force can be broken down into a dissipation-free portion $F_{\mathrm{C,WD}}$ and a dissipativ portion $F_{\mathrm{D,WD}}$. These terms are described by

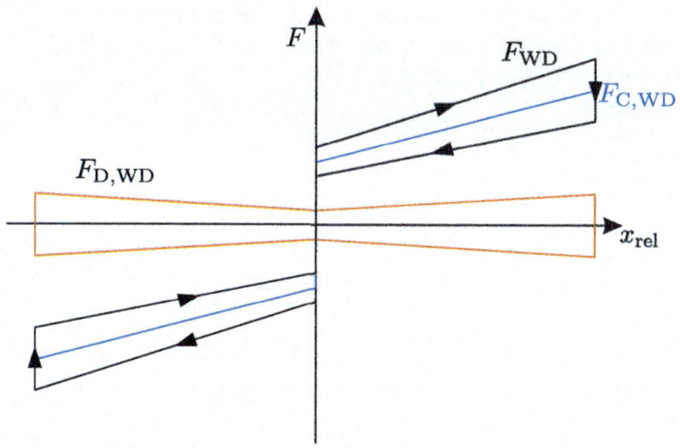

Fig. 6 Break down of the damper force of the prestressed sliding wedge damper with $x_{rel} = x_2 - x_1$

$$F_{WD} = F_{C,WD} + F_{D,WD}, \tag{24}$$

$$F_{C,WD} = 2c_2(2\tan\alpha(x_2 - x_1) + \Delta\ell \operatorname{sgn}(x_2 - x_1))\tan\alpha, \tag{25}$$

$$F_{D,WD} = 2c_2\,(2\tan\alpha|x_2 - x_1| + \Delta\ell)\,\frac{\mu}{\cos^2\alpha}\operatorname{sgn}(\dot{x}_2 - \dot{x}_1). \tag{26}$$

The nonlinearities in the wedge damper are much more pronounced than in the lock-up damper, because for this damper there are discontinuities in its force both as a function of the position and as a function of the velocity. In addition, the dissipative part of the damper force is proportional to the displacement. As a result, the dissipated energy is proportional to the square of the relative amplitude, similar to a viscous damper. Based on this finding, the authors of this work refer to this device as a pseudo-viscous damper.

3.1 Numerical Investigations

The first findings of this system are obtained by numerical simulations. As in Sect. 2, the parameters of the main system are exemplary set to one and the secondary mass is chosen as one tenth of the main mass. However, the prestresses sliding wedge damper offers a higher design flexibility, since the wedge angle α, the coefficient of friction μ, the preload length $\Delta\ell$ and the secondary stiffness c_2 can be considered as design parameters. The numerical studies of this work are limited to two design parameters α and $\Delta\ell$ and to the robustness of the system against the excitation amplitude F. Unless otherwise specified, the following standard parameters are used for the numerical studies

$$m_1 = 1\text{ kg},\ m_2 = 0.1\text{ kg},\ c_1 = 1\text{ N/m},\ c_2 = 0.1\text{ N/m},$$
$$F = F_0 = 0.01\text{ N},\ \alpha = 30°,\ \Delta\ell = 0.01\text{ m},\ \mu = 0.01.$$

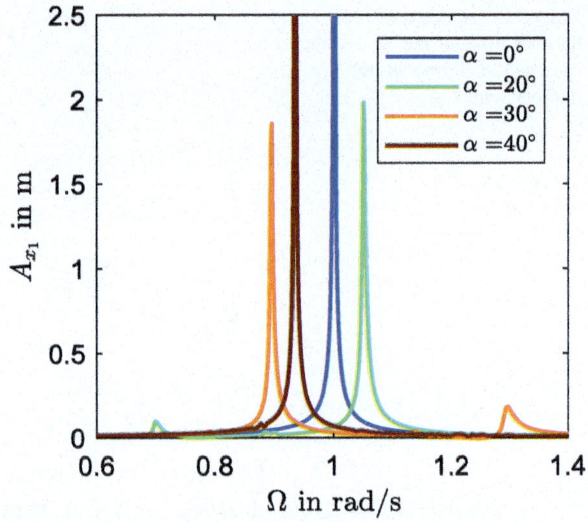

Fig. 7 Parameter study of the prestressed sliding wedge damper for different wedge angles α

The variation of the wedge angle α shows that this parameter has a significant influence on the effective stiffness $c_{2,\text{eff}}$ between the masses. Based on the non-dissipative part of the damper force, the effective stiffness can be derived as the coefficient of relative displacement, $c_{2,\text{eff}} = 4c_2 \tan^2 \alpha$, cf. Eq. 25. This influence is shown in Fig. 7. At small angles, there is a low effective stiffness and thus a peak at low frequencies and a peak near the resonance of the sticking system. Increasing the effective stiffness by the changing angle α causes both resonances to shift to the right, increasing the first peak and decreasing the second one. Additionally, the wedge angle has an influence on the dissipated energy. The higher the wedge angle, the higher the dissipative force at constant relative displacement, cf. Eq. 26.

Similar to the friction force of the lock-up damper, the prestress displacement determines the breakaway frequency at which the system changes to the nonlinear stick-slip range. The parameter study of the prestress displacement $\Delta\ell$ is shown in Fig. 8. A nonexistent prestress ($\Delta\ell = 0$) results in a two degrees of freedom system that does not stick. If the prestress displacement is increased, the sliding range is reduced and two sticking ranges appear at the edges of the amplitude response. The higher the displacement, the larger the sticking ranges become. For $\Delta\ell \to \infty$ the system always sticks. An optimum prestress displacement exists for this system as well. This optimum causes the best possible switching between the two extreme cases and thus achieves lower vibration amplitudes. Furthermore, non-periodic solutions can occur in the system. These can be both quasi-periodic and chaotic. For lowest possible vibration amplitudes over the whole frequency range, this type of solution should be avoided.

The last study of the friction absorber deals with the robustness of the system against a change of the excitation force amplitude. Figure 9 shows the normalized magnification factor of the system for different excitation forces. In contrast to the

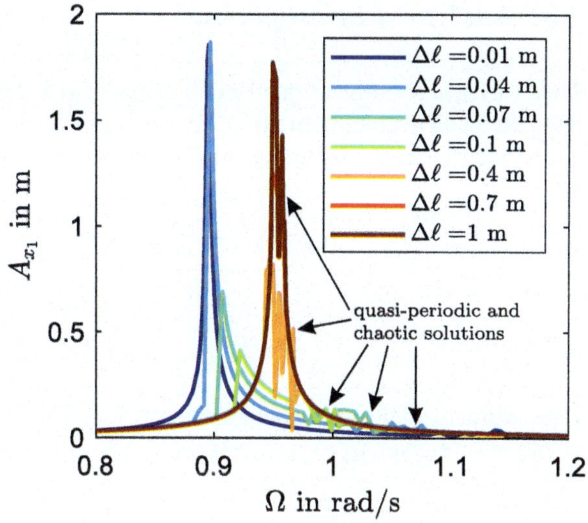

Fig. 8 Parameter study of the prestressed sliding wedge damper for different preload displacements $\Delta\ell$

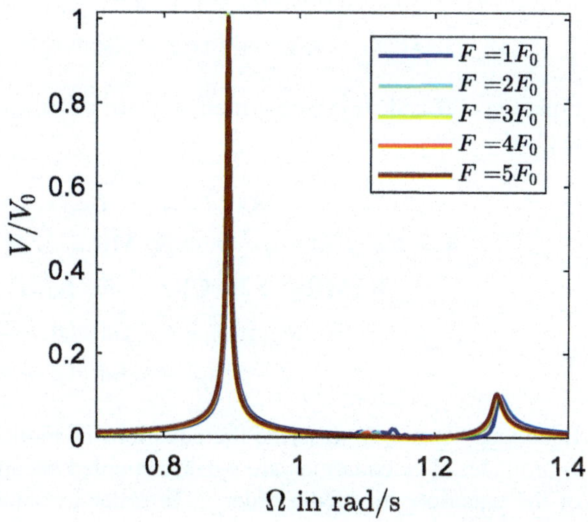

Fig. 9 Parameter study of the prestressed sliding wedge damper for different excitations forces F with $V = A_{x1,F}/F$, $V_0 = A_{x1,F_0}/F_0$

lock-up damper, the amplitudes remain limited. The curves are not directly on top of each other, but the differences are minimal. An increase of the excitation force causes almost no change of the magnification factor. Consequently, the vibration amplitudes of the system are amplified by approximately the same factor. This implies a scalability of the amplitude responses, although an exact scalability can only be observed in linear mechanical systems with viscous damping. These last correlations confirm the pseudo-viscous behavior of the friction damper.

3.2 Analytical Investigations

The analytical solution of the friction damper [4] is derived analogous to the lock-up damper via an averaging method. For this purpose the equations are nondimensionalized and the following dimensionless parameters are introduced

$$\lambda = \frac{m_2}{m_1},\ k^2 = \frac{c_1}{m_1},\ f = \frac{F}{m_1 k^2}, \tau = kt,\ (\cdot)' = \frac{\mathrm{d}}{\mathrm{d}\tau}(\cdot),\ \eta = \frac{\Omega}{k}, \tag{27}$$

$$a = \tan^2\alpha,\ b = \mu\frac{\tan\alpha}{\cos^2\alpha} c = \Delta\ell\tan\alpha,\ d = \frac{\Delta\ell\mu}{\cos^2\alpha}, \tag{28}$$

$$\varepsilon \ll 1, \qquad f,\ \Delta\ell,\ \mu = \mathcal{O}(\varepsilon). \tag{29}$$

If these dimensionless parameters are set in Eqs. 19–22, the dimensionless differential equations are given by

$$\begin{aligned} x_1'' + x_1 + 4\lambda a(x_1 - x_2) &= f\sin\eta\tau \\ &\quad -4\lambda b|x_1 - x_2|\mathrm{sgn}(x_1' - x_2') - 2\lambda c\,\mathrm{sgn}(x_1 - x_2), \end{aligned} \tag{30}$$

$$\lambda x_2'' - 4\lambda a(x_1 - x_2) = 4\lambda b|x_1 - x_2|\mathrm{sgn}(x_1' - x_2') + 2\lambda c\,\mathrm{sgn}(x_1 - x_2). \tag{31}$$

Equations 30 and 31 are rewritten in matrix form and a modal transformation is applied

$$\mathbf{M}\mathbf{x}'' + \mathbf{C}\mathbf{x} = \varepsilon\mathbf{f}_{NL}(\mathbf{x}, \Omega), \tag{32}$$

$$\mathbf{x} = \mathbf{R}\mathbf{z} = \mathbf{R}[p, q]^\top \text{ with } \mathbf{R}^\top\mathbf{M}\mathbf{R} = \mathbf{I},\ \mathbf{R}^\top\mathbf{C}\mathbf{R} = \mathrm{diag}(\eta_i^2), \tag{33}$$

$$\mathbf{R}^\top\mathbf{M}\mathbf{R}\mathbf{z}'' + \mathbf{R}^\top\mathbf{C}\mathbf{R}\mathbf{z} = \varepsilon\mathbf{R}^\top\mathbf{f}_{\mathrm{NL}}(\mathbf{z}) = \varepsilon\tilde{\mathbf{f}}_{\mathrm{NL}}(\mathbf{z}), \tag{34}$$

$$p'' + \eta_{01}^2 p = \varepsilon\tilde{f}_{\mathrm{NL},1}(p, q, \Omega), \tag{35}$$

$$q'' + \eta_{02}^2 q = \varepsilon\tilde{f}_{\mathrm{NL},2}(p, q, \Omega). \tag{36}$$

Equations 35 and 36 describe the equations of motion of the system in the modal coordinates. The equations are weakly coupled because the terms on the right side of the equation are of the order ε. In order to obtain decoupled equations, only the corresponding modal coordinate is considered exclusively near its resonant frequency. Therefore the remaining modal coordinate is neglected [4]. This leads to the decoupled equations of motion in the modal coordinates

$$\eta \approx \eta_{01}:\ p'' + \eta_{01}^2 p = \varepsilon\tilde{f}_{\mathrm{NL},1}(p, 0, \Omega), \tag{37}$$

$$\eta \approx \eta_{02}:\ q'' + \eta_{02}^2 q = \varepsilon\tilde{f}_{\mathrm{NL},2}(0, q, \Omega). \tag{38}$$

In order to derive an analytical solution, a Van der Pol transformation is applied to the system analogous to the lock-up damper and the slowly changing amplitude and phase differences of the system are investigated

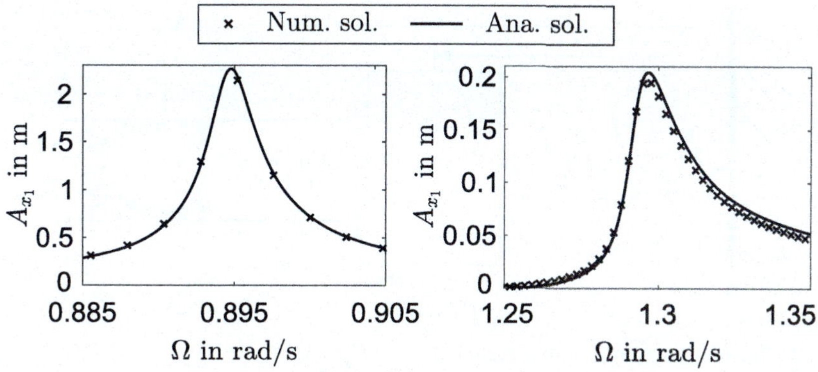

Fig. 10 Analytical solution of the prestressed sliding wedge damper for the standard parameters

$$p = A_1 \sin\varphi_1,\ \ p' = A_1\eta_{01}\cos\varphi_1,\ \ \varphi_1 = \eta\tau + \psi_1, \tag{39}$$

$$q = A_2 \sin\varphi_2,\ \ q' = A_2\eta_{02}\cos\varphi_2,\ \ \varphi_2 = \eta\tau + \psi_2, \tag{40}$$

$$i = \{1, 2\},\ \ \delta_i = \eta_{0i} - \eta, \tag{41}$$

$$A_i' = \varepsilon \tilde{f}_{\mathrm{NL,i}}(A_i, \psi_i, \varphi_i)\cos\varphi_i, \tag{42}$$

$$\psi_i' = \varepsilon\left(\delta_i - \frac{1}{A_i\eta_{0i}}\tilde{f}_{\mathrm{NL,i}}(A_i, \psi_i, \varphi_i)\cos\varphi_i)\right). \tag{43}$$

The stationary solution of the equations for A_i and ψ_i is determined by the averaging procedure and the results are shown in Fig. 10. It can be seen that the analytical solution approximates the numerical results with the accuracy of asymptotic methods. The deviations from the numerical solution are of the order ε, which confirms the validity of the analytic solution.

4 The Friction Damper with Polynomial Contact Geometries

The friction damper with polynomial contact geometries is quite similar to the sliding wedge damper analyzed in Sect. 3. The damper is attached to the main system and is composed of two main elements: a vibration absorber and two contact surfaces with a polynomial geometry. The secondary spring c_2 and secondary mass m_2 make up the absorber portion of the damper and allow it to reduce vibration in the vicinity of the tuned frequency. As with the sliding wedge damper, the contact surfaces are clamped on to the secondary mass via a third spring c_3 and the prestress displacement $\Delta\ell$. The geometry of the contact surfaces are described by the function $y(x_{\mathrm{rel}}) = y(x_2 - x_1) = \gamma\,|x_2 - x_1|^n$ with $n \in \mathbb{N}$. For the sake of simplicity only one polynomial term is introduced in the function y. Additionally, the coefficient μ describes the

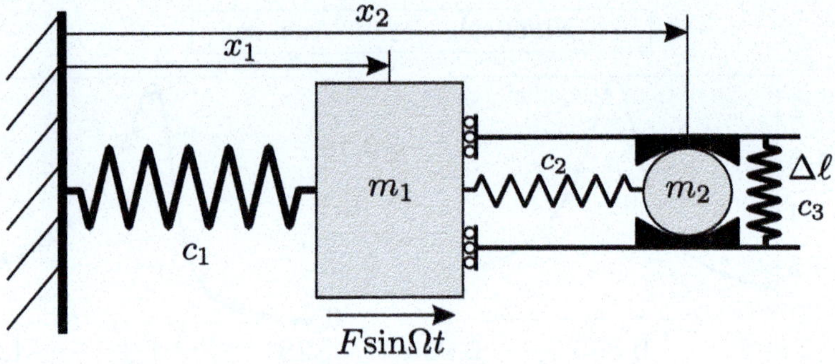

Fig. 11 The friction damper with polynomial contact geometry

relationship between the normal force and the friction force acting on the secondary mass (or on the contact surfaces) (Fig. 11).

As with the sliding wedge damper, the prestress level determines when the system finds itself in the linear sticking range or in the nonlinear stick-slip range. The equations of motion for both ranges are given by

while sticking

$$(m_1 + m_2)\ddot{x}_1 + c_1 x_1 = F \sin \Omega t, \tag{44}$$

$$H = \frac{m_2}{m_1 + m_2}(F \sin \Omega t - c_1 x_1), \tag{45}$$

while sliding

$$m_1 \ddot{x}_1 + c_1 x_1 - c_2(x_2 - x_1) - F_{PD} = F \sin \Omega t, \tag{46}$$

$$m_2 \ddot{x}_2 + c_2(x_2 - x_1) + F_{PD} = 0, \tag{47}$$

$$F_{PD} = 2c_3(2y + \Delta\ell)\frac{y_x + \mu \operatorname{sgn}(\dot{x}_2 - \dot{x}_1)}{1 - \mu \operatorname{sgn}(\dot{x}_2 - \dot{x}_1) y_x}, \tag{48}$$

$$y = \gamma |x_2 - x_1|^n, \quad y_x = n\gamma |x_2 - x_1|^{n-1} \operatorname{sgn}(x_2 - x_1). \tag{49}$$

Analogous to the sliding wedge damper, the force F_{PD} is linearized with respect to the friction coefficient. This linearization allows an insight into the damper force and a practicable separation into a dissipation free $F_{\mathrm{C,PD}}$ and dissipativ portion $F_{\mathrm{D,PD}}$. These quantities are given by

$$F_{\mathrm{PD}} \approx F_{\mathrm{PD,lin}} = F_{\mathrm{C,PD}} + F_{\mathrm{D,PD}}, \tag{50}$$

$$F_{\mathrm{C,PD}} = 4c_3\gamma^2 n(x_2 - x_1)^{2n-1} + 2c_3 \Delta\ell \gamma n |x_2 - x_1|^{n-1} \operatorname{sgn}(x_2 - x_1), \tag{51}$$

$$\begin{aligned} F_{\mathrm{D,PD}} = {} & 4c_3\mu \left(\gamma |x_2 - x_1|^n + \gamma^3 n^2 |x_2 - x_1|^{3n-2}\right) \operatorname{sgn}(\dot{x}_2 - \dot{x}_1) \\ & + 2c_3 \Delta\ell \mu \left(1 + \gamma^2 n^2 |x_2 - x_1|^{2n-2}\right) \operatorname{sgn}(\dot{x}_2 - \dot{x}_1). \end{aligned} \tag{52}$$

As noticed from Eqs. 51 and 52, the contact geometries introduce a dominant nonlinear stiffness of the degree $2n - 1$, as well as a dominant nonlinear damping of the degree $3n - 2$. These terms will mainly determine the behavior of the systems in the resonance regimes and the damping capability of the system.

Since the analytical procedure was validated for the sliding wedge damper, the investigations in this section are limited to analytical considerations of the friction damper with polynomial contact geometries. To this end, the equations of motion are nondimensionalized. In order to ensure a vibration absorption frequency, dominant linear terms are required for low amplitude vibrations. Therefore, a soft spring c_3 is chosen with $c_3 = \varepsilon\tilde{c}_3$ and $\varepsilon \ll 1$. The necessary transformations for the analytical considerations are given by

$$\frac{m_2}{m_1} = \lambda, \quad \frac{c_1}{m_1} = \omega_{01}^2, \quad \tau = \omega_{01}t, \quad \frac{\mathrm{d}(\ )}{\mathrm{d}t} = \omega_{01}\frac{\mathrm{d}(\ )}{\mathrm{d}\tau}, \quad \eta = \frac{\Omega}{\omega_{01}}, \quad p = \frac{c_2}{m_1\omega_{01}}, \tag{53}$$

$$a = \frac{4\tilde{c}_3}{m_1\omega_{01}^2}, \quad b = \frac{4\tilde{c}_3\mu}{m_1\omega_{01}^2}, \quad c = \frac{2\tilde{c}_3\Delta\ell}{m_1\omega_{01}^2}, \quad d = \frac{2\tilde{c}_3\Delta\ell\mu}{m_1\omega_{01}^2}, \quad \varepsilon f = \frac{F}{m_1\omega_{01}^2}. \tag{54}$$

Inserting these transformations in the equations of motion yields

$$x_1'' + x_1 - p^2(x_2 - x_1) = \varepsilon(f\sin\eta\tau + f_{\mathrm{PD}}) = \varepsilon f_1, \tag{55}$$

$$\lambda x_2'' + p^2(x_2 - x_1) = -\varepsilon f_{\mathrm{PD}} = \varepsilon f_2, \tag{56}$$

$$\begin{aligned} f_{\mathrm{PD}} = {} & a\gamma^2 n(x_2 - x_1)^{2n-1} + b\left(\gamma|x_2 - x_1|^n + \gamma^3 n^2|x_2 - x_1|^{3n-2}\right)\mathrm{sgn}(x_2' - x_1') \\ & + c\gamma n|x_2 - x_1|^{n-1}\mathrm{sgn}(x_2 - x_1) + d\left(1 + \gamma^2 n^2|x_2 - x_1|^{2n-2}\right)\mathrm{sgn}(x_2' - x_1'). \end{aligned} \tag{57}$$

Subsequently, a modal coordinate transformation is applied yielding first the weakly coupled differential equations in the modal coordinates. By only considering the corresponding modal coordinate in the respective resonance regime, the fully decoupled differential equations are obtained

$$\eta \approx \eta_{01}: \ p'' + \eta_{01}^2 p = \varepsilon(r_{11}f_1(p, 0) + r_{21}f_2(p, 0)), \tag{58}$$

$$\eta \approx \eta_{02}: \ q'' + \eta_{02}^2 q = \varepsilon(r_{12}f_1(0, q) + r_{22}f_2(0, q)). \tag{59}$$

For the sake of brevity and due to the length of the expressions further equations are omitted. The parameters are chosen as follows

$$m_1 = 1, \ m_2 = 0.1, \ c_1 = 1, \ c_2 = 0.1, \ c_3 = 0.01,$$
$$\Delta\ell = 0.1, \ \gamma = 1, \ \mu = 0.1, \ F = 0.01.$$

The first consideration focuses on the amplitude response of the polynomial damper for different polynomial degrees, c.f. Fig 12. As expected, multiple solution branches exist for $n > 1$ due to the nonlinear stiffness terms. Here the advantages and disadvantages of the damper are noted. Due to the nonlinear dissipative terms the maximal amplitude of the system is greatly reduced. For example a polynomial degree of $n = 4$ leads to a maximum amplitude reduction of 72, 9% in comparison to a polynomial contact surface with $n = 1$. This amplitude reduction comes at the price of multiple branch solutions, thus leading to higher possible amplitudes within the multiple

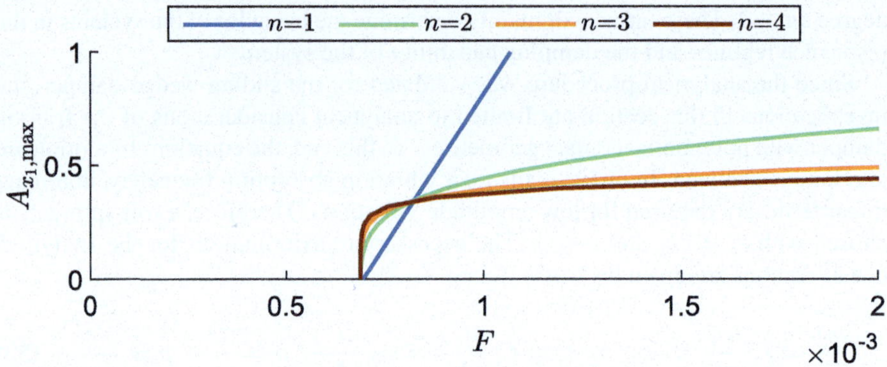

Fig. 12 Amplitude response of the friction damper with polynomial contact geometries for different polynomial degrees

solution range and amplitude jump when leaving said range. Especially the jumps in amplitude could prove detrimental to the function and life span of the main system [3]. Furthermore, the multiple solution range increases with the excitation leading to an amplitude rise over a wider range. However, if the excitation amplitude is known and a system overload can be excluded, the polynomial contact surface can be designed in order to avoid multiple solutions and effectively reduce amplitude vibrations.

An additional analytical consideration evaluates the relationship between the excitation force and the maximal amplitude in the system's response. Figure 13 shows that for low excitation forces the contact surface with lower polynomial degree result in lower maximal amplitudes. This is due to the relationship between the dissipated energy and the relative displacement, which is approximately described by $E_D \sim (A_{\text{rel}}/K)^{3n-1}$. As is seen from this relationship, low values of A_{rel} caused by

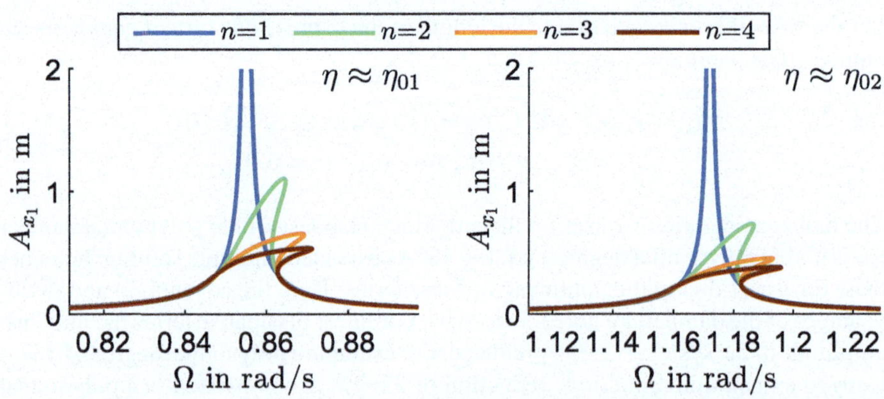

Fig. 13 Excitation vs maximal amplitude relationship for different polynomial degrees

low excitation forces lead to a significantly lower dissipated energy for $n > 1$. Furthermore, all curves cross the same point where the exponent has no influence on the dissipated energy, c.f. Fig 12.

5 Experiments

Based on the described analytic results, an experimental setup has been developed. The first experiments were performed together with the group of Professor Sattel at the TU Ilmenau, see Fig. 14. This setup allows for the investigation of different add-on damper systems at comparatively low frequencies (10–60 Hz) with large amplitudes (0.1–2 mm). The results confirm the theoretically predicted behavior, c.f. Figs. 2, 3 and 15. In Fig. 15a, the curves represent different preloads of the friction contact. The blue curves correspond to a zero breakaway force (no preload). The system behaves as a tuned mass damper, demonstrating two prominent peaks and the strong suppression of vibrations at the tuning frequency. The red curves correspond to very high preload, which ensures permanent sticking in the friction contact. The system's behavior in the last case corresponds to a one degree of freedom oscillator. The curves in between demonstrate that the lock-up damper is able to damp the peak around the first resonance. Furthermore, the systems sensitivity with respect to the excitation amplitude was also validated, c.f. Fig 15b. A comparison between the dry friction lock-up damper and a magneto-electro rheological damper is presented in [12].

In order to validate the numerical and analytical predictions of the performance of the wedge damper, a second setup was designed, see Fig. 16. To a certain extent, this setup uses the same parts as the setup for the lock-up damper and is also designed for low frequencies and large amplitudes. The experimental results confirm the numerical and analytical simulations. The prestress level on the damper determined how the

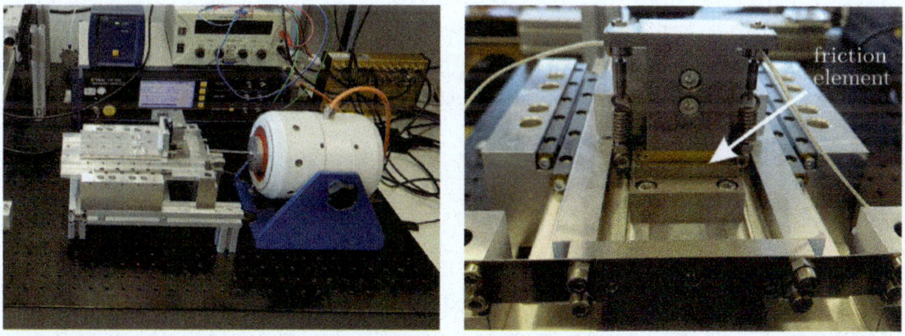

a) b)

Fig. 14 **a** Experimental setup for testing of the lock-up mass damper. **b** Detailed view on the lock-up element

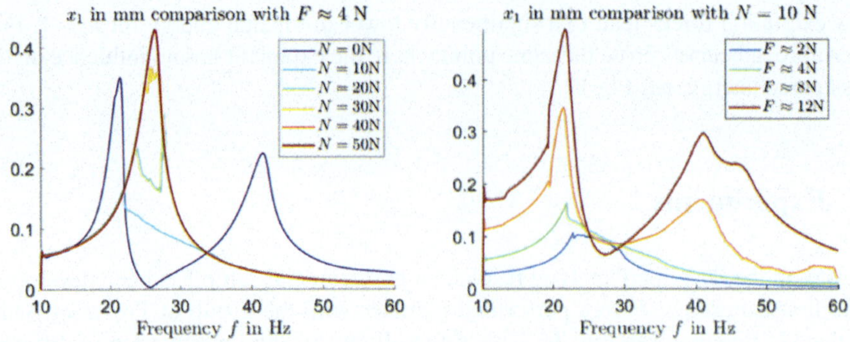

a) b)

Fig. 15 **a** Prestress variation results. **b** Excitation amplitude variation results

a) b)

Fig. 16 **a** Experimental setup for testing of the wedge damper. **b** Detailed view on the wedge element

system behaves, analogously to the breakaway force of the lock-up damper. If the prestress level is too high, the system behaves as a one degree of freedom system, whereas if it is set to zero, a system with two degrees of freedom is observed, c.f. Fig. 16a. Additionally, Fig. 16b shows the magnification factor V for different excitations amplitudes and proves the damper's robustness, i.e. its scalability (Fig. 17).

6 Conclusions

This work investigated three friction based dampers: the lock-up damper, the sliding wedge damper, and the friction damper with polynomial contact surfaces. It can be stated that the friction-based dampers demonstrate their ability to diminish forced vibrations tightly focused on the resonance ranges. Furthermore, the contact geometry significantly determines the characteristics of the damper. The planar contact

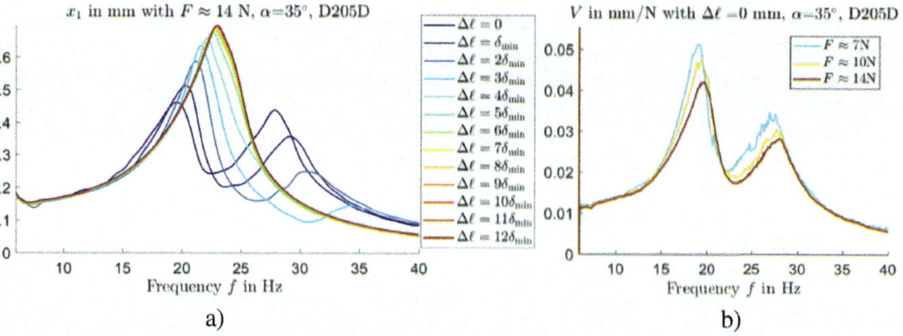

Fig. 17 **a** Prestress variation results $6\delta_{\min} = 0.7$ mm. **b** Magnification factor results for different excitation amplitudes

geometry of the dry lock-up damper has a limited robustness and therefore a reduced applicability. An increase in robustness is observed in the prestressed wedge damper, which has a linear varying contact geometry. In order to avoid unnecessary energy losses, the wedge damper can be prestressed and forced into the stick-phase as long as the vibrations remain sufficiently small. These passive dampers are either optimal in the passage through resonance or at a certain tuned frequency, but not in both. The friction based damper with polynomial contact surfaces addresses this limited applicability to a certain extent. Namely, when the range of the excitation force amplitude is known a priori.

The developed analytical approaches enable to predict the behavior of such systems and to make the reasonable parameter choice for the design of real devices. These analytical results are verified via experiments for the lock-up damper and the sliding wedge damper. Comparing the analytical results, it is noted that the developed approach based on the averaging technique enables accurate prediction of the dynamics of such devices.

Acknowledgements The authors would like to express their gratitude towards the German Research Foundation, Deutsche Forschungsgemeinschaft (DFG) for the financial support within the priority program SPP 1897, 'Calm, Smooth, and Smart - Novel Approaches for Influencing Vibrations by Means of Deliberately Introduced Dissipation'.

References

1. Den Hartog, J.P.: Mechanical Vibrations. Dover Publications, New York (1985)
2. Dixon, J.C.: The Shock Absorber Handbook. Wiley, New York (2008)
3. Dresig, H., Fidlin, A.: Schwingungen mechanischer Antriebssysteme. Springer, Berlin (2006)
4. Fidlin, A., Gafur, N.: On the dynamics of friction based tuned mass dampers. In: Proceedings of ENOC the 9th EUROMECH Nonlinear Dynamics Conference (2017)
5. Fidlin, A., Lobos, M.: On the limiting of vibration amplitudes by a sequential friction-spring element. J. Sound Vib. **333**(23), 5970–5979 (2014)

6. Fidlin, A., Aramendiz, J.: Study on the dynamics of a lockup mass damper: asymptotic analysis and application limits. Nonlinear Dyn. **97**, 1867–1875 (2018)
7. Fleischer, G., Wachter, K.: Konstruktionslehre für Maschineningenieure. Verl. Technik, Berlin (1987)
8. Hausner, M., Hässler, M.: Kupplungsscheibe mit frequenztilger gegen rupfschwingungen. ATZ-Automobiltechnische Zeitschrift **114**(1), 64–69 (2012)
9. Khoo, H.H., Clifton, C., Butterworth, J., MacRae, G., Gledhill, S., Sidwell, G.: Development of the self-centering Sliding Hinge Joint with friction ring springs. J. Constr. Steel Res. **78**, 201–211 (2012)
10. Panning, L.; Sextro W.; Popp K.: Optimization of interblade friction damper design. In: Proceedings of the ASME Turbo Expo 2000: Power for Land, Sea, and Air (2000)
11. Sun, J., Jolly, M.R., Norris, M.: Passive, adaptive and active tuned vibration absorbers-a survey. J. Mech. Design **117**(B), 234–242 (1995)
12. Tan, A.S., Aramendiz, J., Ross, K.H., Sattel, T., Fidlin, A.: Comparative study between dry friction and electrorheological fluid switches for tuned vibration absorbers. J. Sound Vibrat. 114874 (2019)
13. Weber, F., Høgsberg, J., Krenk, S.: Optimal tuning of amplitude proportional Coulomb friction damper for maximum cable damping. J. Struct. Eng. **136**(2), 123–134 (2002)
14. Wu, Q., Cole, C., Spiryagin, M., Sun, Y.Q.: A review of dynamics modelling of friction wedge suspensions. Veh. Syst. Dyn. **52**(11), 1389–1415 (2014)

Simulation-Based Design of Hybrid Particle Dampers with Application to Flexible Multibody Systems

Niklas Meyer and Robert Seifried

1 Introduction

Passive damping techniques are often used to reduce structural vibrations. Classical liquid dampers are mostly seen for these applications. These dampers are well studied and mathematically easy to describe. However, liquid dampers perform insufficiently under extreme temperatures, due to the change of viscous properties, and do need an anchor point. Hence, for applications where liquid dampers are not suitable, particle dampers are becoming more and more popular.

Particle damping (PD) technology is classified as an auxiliary-mass type vibration technique [1]. Either containers attached to the vibrating structure or holes within the vibrating structure are filled with a granular material of convex or non-convex shape. Various different materials like steel, tungsten, carbide, polymers and many more can be used. The particle size normally ranges from the micrometer scale to the millimeter scale. Thus, from several dozens up to millions of particles might be included in a single particle damper. By structural vibrations, momentum is transferred to the granular material causing granular movement. By particle interactions, energy dissipation occurs due to inelastic normal collisions and frictional losses. This reduces the structural vibrations.

Compared to other damping techniques, particle dampers show various advantages due to their conceptual simplicity and passive nature. They are cost-efficient devices, do not need an anchor point, and do not often degrade in time [2]. Using appropriate particles, e. g. steel or tungsten, particle dampers can even operate in extreme environmental conditions [1, 3, 4]. Furthermore, particle dampers add only

N. Meyer (✉) · R. Seifried
Institute of Mechanics and Ocean Engineering, Hamburg University of Technology, Eißendorfer Straße 42, 21073 Hamburg, Germany
e-mail: n.meyer@tuhh.de

R. Seifried
e-mail: robert.seifried@tuhh.de

P. Eberhard (ed.), *Calm, Smooth and Smart*, Lecture Notes in Applied and Computational Mechanics 102, https://doi.org/10.1007/978-3-031-36143-2_3

little mass to the primary system [2] causing no significant change in its mass and stiffness [5]. It is also reported that particle dampers can be applied to a wide frequency range [6].

Although particle dampers show huge potential, they are so far only rarely used in technical applications. The major reason for this is that currently there exists no easy design guideline, which is due to their nonlinear behavior and the variety of influence parameters. These influence parameters strongly affect the particle motion, also called state of matter or *motion mode* (of the rheology behavior), which correlates in a non-trivial way with the damper's energy dissipation. These correlations are often only poorly understood, leading to a trial and error-based design process of particle dampers even nowadays. In [7], a systematic multiscale design methodology in form of a *toolchain* is developed, which is based on computational models as well as models derived from experiments. These models are extremely useful for supporting the damper design and damper integration on a structure. In addition, they also provide useful insights into the complex processes, the nonlinear effects, and the design parameters influencing the efficiency of particle dampers.

Within this paper, the toolchain is used to develop single particle damper units with predefined characteristics, which do not rely on a specific application. Different hybrid damper approaches are utilized to increase the efficiency of the individual particle dampers. Multiple container layers with different filling ratios are used to obtain an optimized damping behavior for transient vibrations. Inner structures within the dampers show a more robust behavior and buffering the damper's walls with a polymer leads not only to a considerable noise reduction, but leads also to an increased efficient range of operation.

This paper summaries the research on the DFG projects SE1685/5-1 and 424825162 within the SPP1897. This paper is based on the publications [7–16] which originated from the two project phases and is organized in the following way: In Sect. 2 the systematic multiscale design methodology in form of a toolchain is presented. In the following sections the design toolchain is applied to design particle dampers for horizontal free and forced vibration systems of low intensity, see Sect. 3 and horizontal and vertical forced vibration systems of high intensity, see Sect. 4. Finally, the conclusion is given in Sect. 5.

2 Design Toolchain

Particle dampers show a nonlinear dynamical behavior, starting at the micromechanical effects during single particle impacts and sliding contacts, continuing with the energy dissipation inside the particle container, and ending at the interaction within a structure. For a better understanding, and thus also for the design of particle dampers, investigations on these different scales or levels, respectively, are necessary. A systematic multiscale design methodology in form of a toolchain is therefore developed, see Fig. 1 and [7]. Insights made on one level can be used on the next level to better understand the dynamical properties and shorten the

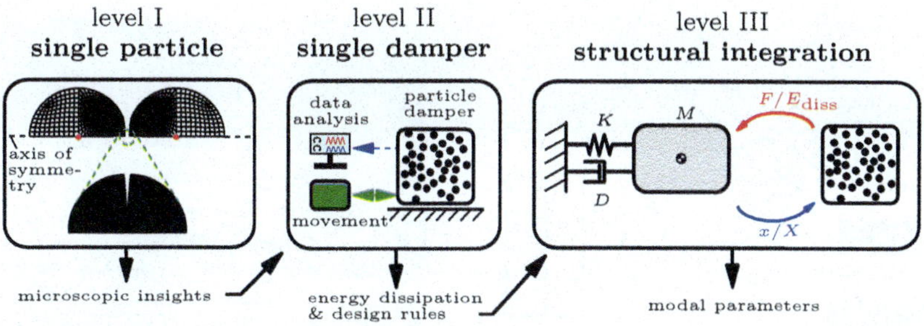

Fig. 1 Toolchain for the analysis of particle dampers

overall design process of particle dampers. On all levels, numerical models, as well as experimental tests, might be used for analysis.

On the **first level** of the toolchain, the micro-mechanical behavior of single particle–particle and particle–wall interactions is analyzed. This provides important input for the second level. The **second level** represents investigations into an isolated particle damper subjected to a harmonic motion and the determination of the damper's energy dissipation for a given excitation frequency range and amplitude range. First design rules and, if applicable, analytical formulas for the energy dissipation are derived here. Finally, the **third level** represents the integration of the particle damper in a vibrating structure to evaluate the overall damping effect.

2.1 Level I

The **first level** of the toolchain, see Fig. 1-single particle, is completely independent of the other two levels. On this level, the micro-mechanical behavior during a normal impact of two bodies of macroscopic size, like particle–particle or particle–wall, is analyzed. Here, a particle collides in a defined manner with the collision partner. Thereby, the energy dissipation during the impacts is of major interest and can mainly be characterized by the *coefficient of restitution* (COR) ε describing the velocity change during contact [17, 18]. The COR designates the quotient between the velocities right after (1) and before (0) the impact for the bodies I and II as $\varepsilon = -\left(v_{\mathrm{I}}^{1} - v_{\mathrm{II}}^{1}\right) / \left(v_{\mathrm{I}}^{0} - v_{\mathrm{II}}^{0}\right)$. For $\varepsilon = 1$ the impact is called elastic. For $\varepsilon = 0$ it is called fully inelastic and both collision partners move with the same velocity afterward.

In *discrete element method* (DEM) simulations, which are used on **level II**, often a constant COR is applied. However, the impact velocity of the collision partners has a big influence on the COR and should thus be considered. Numerical and experimental models can be used to determine the COR [19–21]. For example, Fig. 2 shows the experimental setup for the impact of a steel sphere against different planar wall materials, as detailed discussed in [14, 16]. The testbed consists of a steel sphere

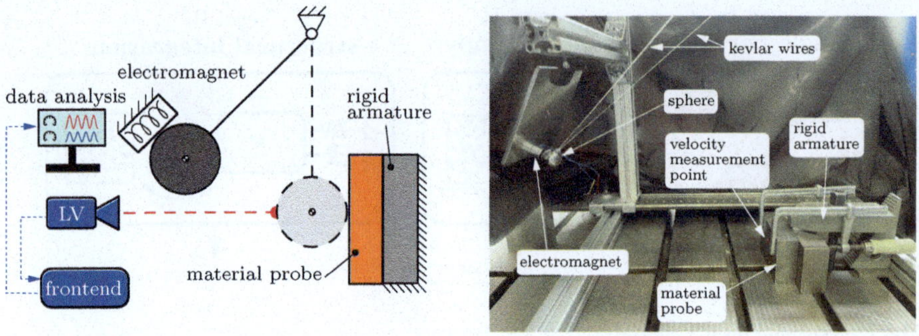

(a) Schematic representation of setup. (b) Picture of test bed.

Fig. 2 Test bed to determine the COR for a sphere–wall contact

of 15 mm radius which is suspended by thin wires impacting the wall material probe glued or fixed to a rigid steel block.

The sphere is held in the deflected state by an electromagnet. As the position of the electromagnet is variable, different impact velocities are achieved. After release, the velocity of the sphere is measured by the laser vibrometer (LV) PSV-500 from POLYTEC with a sampling frequency of 250 kHz.

Alternatively, numerical investigations using the *finite element method* (FEM) can be performed for determining the COR [20, 22]. A schematic representation of the sphere–sphere FEM model is shown in Fig. 1–**level I**. Here, for example, the spheres have an initial radius of 5 mm, which can be scaled to different sizes. Each sphere consists of 6093 axis symmetric 2D elements, called CAX4R in ABAQUS. The element size varies between 0.015 mm to 0.5 mm and both spheres are assigned with half the impact velocity with opposed signs [14, 16].

The analyzed steel–steel and steel–aluminum contacts show a high dependency on impact velocity [14, 16]. Exemplary, in Fig. 3a the experimental and numerical COR results for steel sphere–steel wall impacts are shown. These start at high COR values for low impact velocities and show a digressive decay towards higher impact velocities. Due to plastic deformations, repeated impacts onto the same spot show a much higher COR [22]. The FEM simulations are capable to reproduce the quantitative progression of the COR observed in experiments. Only small quantitative differences remain.

Besides the metal–metal impacts, steel-polymers combinations are investigated. Exemplary, in Fig. 3b the experimental and numerical COR results for steel sphere–polymer wall impacts are shown. The steel–polymer contacts show only little dependency on impact velocity. For $v_{\mathrm{I}}^0 > 0.1\,\frac{m}{s}$, only a little linear decrease of COR with impact velocity is observed. The effect of repeated impacts vanishes [16], as no plastic deformations in the contact zone occur. Numerically, a good agreement with the experimental measurements is achieved for low impact velocities, i. e. $v_{\mathrm{I}}^0 < 1\,\frac{m}{s}$. At high impact velocities, bigger differences are seen.

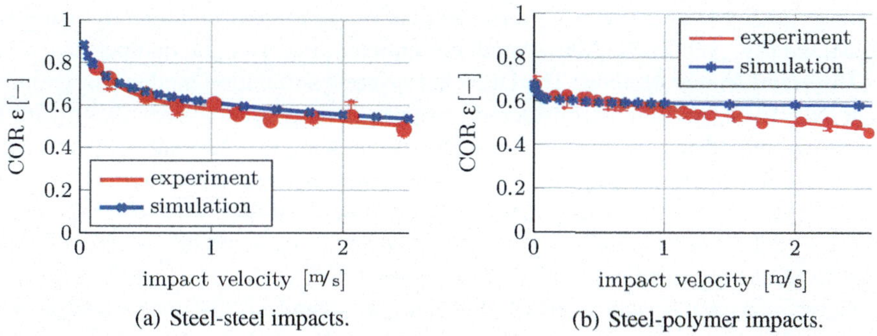

Fig. 3 Experimental and numerical COR results for sphere-wall impacts of different materials

Using the validated numerical models, the CORs for various sphere diameters are calculated, which can be used for later DEM simulations. Indeed, it turns out that the dependency on impact velocity is much bigger than on the sphere's diameter. See [16] for further details.

2.2 *Level II*

The **second level** of the toolchain, see Fig. 1-single damper, represents investigations of an isolated particle damper subjected to a defined horizontal or vertical vibration. The energy dissipation of the particle damper is determined for the given excitation frequency range and excitation amplitude range. Additional, insights about the movement of the particle bed, called *motion mode*, are gained. The particle container is excited by a harmonic motion using a rheonomic constraint $x_c = X \sin(\Omega t)$, with container amplitude X and angular frequency $\Omega = 2\pi f$. The corresponding container velocity and acceleration follow as $\dot{x}_c = V \cos(\Omega t)$ and $\ddot{x}_c = -A \sin(\Omega t)$ with $V = X\,\Omega$ and $A = X\,\Omega^2$. The dimensionless excitation intensity is defined as $\Gamma = A/g$ with g as gravity constant. From the velocity of the particle container and the excitation force, the *complex power* can be determined. Using the complex power, the energy dissipation and the reduced loss factor are obtained, displayed as characteristic diagrams over the excitation frequency and excitation amplitude. These characteristic diagrams are also called *effective fields* in the following. The calculated effective fields can be stored and used on the third level of the toolchain for the integration process of the particle dampers in a vibrating structure. If applicable, the effective fields might be approximated by an analytical formula, which can also be integrated on the next level. All these aspects help to shorten the design phase on **level III**.

Investigations can be performed experimentally using a closed-loop controlled shaker setup or a linear drive. This depends on the excitation frequency and amplitude, see e. g. [12, 14, 15]. Investigations can also be performed numerically using the DEM, see e. g. [7]. Such a DEM model is shown in Fig. 1-single damper.

Discrete Element Method: The DEM is a discrete simulation method for granular materials. Every particle is considered as an unconstrained moving body only influenced by applied forces, e. g. the particle–particle and particle–wall contact forces. The dynamics are described by *Newton's and Euler's* equation of motion for every particle [23]. This results in general in a coupled nonlinear differential equation with $6\,n_\mathrm{p}$ degrees of freedom for 3D simulations with n_p being the number of particles. Here, the algorithms presented in [11] are used.

Complex power: To analyze the energy dissipation and the efficiency of the particle damper, the complex power method, introduced by Yang [24], is used. The complex power is determined to $P = 0.5\, F^{*\top}\, \bar{V}^*$. Hereby, F^* denotes the complex amplitude calculated by the *fast Fourier transform* (FFT) of the driving force signal acting on the container and $\bar{V}^*$ is the conjugate complex amplitude by FFT of the velocity signal of the container motion. The dissipated energy per cycle $\widetilde{E}_\mathrm{diss}$ follows from the complex power to $\widetilde{E}_\mathrm{diss} = 2\,\pi\, E_\mathrm{diss} = 2\,\pi\, \mathrm{Real}(P)/\Omega$. To judge about the damper's efficiency the reduced loss factor η^* [14] is utilized. It is calculated by a scaling of the dissipated energy with the kinetic energy of the particle system $E_\mathrm{kin} = 0.5\, m_\mathrm{bed}\, |V^*_\Omega|^2$ using the mass of the particle bed m_bed, i. e. the mass of all particles, to $\eta^* = E_\mathrm{diss}/E_\mathrm{kin}$ with V^*_Ω being the complex amplitude of the velocity signal at driving frequency. As consequence, the reduced loss factor is independent of the container and particle mass and enables the comparison of different particle settings.

2.3 Level III

The **third level** of the toolchain, see Fig. 1-structural integration, represents the integration of particle dampers in a vibrating structure to evaluate their overall damping effect. This can be done experimentally or numerically. For simplicity on this level, only 2D motion is considered in this project, i. e. either horizontal or vertical vibrations. For the numerical investigations, three different approaches have been developed. In the first approach, the complete DEM model describing the particle damper used on **level II** is coupled with the dynamical model of the structure. The flexible structure might be described by a reduced linear finite element model or a flexible multibody system. While this approach is very accurate and useful for verification purposes, it is computationally very expensive. Alternatively, the flexible structure model can be coupled with the effective fields or the analytical formulas previous determined in **level II**. This coupling is easily implemented and leads to accurate results and short computation times, see [7, 12] for detailed information about the coupling process. If an analytical formula is present, it even might be used for a damper optimization.

In the following, the toolchain is applied to design different particle damper systems for various application fields and their results are presented.

3 Horizontal Vibration Systems of Low Vibration Intensity

Low vibration intensity is defined here as a vibration with particle damper acceleration amplitudes below the gravitational acceleration $A < g$ and frequencies/ eigenfrequencies of $f_0 < 5\,Hz$. Often, only small energy dissipation rates are obtained for such systems so far, due to sticking of particles. Hence, a new and more efficient design of particle dampers is necessary for these applications, whereby the focus is on horizontal vibrations. The proposed design makes use of the rolling property of spheres inside particle containers with flat bases, see [15]. The presentation of this chapter is based on [7, 13, 15].

3.1 *Level II—Considerations*

The experimental setup to analyze the particle damper's motion modes and energy dissipation under horizontal forced vibration of low intensity is shown in Fig. 4-*top*. For this purpose, a linear drive is used.

The cuboid container is filled with 36 spherical steel particles of 5 mm radius with clearance h_{roll} to the other container side. The container is mounted via a force transducer on a linear drive. Thus, the excitation force acting on the particle container is measured. During the conducted experiments, two different motion modes of the particle bed are observed. Trajectories of these motion modes, obtained from DEM simulations, are shown in Fig. 5. The reduced loss factor, i. e. the efficiency factor of the particle bed, is shown in Fig. 4-*bottom*. For low excitation amplitudes $X < X_{\text{rol}}^{\text{opt}}$, the system is in the so-called *scattering* motion mode. No regular or synchronous motion of the particles is observed, see also Fig. 5a. Hence, only a little amount of energy dissipates, resulting in a low reduced loss factor. When the container amplitude reaches a certain threshold amplitude $X_{\text{rol}}^{\text{opt}}$ the system turns suddenly into the *rolling collect-and-collide* motion mode, i. e. for $X > X_{\text{rol}}^{\text{opt}}$. Here, the particle bed stays together as one particle block and slides and rolls over the container base. The collisions with the container walls are inelastic, i. e. after impact, the particle bed adopts the container's velocity and does not rebound from it. This happens due to multiple inter-particle collisions during impact, see [25, 26] for further details. Hence, a synchronous particle motion with the container is achieved, see Fig. 5b. This leads to a high energy dissipation at $X = X_{\text{rol}}^{\text{opt}}$ with a slight decrease to higher excitations amplitudes. Equipping the container with buffered walls, making the particle damper hybrid, does not influence its energy dissipation but leads to a considerable noise reduction for this excitation regime.

The numerical DEM results for the reduced loss factor are also pictured in Fig. 4. For the scattered motion mode, i. e. $X < X_{\text{rol}}^{\text{opt}}$, the results are on the same scale as the experimental results. However, neither a qualitative nor quantitative agreement of the observed curves for this area is achieved. For the rolling collect-and-collide

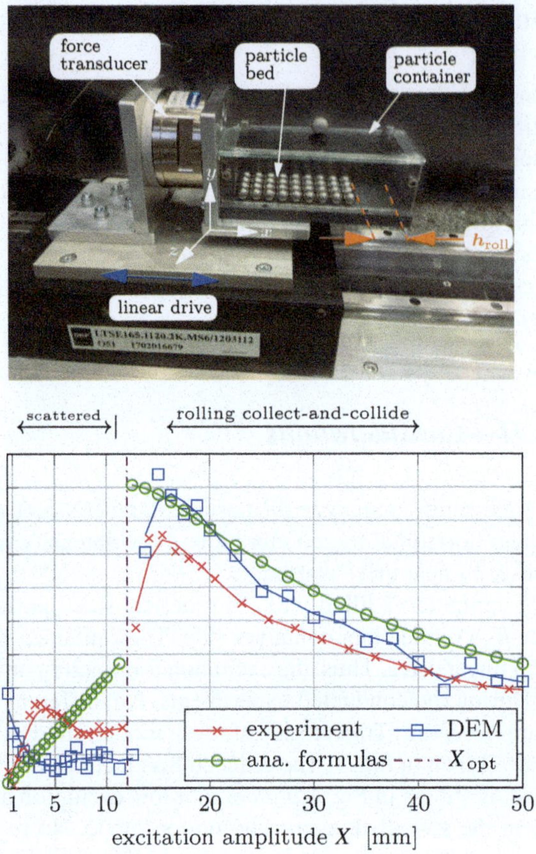

Fig. 4 Horizontal vibrations analysis of low vibration intensities with experimental setup (*top*) and corresponding reduced loss factor for an excitation frequency of $f = 2\,\mathrm{Hz}$ (*bottom*)

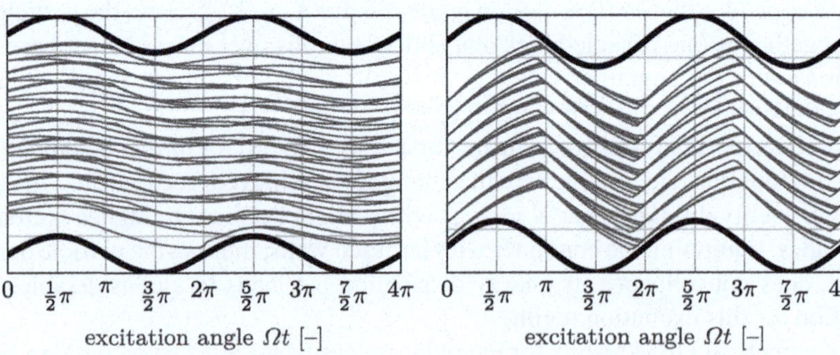

(a) Scattered state $X < X_{\mathrm{rol}}^{\mathrm{opt}}$.

(b) Rolling collect-and-collide $X > X_{\mathrm{rol}}^{\mathrm{opt}}$.

Fig. 5 Particles trajectories obtained from DEM simulations of particle setting shown in Fig. 4 for different container strokes

motion mode, i.e. $X > X_{\mathrm{rol}}^{\mathrm{opt}}$, a good qualitative agreement with the experiments is obtained.

For the scattered state, an empirical formula describing the energy dissipation is found, yielding a rough approximation. For the rolling collect-and-collide regime, instead, an deterministic equation for the energy dissipation is derived. For this analytical equation, the curve progression of the reduced loss factor agrees well with the experiments, see Fig. 4. However, the obtained reduced loss factor values are above the experimentally measured ones for all excitation amplitudes. Additionally, a simple expression for the optimal stroke is achieved to $X_{\mathrm{rol}}^{\mathrm{opt}} \approx 0.4\, h_{\mathrm{roll}}$, see also Fig. 4. This formula is in great agreement with experimental results [15] and enables a quick and reliable damper design for a structural integration.

Additionally, intensive sensitivity analyses are performed experimentally and numerically in [15]. Most of the particle properties, like Young's modulus, density, coefficient of restitution, or particle number have a negligible influence on the damper's efficiency. However, it turns out that a low friction coefficient and a high particle radius are beneficial. Also, a tilt around the damper's axis are studied. Hereby, a little tilt around the dampers yaw axis is showing only little influence on the damper's efficiency. Indeed, a tilt around its longitudinal or pitch axis might significantly decrease the efficiency of the rolling collect-and-collide motion mode. Finally, the container shape is analyzed. The cuboid shape is replaced by a cylindrical shape heading against gravity. While the efficiency of the damper is only a little reduced, this cylindrical shape is showing the great advantage of applying to vibrations in the whole horizontal plane. Thus, this new efficient damper design for low acceleration vibrations opens a completely new area of applications for particle dampers in mechanical and civil engineering, like the damping of high-rise buildings or wind power plants [15].

3.2 Level III—Considerations

To show the applicability of the derived damper design, it is applied to free and forced vibrations in the following.

3.2.1 Free Vibrations

The system to be damped for these vibration intensities is a parallel lightweight manipulator with highly elastic links [7], see Fig. 6-*top*. The lightweight manipulator consists of two linear motors set up in a "T–configuration". Elastic links, made of spring steel, are mounted via revolute joints on both sliders and are connected via a third revolute joint forming a parallel robot. At the end of link II, the end-effector is mounted, which consists of the hybrid particle damper of multiple layers. The elastic deformation in link II is introduced via the rigid body motion, i.e. when the linear drives are moving, and is dominated by the first (bending) eigenmode.

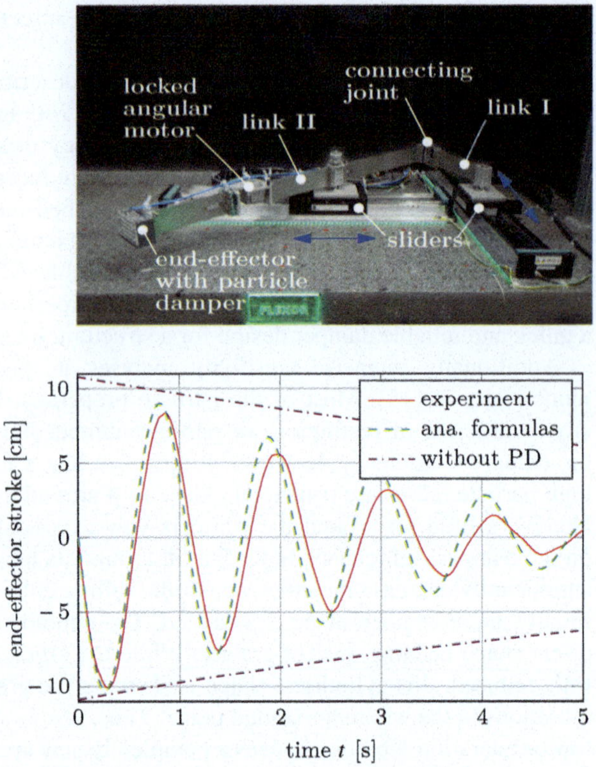

Fig. 6 Parallel lightweight manipulator used for harmonic vibration analyses with overview of the system (*top*) and comparison of the end-effector movement for the optimized particle damper (*bottom*)

For this system, all three particle damper models, i. e. full DEM model, effective fields, and analytical formulas, are coupled to the modal reduced model of the system. All coupled models are showing a good agreement to experimental measurements [7]. Finally, the analytical formulas are used to optimize the filling ratio of the different particle damper layers for an initial deflection of link II. The effectiveness of the optimized design is demonstrated experimentally as shown in Fig. 6-*bottom*. Only a small difference between experiment and analytical formulas is found and the system is greatly damped compared to the undamped case.

3.2.2 Forced Vibrations

For the forced vibrations, a simple beam-like structure is used as an application example, see Fig. 7-*top*. Its base point is subjected to a harmonic motion of variable frequency using a linear drive. The particle damper is mounted at the tip of the beam and its velocity is measured using a laser scanning vibrometer. Thus, the *frequency response function* (FRF) $|H^*(f)| = |X_c^*(f)/U^*(f)|$, with complex particle damper amplitude X_c^* and complex excitation amplitude U^* is obtained

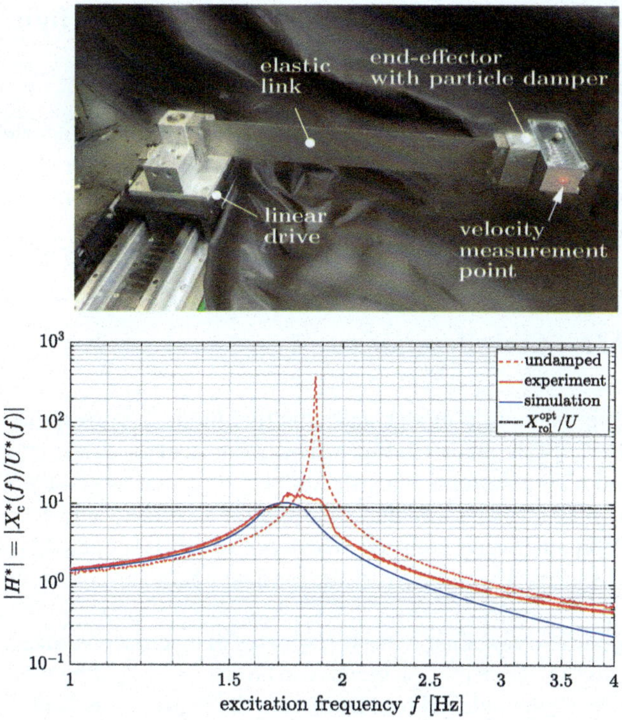

Fig. 7 Simple beam-like structure setup for forced vibration analyses with overview of system (*top*) and FRF's of optimized particle damper for an excitation of $U = 1.33$ mm and 36 steel particles of 5 mm radius (*bottom*)

experimentally. Numerically, the equations describing the energy dissipation of the particle damper, which are shown in the normalized form in Fig. 4-*bottom*, are coupled to a modal reduced model of the structure. A good agreement between analytical and experimental obtained frequency response function is achieved for various excitation intensities, validating the presented approach, see [13] for a detailed discussion.

Finally, the coupled model is used to calculate the design parameter of the particle damper to operate it at its maximum efficiency, i. e. at $X_{\text{rol}}^{\text{opt}}$ see Fig. 4-*bottom*. A simple analytical expression is obtained. Its accuracy is proven by comparison to an experiment as shown in Fig. 7-*bottom*. Simulation and experiment are only slightly crossing the optimal normalized amplitude of $X_{\text{rol}}^{\text{opt}}/U$, with U being the excitation amplitude. However, due to uncertainties within the experimental setup, some differences remain in the obtained FRF's. Still, the derived formula provides a powerful tool to design particle damper for applications of low acceleration intensity, see the discussion in [7].

4 Horizontal and Vertical Vibration Systems of High Vibration Intensity

High intensity vibrations are defined here as vibrations with an acceleration amplitude $A \gg g$ and (eigen)frequencies $f_0 \gg 10\,\text{Hz}$. For such vibrations, the particle dynamics completely change compared to the previous two sections, but the influence of the gravitational acceleration becomes less.

4.1 Level II—Considerations

For the analysis of motion modes and effective fields, a corresponding testbed is developed [9–11, 14] as shown in Fig. 8. The cylindrical particle container is excited by a controlled harmonic force via a shaker perpendicular to gravity. The excitation force is controlled in such a way that the vibration frequency and acceleration magnitude of the container stays constant. The force sensor is mounted between particle container and shaker. The velocity of the particle container is measured via a laser vibrometer.

Filling the particle container with steel spheres of macroscopic size, i. e. from 0.3–5 mm radius, five different motion modes can be observed, as shown as velocity fields in Fig. 9. In Fig. 10 the reduced loss factor and the distribution of motion modes for such a setting are plotted. The solid-like state, see Fig. 9a, is characterized by almost no relative motion between particles and container. This causes the granular matter to look like an added block, staying at the container base and moving with the same velocity as the container. Hence, only little dissipation rates are obtained, see Fig. 10. For the local fluidization, see Fig. 9b, particles located at the top surface become fluidized. Here small to medium dissipation rates are achieved, see Fig. 10. When the whole particle system gets fluidized, the global-fluidization is obtained, as

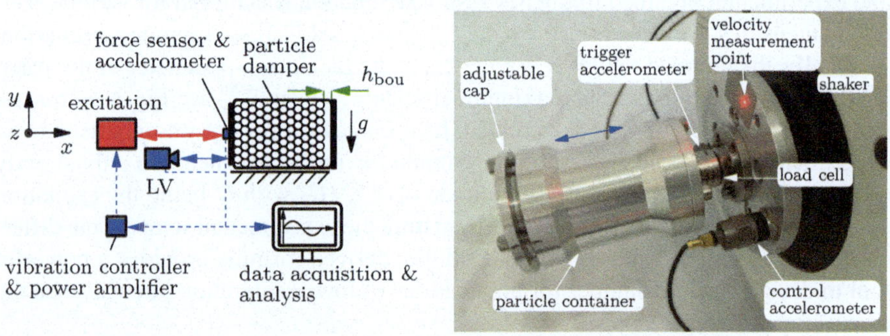

Fig. 8 Testbed for determination of effective fields of particle damper for high vibration intensities with schematic representation (*left*) and picture (*right*)

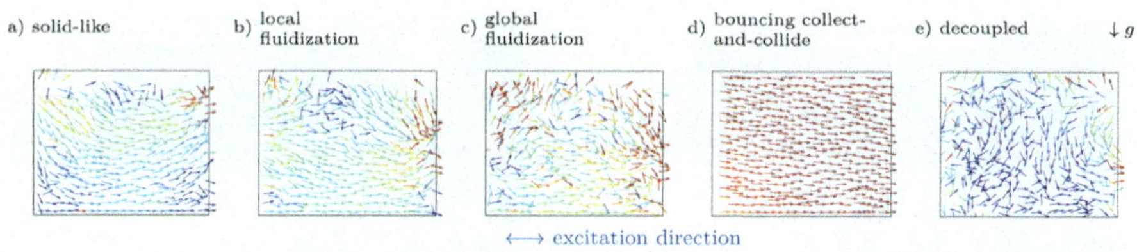

Fig. 9 Velocity fields of motion modes for high excitation intensities for particle damper shown in Fig. 8. The colors show the magnitude of the in-plane particle velocity normed by the container velocity V from low (blue) to high (red)

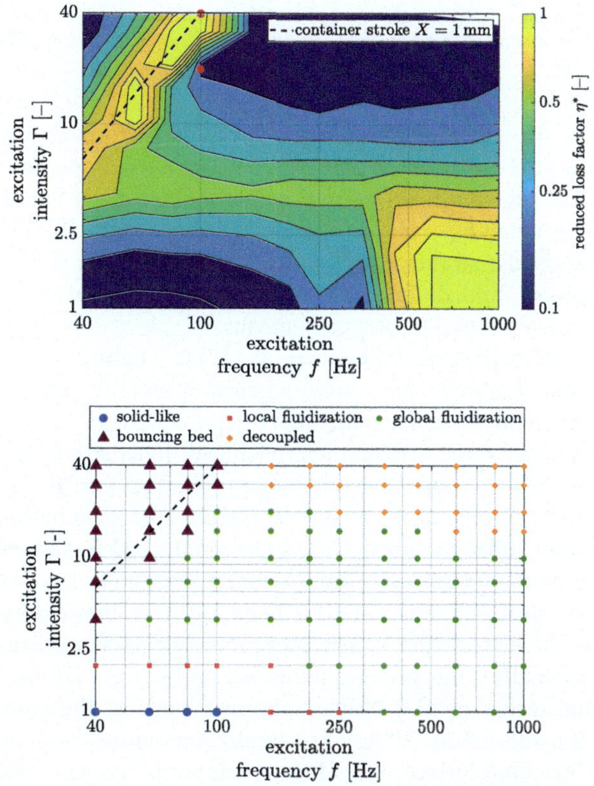

Fig. 10 Experimentally determined reduced loss factor (*top*) and numerically obtained motion modes (*bottom*) of particle container for horizontal high excitation intensities shown in Fig. 8

shown in Fig. 9c. Medium to high reduced loss factors are seen. Within the bouncing collect-and-collide motion mode, see Fig. 9d, the particles move as one single particle block synchronously with the driven particle container and collide in elastically with the container walls, leading to medium to high reduced loss factors. The decoupled motion mode is shown in Fig. 9e. It is characterized by a very small absolute particle velocity compared to the velocity of the container. Thus, the granular matter appears to be decoupled from the container and results in small dissipation rates.

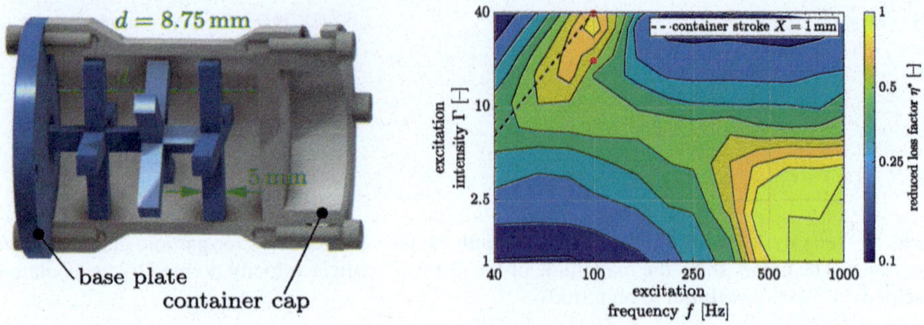

Fig. 11 Inner structures used for high excitation vibrations with schematic representation (*left*) and result for same particle setting as used in Fig. 10 equipped with inner structure (*right*)

An intensive sensitivity analysis is performed experimentally and numerically in [14] on different particle and container properties affecting the motion modes. The bouncing collect-and-collide motion mode is rather insensitive except for the clearance h_{bou}. High reduce loss factor values are again obtained along a constant container stroke of $X_{\mathrm{bou}}^{\mathrm{opt}} = h_{\mathrm{bou}}/\pi$, see also Fig. 8 for the definition of h_{bou}. For the global fluidization, a high filling ratio of the particle container and a small particle size are found to be advantageous.

Hybrid Particle Dampers: In [27] two hybrid particle damper approaches are presented to make the damping behavior around high reduced loss factor values more robust and efficient compared to a pure particle damper. These findings [27] are summarized here. The first approach uses inner structures inside the particle container, see Fig. 11-*left* for a schematic representation. In Fig. 11-*right* the corresponding reduced loss factor is shown. Hereby, the same particle settings as for Fig. 10 are used. Inner structures lead to lower reduced loss factors of the bouncing collect-and-collide motion mode, but to a more robust (wider) behavior in this excitation area, compare red dots in Figs. 10 and 11. Also, they lead to higher reduced loss factors at high frequencies. The more inner structures are used and the smaller the particle radius, the stronger the effect on the reduced loss factor. The second approach presented in [27] utilizes buffered walls of the particle container. The approach aims to influence the local fluidization mode, such that a similar motion as in the bouncing collect-and-collide is achieved. Instead of taking off the container base and flying through the container as for the bouncing collect-and-collide motion mode, the particle bed penetrates the buffered wall material. When the buffered material is completely compressed, an inelastic collision with the container's wall occurs and the relative velocity between particles and wall vanishes. This produces a new particle motion, also called compression collect-and-collide. In Fig. 12-*left*, the corresponding velocity field obtained from DEM simulations is shown. In contrast to the local fluidization mode, a much higher particle velocity is achieved, compare with Fig. 9b. An analytical formula based on Hertz impact theory has been derived, enabling a fast dimensioning of the buffered wall material, see [27]. In Fig. 12-*right*, the reduced loss factor is shown for the particle setting of Fig. 10 using buffered walls

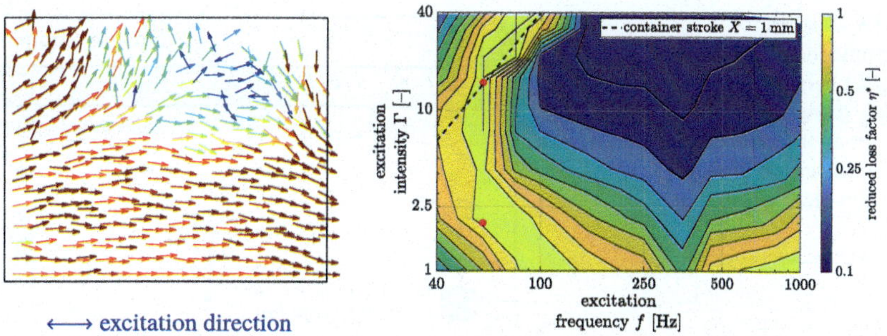

Fig. 12 Buffered walls used for high excitation vibrations with particle damper's velocity field (*left*) and reduced loss factor for same particle setting as used in Fig. 10 equipped with buffered walls designed for $f = 60$ Hz (*right*)

designed for an excitation frequency of $f = 60$ Hz. Here, a very robust behavior of the reduced loss factor concerning the excitation intensity at the desired excitation frequency is achieved.

4.2 Level III—Considerations

In the next step, the particle container of **level II** is coupled to an underlying structure as presented in detail in [11, 12]. The previous experimentally and numerically determined effective fields or the full DEM model are used to predict the overall damping of the system. As an application example, a beam-like structure with free–free boundary condition is used. A picture and a schematic representation of the beam and testbed are shown in Fig. 13. The testbed consists of a flexible beam with a hollow profile supported by three soft cables. The beam is exited in the transverse direction with a variable force by a shaker at its free end using a sine sweep excitation. The particle container can be placed at any desired position on the beam. For the investigated analysis here, the container is placed at the free, not excited end, as shown in Fig. 13. Later on, the position of the particle damper can be varied.

Coupling the effective fields of the particle damper to a modal reduced model of the beam, various investigations have been conducted to show its qualitative accuracy and efficiency [11, 12]. In Fig. 14 an exemplary result of this coupling procedure is shown. While a perfect quantitative fit is not obtained, so gives the qualitative results useful guidelines during the particle damper design process. The position of the particle damper plays an important role. Placing the particle damper at an antinode of the shape function a good agreement between damping prediction and the experimental result is achieved. By placing the damper at a position where the shape function exhibits an additional rotation, the damping prediction is still acceptable. Although, in some cases this greatly reduces the energy dissipation of the damper, i. e. for the bouncing collect-and-collide motion mode. Even multiple eigenmodes

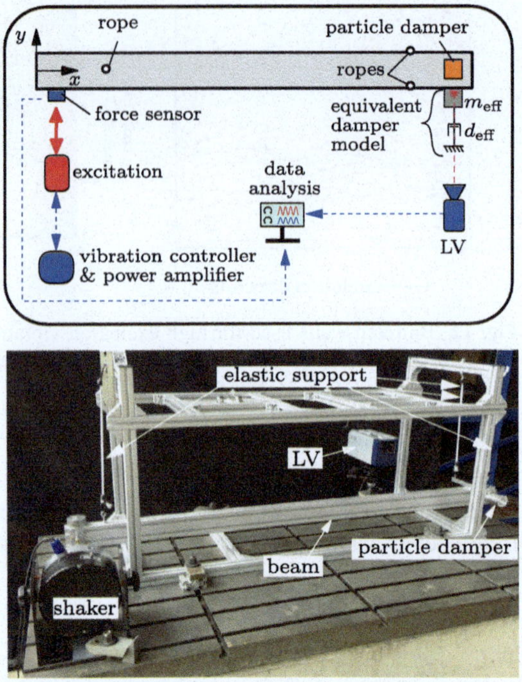

Fig. 13 Testbed to determine the overall damping behavior of the particle damper for high excitation vibrations with schematic representation (*top*) and picture (*bottom*)

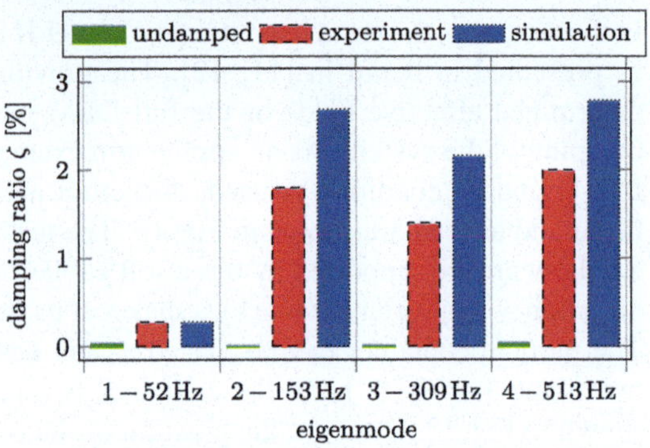

Fig. 14 Damping ratios of beam-like structure equipped with particle damper of Fig. 8 for high excitation intensities

can be damped efficiently if the particle damper is placed at a position, where these modes have a high shape function value. For further details on the different coupling methods of the numerical models see [11, 12].

5 Conclusion

So far particle dampers have been mostly developed by time-consuming experimental-based trial and error strategies for very specific applications, where the adaption to other systems is extremely limited. This might be due to the fact, that the processes in the particle dampers are highly nonlinear and depend on a variety of different influence parameters, like the coefficient of restitution, the coefficient of friction, the excitation frequency, and the vibration amplitude. Due to the lack of understanding of these processes in the dampers and missing systematic design approaches, particle damper's application is so far limited.

The goal of this project is the development of a new design methodology in form of a toolchain for passive vibration damping of lightweight structures and machines using particle dampers. Thereby, using simulations that are verified by experiments, also a deeper understanding of the micro-mechanical processes in the dampers are obtained. This is crucial in the systematic design of particle dampers using numerical methods. By this new design methodology, which is in parts independent of the specific application, it is possible to extend particle dampers to a variety of very different applications, which has been shown at multiple examples.

Using the developed toolchain single particle damper units with predefined characteristics are developed which do not rely on a specific application. Hybrid approaches are used to increase the efficiency of the individual particle dampers. Multiple layers with different filling ratios might be used to obtain an optimized damping behavior for transient vibrations. Inner structures within the dampers cause a more robust behavior and buffering the dampers walls with a polymer leads not only to a considerable noise reduction but can also be used to extend the damper's efficient range of operation for high intensity applications. These individual particle dampers finally form an assembly set, which is ultimately used in the overall damping concept for specific applications.

Acknowledgements The authors would like to thank the German Research Foundation (DFG) for the financial support of the project 352324024 during its first project phase, being an associated project to the SPP1897 and for the financial support of the project 424825162 as part of the SPP 1897 during the second project phase.

The authors would also like to thank Dr.-Ing. Marc-André Pick, Dipl.-Ing. Riza Demir, Dipl.-Ing. Norbert Borngräber-Sander and Wolfgang Brennecke for helping to design and realize the experimental rigs.

References

1. Lu, Z., Wang, Z., Masri, S.F., Lu, X.: Particle impact dampers: past, present, and future. Struct. Control Health Monitor. **25**(1), e2058 (2017)
2. Johnson, C.D.: Design of passive damping systems. J. Mechan. Design **117**(B), 171–176 (1995)
3. Panossian, H.: Non-obstructive particle damping experience and capabilities. Proc. SPIE - Int. Soc. Opt. Eng. **4753**, 936–941 (2002)
4. Simonian, S.S.: Particle beam damper. Proc. SPIE - Int. Soc. Opt. Eng. **2445**, 149–160 (1995)

5. Saeki, M.: Impact damping with granular materials in a horizontally vibrating system. J. Sound Vib. **251**(1), 153–161 (2002)
6. Chen, T., Mao, K., Huang, X., Wang, M.: Dissipation mechanisms of nonobstructive particle damping using discrete element method. Proc. SPIE - Int. Soc. Opt. Eng. **4331**, 294–301 (2001)
7. Meyer, N., Schwartz, C., Morlock, M., Seifried, R.: Systematic design of particle dampers for horizontal vibrations with application to a lightweight manipulator. J. Sound Vibrat. **510**, 116319 (2021)
8. Meyer, N.; Seifried, R.: Investigation of the Influence of Parameters on Particle Dampers. Computational Structures Technology 2018, Sitges, Extended Abstract (2018)
9. Meyer, N., Seifried, R.: Experimental and numerical investigations on parameters influencing energy dissipation in particle dampers. In: VI International Conference on Particle-Based Methods - Fundamentals and Applications, vol. 6, pp. 260–271 (2019)
10. Meyer, N.; Seifried, R.: An experimental model for the analysis of energy dissipation in particle dampers. PAMM **19**(1) (2019)
11. Meyer, N., Seifried, R.: Numerical and experimental investigations in the damping behavior of particle dampers attached to a vibrating structure. Comput. Struct. **238**, 106281 (2020)
12. Meyer, N., Seifried, R.: Damping prediction of particle dampers for structures under forced vibration using effective fields. Granul. Matter **23**(3), 64 (2021)
13. Meyer, N.; Seifried, R.: Optimal design of particle dampers for structures with low first eigenfrequency under forced vibration. In: 7th Edition of the International Conference on Particle-Based Methods, vol. 7 (2021)
14. Meyer, N., Seifried, R.: Toward a design methodology for particle dampers by analyzing their energy dissipation. Comput. Particle Mech. **8**(4), 681–699 (2021)
15. Meyer, N.; Seifried, R.: Energy dissipation in horizontally driven particle dampers of low acceleration intensities. Nonlinear Dyn. (2022)
16. Meyer, N., Wagemann, E.L., Jackstadt, A., Seifried, R.: Material and particle size sensitivity analysis on coefficient of restitution in low-velocity normal impacts. Comput. Particle Mech. (2022)
17. Stronge, W.J.: Impact Mechanics, 2 edn. Cambridge University Press, Cambridge (2018)
18. Pöschel, T., Brilliantov, N.V.: Extremal collision sequences of particles on a line: optimal transmission of kinetic energy. Phys. Rev. E **63**(2), 1–9 (2001)
19. Goldsmith, W.: Impact: The Theory and Physical Behavior of Colliding Solids. Edward Arnold Publishers, London (1960)
20. Seifried, R., Minamoto, H., Eberhard, P.: Viscoplastic effects occurring in impacts of aluminum and steel bodies and their influence on the coefficient of restitution. J. Appl. Mech. **77**(4) (2010)
21. Tatara, Y., Moriwaki, N.: Study on impact of equivalent two bodies: coefficients of restitution of spheres of brass, lead, glass, porcelain and agate, and the material properties. Bull. JSME **25**(202), 631–637 (1982)
22. Minamoto, H., Seifried, R., Eberhard, P., Kawamura, S.: Analysis of repeated impacts on a steel rod with visco-plastic material behavior. Eur. J. Mech. A. Solids **30**(3), 336–344 (2011)
23. Pöschel, T.: Computational Granular Dynamics: Models and Algorithms. Springer, Berlin (2005)
24. Yang, M.Y., Lesieutre, G.A., Hambric, S., Koopmann, G.: Development of a design curve for particle impact dampers. Noise Control Eng. J. **53**, 5–13 (2005)
25. Sanchez, M., Rosenthal, G., Pugnaloni, L.: Universal response of optimal granular damping devices. J. Sound Vib. **331**(20), 4389–4394 (2012)
26. Bannerman, M.N., Kollmer, J.E., Sack, A., Heckel, M., Mueller, P., Pöschel, T.: Movers and shakers: granular damping in microgravity. Phys. Rev. E **84** (2011)
27. Meyer, N., Seifried, R.: Design of Robust Particle Dampers Using Inner Structures and Buffered Walls. Preprint series of the SPP 1897 Calm, Smooth, Smart, vol. 35, p. 24 (2022)

A Systematic Approach to Smart Damping of Mechanical Systems Based on Inductive Electro-Mechanical Coupling

Mitja Rosenboom and Hartmut Hetzler

1 Introduction

The field of inductive damping of structural vibrations is best described from an energetic point of view. The kinetic energy of the structural vibration is converted into electric energy by electromagnetic induction. The electric energy is then dissipated by ohmic resistors and is thus extracted from the mechanical system. In order to convert kinetic energy into electric energy, the magnetic flux through some conductive material has to be modulated. This process can be divided into four basic functionalities: source of magnetic flux, transport of magnetic flux, modulation of magnetic flux and induction of electric current.

For each of these functionalities different realizations are possible. The source of the magnetic flux can either be a permanent magnet or an electromagnet, the transport of the magnetic flux can be guided through the structure by use of high permeability iron cores or can be unguided. The modulation of the flux in a conductive material may be due to a change of the magnetic flux itself or due to a movement of a conductive material relative to a magnetic field. The induction can either occur in a coil (as a lumped element of the system) or in form of eddy currents, distributed over a part of the structure. Figure 1 shows a matrix which gives an overview on the different functionalities as well as symbolic design examples.

Based on this matrix, damping devices may systematically be created by (rather) freely combining different alternatives to implement the basic functionalities. For example, an inductive damping device may be constructed by combining a guided transport of the magnetic flux, a modulation of the flux by varying an air gap, an inductive coupling by means of a coil and providing dissipation using an ohmic resistor. For the source of the magnetic flux, basically two options are available:

M. Rosenboom (✉) · H. Hetzler
Institute of Mechanics, Engineering Dynamics Group, University of Kassel, Mönchebergstraße 7, 34125 Kassel, Germany
e-mail: rosenboom@uni-kassel.de

P. Eberhard (ed.), *Calm, Smooth and Smart*, Lecture Notes in Applied and Computational Mechanics 102, https://doi.org/10.1007/978-3-031-36143-2_4

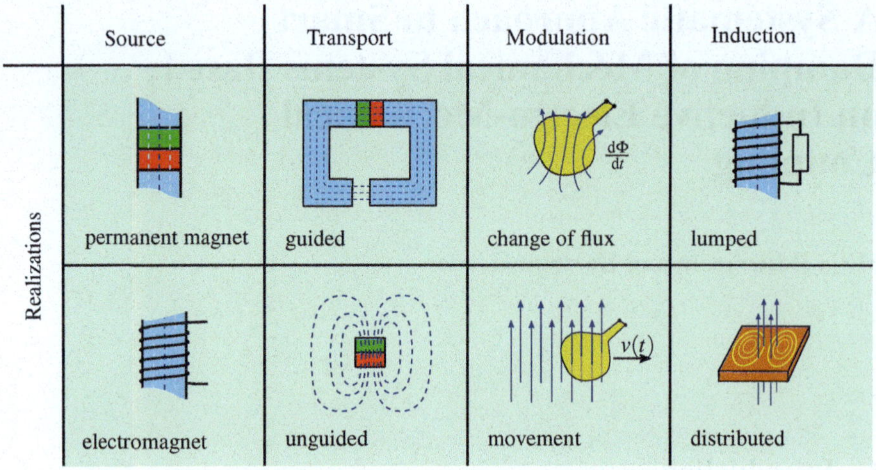

Fig. 1 Overview on functional elements of electromagnetic damping devices

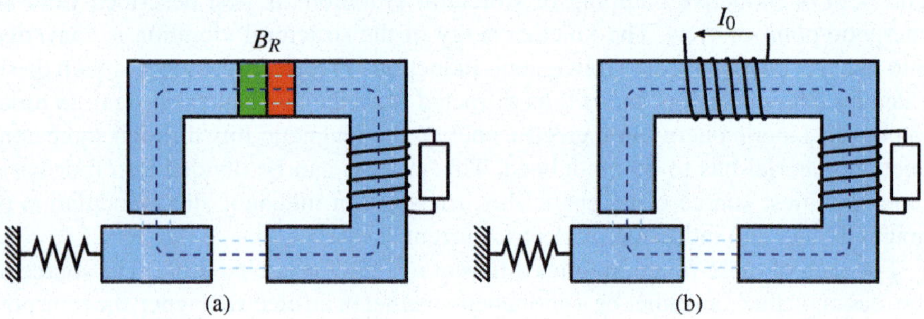

Fig. 2 Examples built from the matrix of basic functionalities using different realizations of the source of the magnetic flux: **a** permanent magnet, **b** electromagnet

it may either be produced by a permanent magnet with a remanence of B_R (cf. Fig. 2a) or by means of an electromagnet fed with a constant current I_0 (cf. Fig. 2b).

In the past decades several investigations on inductive damping have been published. Behrens et al. [4] have introduced electromagnetic shunt damping. They proposed a plunger, consisting of permanent magnets, that is moving in a coil, which is connected to an impedance network.

Przybylowicz and Szmidt [8, 9] theoretically investigated a mechanical oscillator between two electromagnets. The magnetic flux is guided with iron cores through the mechanical oscillator and builds two independent magnetic circuits with an air gap. The length of the air gap is modulated by the mechanical movement and thus, eddy currents are induced in the iron core. The investigated model shows strongly nonlinear behavior.

Bae et al. [1, 2] studied the behavior of a cylindrical permanent magnet moving in a conductive tube. Sodano et al. [12, 13] investigated a model consisting of a cantilever beam with a conducting plate, that is moving in the magnetic field of a permanent magnet. Later on Sodano and Inman [14] proposed an active damping device where they used again a cantilever beam with a conductive plate. This time an electromagnet generates the magnetic field and a feedback control system is used to control the oscillations of the structure. Laborenz et al. [5, 6] experimented with eddy current damping to reduce the oscillations of steam turbine blades. They, as well, used a copper plate oscillating in the magnetic field of a permanent magnet.

Bae et al. [3] studied the use of an eddy current damper as a magnetically damped tuned mass damper to reduce oscillations of a beam structure. They showed, that the resonance amplitudes of the structure were decreased by applying eddy current damping to the tuned mass damper. Lian et al. [7] proposed an eddy current-tuned mass damper for wind turbines.

The objective of this contribution is to systematically analyze different realizations of inductive damping elements. Therefore models using different elements of the basic functionalities shown in Fig. 1 will be investigated. Furthermore, the possibility to modify inductive damping systems with additional nonlinearities to show a specific behavior is presented.

2 Analysis of Models Based on Magnetic Circuits

In this section the equations of motions for the proposed models shown in Fig. 2 will be derived and the static and dynamic behavior will be analyzed. The derivation of the equations of motion is exemplary shown for the system with permanent magnet, illustrated in Fig. 2a.

To describe an inductive damping model mathematically, the system can be divided into three subsystems, as shown in Fig. 3. Here, the electrical and the mechanical subsystem do not interact directly but will be coupled by the magnetic field which acts as a mediator.

The mechanical system in this case is a simple single degree of freedom (DoF) oscillator with mass m and stiffness k (cf. Fig. 4a). The position of the mass is described by the coordinate x. The mass is excited harmonically by an external force $F(t)$. Furthermore the magnetic force F_{mag} acts on the mass. Another static force F_0 is introduced in order to compensate static magnetic forces and thus to ensure, that the system will have a static resting position at $x = 0$. For instance, such a force could easily be realized by preloading the spring. Eventually, the equation of motion for the mechanical system follows from Newton's law and is given by

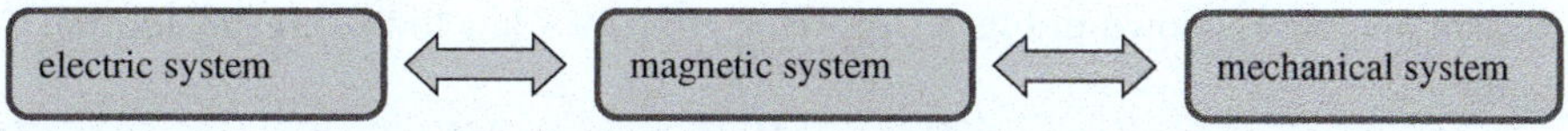

Fig. 3 Separate physical sub-domains involved in an inductive damping device

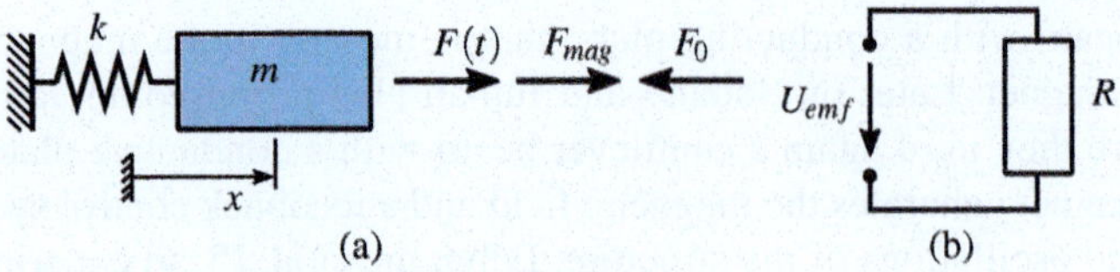

Fig. 4 **a** Mechanical subsystem. **b** Electrical subsystem

$$m\ddot{x} + kx = F(t) + F_{mag} - F_0. \tag{1}$$

The electric subsystem is a simple electrical network, featuring an ohmic resistor R and the induced voltage U_{emf}. The current flowing in the circuit is denoted by I. The equation of motion follows from Kirchhoff's law and reads

$$U_{emf} + RI = 0. \tag{2}$$

Assuming magnetostatic conditions the magnetic problem can be described by the simplified form of Ampère's law [16], reading

$$\oint_\Gamma \mathbf{H} \cdot \mathrm{d}\mathbf{l} = I_{enc}, \tag{3}$$

where $\mathbf{H}$ is the magnetic field, I_{enc} is the current enclosed by the loop Γ and $\mathrm{d}\boldsymbol{\ell}$ is an infinitesimal element of the curve Γ. Furthermore, the conservation law of the magnetic flux holds according to

$$\oint_S \mathbf{B} \cdot \mathrm{d}\mathbf{a} = 0, \tag{4}$$

where $\mathbf{B}$ is the magnetic flux density and $\mathrm{d}\mathbf{a}$ is the outer normal unit vector of the closed surface S. The magnetic flux density $\mathbf{B}$ is connected to the magnetic field $\mathbf{H}$ by the constitutive relation

$$\mathbf{H} = \mathbf{H}(\mathbf{B}). \tag{5}$$

Furthermore, the magnetic flux through a surface S is defined by

$$\Phi = \int_S \mathbf{B} \cdot \mathrm{d}\mathbf{a}. \tag{6}$$

The magnetic subsystem for the proposed inductive damping device with a permanent magnet is shown in Fig. 5. Applying Ampère's law to the illustrated loop Γ yields

$$H_m \ell_m + H_{fe} \ell_{fe} + H_d(\ell_d - x) = NI, \tag{7}$$

Fig. 5 Magnetic subsystem

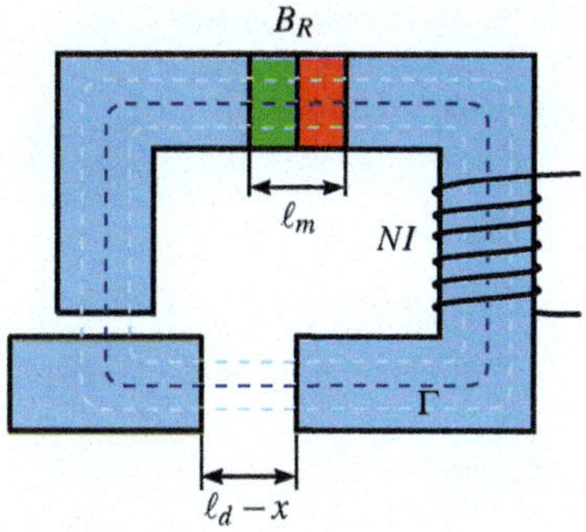

where H_m is the magnetic field in the permanent magnet, ℓ_m is the length of the permanent magnet, H_{fe} is the field in the iron core, ℓ_{fe} is the length of the iron core, H_d is the field in the air gap and ℓ_d is the nominal length of the air gap, N is the number of turns of the coil and I is the current flowing through the coil. From the conservation of the magnetic flux, it may be followed that all individual parts of the structure are crossed by the same flux and thus

$$\Phi_m = \Phi_{fe} = \Phi_d = \Phi \tag{8}$$

holds. Assuming equal cross sections A of the individual parts yields

$$\Phi = BA, \tag{9}$$

which states that the magnetic flux density in all parts is equal. The constitutive relations for the magnetic fields in the permanent magnet (H_m), the air gap (H_d) and the iron core (H_{fe}) are given by

$$H_m = \frac{1}{\mu_0}(B - B_R), \quad H_d = \frac{1}{\mu_0}B \quad \text{and} \quad H_{fe} = \frac{1}{\mu_{fe}(B)}B, \tag{10}$$

where μ_0 is the magnetic permeability of free space and $\mu_{fe}(B)$ is the permeability of the iron. The permeability of the magnet is assumed to be μ_0, as this is approximately the case for neodymium magnets. For this study hysteresis losses of the B–H characteristic are neglected because most of the energy storage of the magnetic field occurs in the air gap, and thus hysteresis losses are assumed to be small [15]. The assumed B–H characteristic is shown in Fig. 6: the slope of the characteristic decreases sharply after the magnetic flux density reaches a magnitude B_{sat}, as the material saturates.

Eventually, the magnetic flux of the system is described by

$$\left((\ell_m + \ell_d - x) + \frac{\mu_0}{\mu_{fe}(B)}\ell_{fe}\right)\Phi = \mu_0 NAI + \ell_m \Phi_R, \tag{11}$$

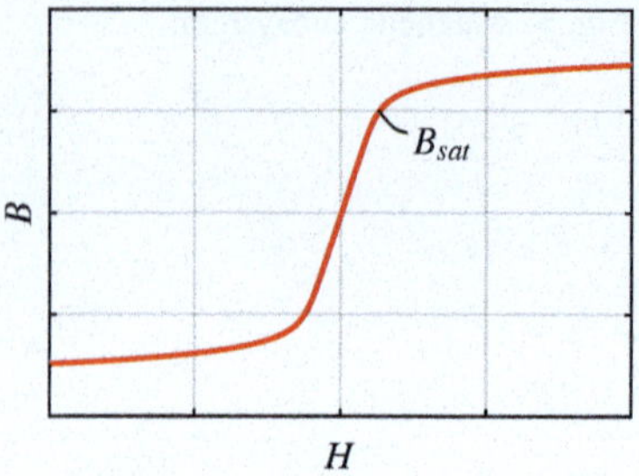

Fig. 6 *B–H* characteristic of the material for the iron core (adapted from [15])

where $\Phi_R = B_R A$ is the magnetic flux of the remanence. The link of the mechanical and the magnetic subsystems may be expressed using Maxwell's stress tensor

$$\mathbf{T} = \mathbf{B} \otimes \mathbf{H} - \frac{1}{2}(\mathbf{B} \cdot \mathbf{H})\mathbf{I}, \tag{12}$$

where $\mathbf{I}$ is the unit dyadic tensor [11]. Accordingly, the total magnetic force acting on an object reads

$$\mathbf{F} = \oint_S \mathbf{T} \cdot \mathrm{d}\mathbf{a}, \tag{13}$$

where S is the surface of the object and $\mathrm{d}\mathbf{a}$ is the outer normal unit vector of the surface. For the considered model, the magnetic force results in

$$F_{mag} = A\frac{B^2}{2\mu_0} = \frac{\Phi^2}{2A\mu_0}. \tag{14}$$

The relation between the electric and the magnetic subsystem is given by the flux linkage of the coil, and thus

$$U_{emf} = \frac{\mathrm{d}\Psi}{\mathrm{d}t}, \quad \text{where} \quad \Psi = N\Phi. \tag{15}$$

For the proposed model, Ψ is not explicitly time-dependent and therefore

$$U_{emf} = N\frac{\mathrm{d}\Phi}{\mathrm{d}t} = N\left(\frac{\partial\Phi}{\partial x}\dot{x} + \frac{\partial\Phi}{\partial I}\dot{I}\right) \tag{16}$$

applies. Summarizing, the equations of motion for the system in Fig. 2a read

$$m\ddot{x} + kx - \frac{\Phi^2}{2A\mu_0} = F(t) - F_0 \tag{17}$$

$$N\left(\frac{\partial\Phi}{\partial x}\dot{x} + \frac{\partial\Phi}{\partial I}\dot{I}\right) + RI = 0 \tag{18}$$

$$\left((\ell_m + \ell_d - x) + \frac{\mu_0}{\mu_{fe}(\Phi/A)}\ell_{fe}\right)\Phi = \mu_0 N A I + \ell_m \Phi_R. \tag{19}$$

Using the mechanical eigenfrequency

$$\omega_0 = \sqrt{\frac{k}{m}} \tag{20}$$

for vanishing electro-magnetic coupling (i.e. $\Phi = 0$), the time t is re-scaled to the dimensionless time $\tau = \omega_0 t$. Consequently, the differential operators transform according to $\dot{(\cdot)} = \omega_0 (\cdot)'$, where $(\cdot)'$ denotes the derivation with respect to τ. Furthermore the dimensionless parameters

$$\kappa = \frac{\ell_m}{b}, \qquad \delta_0 = \frac{\ell_d}{b}, \qquad \beta = \frac{\ell_{fe}}{b}, \tag{21}$$

$$\nu = \frac{N \Phi_R \omega_0}{R I_0}, \qquad \gamma = \frac{\Phi_R^2}{2\ell_d k A \mu_0}, \qquad h(\varphi) = \frac{\mu_0}{\mu_{fe}\left(\Phi_R \varphi / A\right)}, \tag{22}$$

$$\rho = \frac{b \mu_0 N I_0}{\Phi_R}, \qquad f(\tau) = \frac{1}{\ell_d k} F\left(\tau/\omega_0\right), \qquad f_0 = \frac{1}{\ell_d k} F_0 \tag{23}$$

as well as the scaled coordinates

$$\text{displacement: } \xi = \frac{x}{\ell_d}, \qquad \text{flux: } \varphi = \frac{\Phi}{\Phi_R}, \qquad \text{current: } \iota = \frac{I}{I_0} \tag{24}$$

are introduced, where b is the width of the iron core and I_0 is some reference current. Inserting these parameter into the Eqs. (17)–(19) results in the dimensionless equations of motion of the *system with permanent magnet* (cf. Fig. 2a)

$$\xi'' + \xi - \gamma \varphi^2 = f(t) - f_0 \tag{25}$$

$$\nu \left(\frac{\partial \varphi}{\partial \xi} \xi' + \frac{\partial \varphi}{\partial \iota} \iota' \right) + \iota = 0 \tag{26}$$

$$(\kappa + \rho \iota) - \big(\kappa + \delta_0 (1 - \xi) + \beta h(\varphi)\big)\varphi = 0. \tag{27}$$

Apart from the difference in the source term of the magnetic flux, the mechanical and electrical subsystems for the damping device using an electromagnet as depicted in Fig. 2b are equivalent. For this case, Ampère's law yields

$$H_{fe}\ell_{fe} + H_d\left(\ell_d - x\right) = N I + N_0 I_0, \tag{28}$$

where N_0 is the number of turns of the electromagnet and I_0 is a constant current feeding the electromagnet. Following the same steps as for the system with a permanent magnet, the dimensionless equation of motion of the *system with electro-magnet* (cf. Fig. 2b) are found as

$$\xi'' + \xi - \gamma\varphi^2 = f(t) - f_0 \tag{29}$$

$$\nu\left(\frac{\partial\varphi}{\partial\xi}\xi' + \frac{\partial\varphi}{\partial\iota}\iota'\right) + \iota = 0 \tag{30}$$

$$(\rho_0 + \rho\iota) - (\delta_0(1-\xi) + \beta h(\varphi))\,\varphi = 0, \tag{31}$$

where ρ_0 is the source term of the magnetic flux, given by

$$\rho_0 = \frac{b\mu_0 N_0 I_0}{\Phi_r}. \tag{32}$$

2.1 Static Analysis

In Fig. 7 the magnetic flux φ of the considered models is shown as a function of the relative static displacement ξ.[1] Since a static solution is assumed ($\xi' = 0$) no currents are induced and therefore, $\iota = 0$ holds. Notice that a relative displacement of $\xi = 1$ means, that the air gap is closed. The figure compares solutions with as well as those without accounting for magnetic saturation. For the system with permanent magnet, the influence of saturation is very small, as the magnetic flux is restricted due to the remanence of the permanent magnet. In contrast, the system with electromagnet shows a strong influence of saturation as the flux increases for small air gaps. Therefore, for the system with an electromagnet, saturation effects must be taken into account, if small air gaps occur.

As the magnetic forces act equivalent to a nonlinear spring with negative stiffness, the static solution may become unstable for certain parameters. Figure 8 shows stability charts for both models for parameter variations of the source term of the magnetic flux (κ respectively ρ_0) and the coupling parameter of the mechanical and the magnetic subsystem γ. While the system with a permanent magnet only shows instabilities for higher values of γ, the system with an electromagnet becomes unstable

Fig. 7 Magnetic flux of models with permanent magnet (pm) and electromagnet (em) with and without the effect of saturation

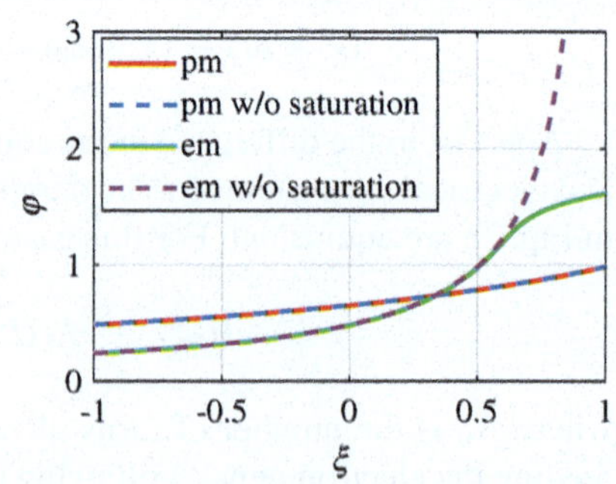

[1] The parameters used for the analysis are $\kappa = 2$, $\delta_0 = 1$, $\beta = 10$, $\rho_0 = 0.5$, $\rho = 0.5$, $\gamma = 0.5$, $\nu = 1$, $B_r = 1.2$.

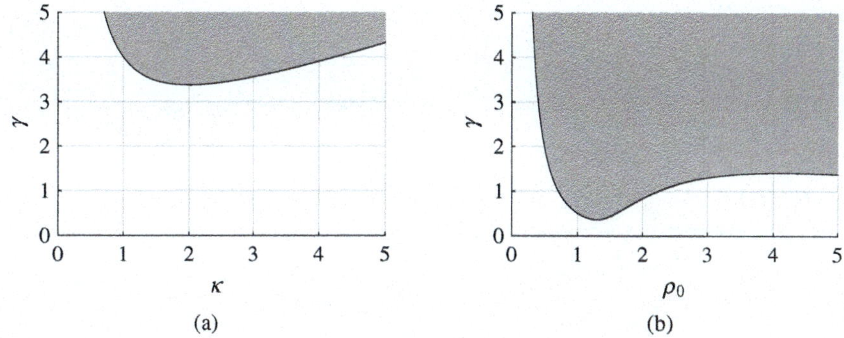

Fig. 8 Stability charts of magneto-mechanical coupling against flux source for **a** system with permanent magnet, **b** system with electromagnet [10]

already for much lower values. As the stiffness of the magnetic force is dependent on the slope of the magnetic forces against ξ, the system becomes unstable if $\gamma \frac{\mathrm{d}\varphi^2}{\mathrm{d}\xi} > 1$ applies [10].

2.2 *Dynamic Analysis*

For a harmonic excitation with $f(t) = \hat{f}\sin(\eta\tau)$, a dynamic analysis of the system is carried out using a simple shooting method. The maximal amplitudes of the occurring oscillations are shown in Fig. 9a. Both systems—the one with permanent magnet as well as the system with electromagnet—show a shift of the resonance frequency. This is caused by the magnetic forces acting as a spring with negative stiffness. Furthermore, in both cases resonance amplitudes are limited. The resonance amplitudes of the system featuring an electromagnet are much lower and thus, the damping is higher. While for the system with a permanent magnet higher amplitudes of the excitation force will lead to system failure, as the deflection reaches the length of the air gap, the frequency responses of the system with an electromagnet for higher excitation force levels are depicted in Fig. 9b. The system shows strongly nonlinear behavior for higher amplitudes. Even higher amplitudes of the force lead to instabilities and therefore are not computable by the used simple shooting method.

2.3 *Numerical Validation*

To validate the obtained results, a numerical model has been set up. To compensate the one-sided magnetic pull and to get a more efficient damping, the model has been expanded to a symmetric model, as depicted in Fig. 10a.

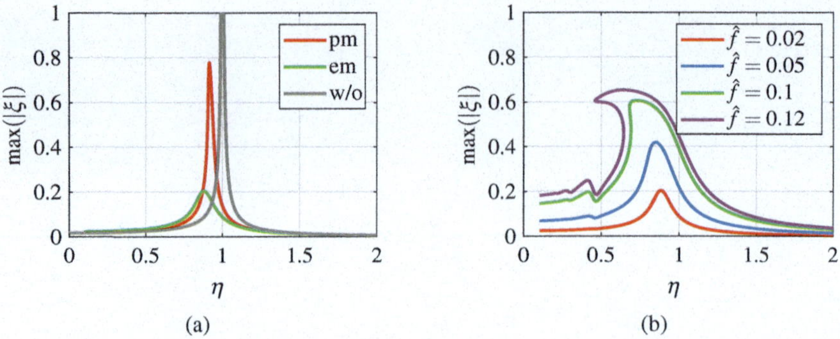

Fig. 9 **a** Frequency responses of the models with excitation force amplitude $\hat{f} = 0.02$. Red: system with permanent magnet—Green: system with electromagnet—Gray: system without magnets. **b** Frequency response of the system with electromagnet for different excitation force amplitudes $\hat{f}$

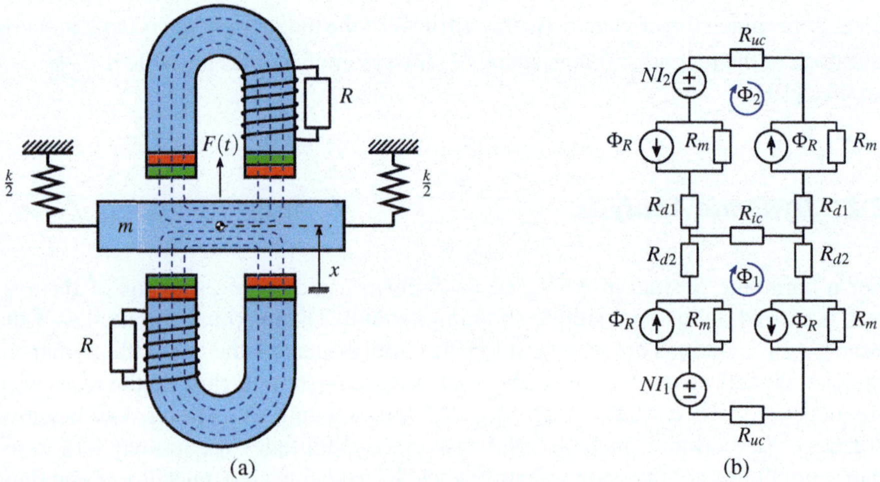

Fig. 10 **a** Model for numerical validation. **b** MEC-network of the model

To describe the model analytically, the method of magnetic equivalent circuits (MEC) is used. The method converts the continuous model into lumped parameters connected by nodes and thus, forming a network, similar to an electric network. The method is equivalent in assumptions to the one presented in the previous subsection, but it brings the advantage of being able to analyze more complex systems with multiple connected magnetic circuits. To transform the continuous model into a MEC, Ampère's law (Eq. (3)) is split into individual sections

$$\oint_{\Gamma} \mathbf{H} \cdot \mathrm{d}\boldsymbol{\ell} = \sum_{d \in D_{\Gamma}} \int_{\ell_d} \mathbf{H} \cdot \mathrm{d}\boldsymbol{\ell} = \sum_{d \in D_{\Gamma}} F_d, \tag{33}$$

where D_Γ is the set of names of the individual parts. F_d is commonly referred to as magnetomotive force (MMF) drop. Accordingly the currents inside the loop Γ are called MMF sources and are denoted as

$$I_{enc,\Gamma} = \sum_{s \in S_\Gamma} N_s I_s = \sum_{s \in S_\Gamma} F_s, \tag{34}$$

where S_Γ is a set, containing the names of the MMF sources. Ampère's law in the theory of MECs is referred to as Kirchhoff's MMF law, which states that the sum of MMF drops in a closed loop equals the sum of MMF sources, i.e.

$$\sum_{d \in D_\Gamma} F_d = \sum_{s \in S_\Gamma} F_s. \tag{35}$$

Additionally the flux conservation law (Eq. (4)) was used in the previous chapter. This is found in Kirchhoff's flux law, which states that the sum of fluxes into respectively out of any node must vanish. It is left to define Ohm's law for magnetic equivalent circuits, which is found by manipulating the MMF drops to a form

$$F_d = R_d \Phi_d, \tag{36}$$

where R_d is called a reluctance of the associated section of the structure. A detailed description of the method is provided in [15].

For the presented model, the MEC network is shown in Fig. 10b. As the previous study showed, that for the considered system with permanent magnet saturation doesn't have an influence on the magnetic flux, saturation is not taken into account for this study. Thus, the MEC model provides a linear algebraic equation to calculate the magnetic flux, given by

$$\mathbf{R}_N \mathbf{\Phi}_N = \mathbf{F}_N. \tag{37}$$

In Eq. (37), $\mathbf{R}_N$ is the network reluctance matrix, $\mathbf{F}_N$ is a column matrix containing the external MMF sources and $\mathbf{\Phi}_N$ is the column matrix of the network fluxes. The equations of motion for the mechanical and the electric system read

$$m\ddot{x} + kx - \frac{1}{\mu_0 A}(\Phi_2^2 - \Phi_1^2) = F(t) \tag{38}$$

$$\mathbf{C}\dot{x} + \mathbf{L}\dot{\mathbf{I}} + \mathbf{R}\mathbf{I} = 0, \tag{39}$$

where m is the oscillating mass, k is the stiffness of the spring, $F(t)$ is a harmonic force, μ_0 is the vacuum permeability, A is the cross section of the iron cores. The matrices $\mathbf{C}$ (coupling), $\mathbf{L}$ (inductance) and $\mathbf{R}$ (electric resistance) are calculated with

$$\mathbf{C} = N\frac{\mathrm{d}\mathbf{\Phi}}{\mathrm{d}x} \quad \text{and} \quad \mathbf{L} = N\frac{\mathrm{d}\mathbf{\Phi}}{\mathrm{d}\mathbf{I}} \quad \text{and} \quad \mathbf{R} = \begin{bmatrix} R & 0 \\ 0 & R \end{bmatrix}, \tag{40}$$

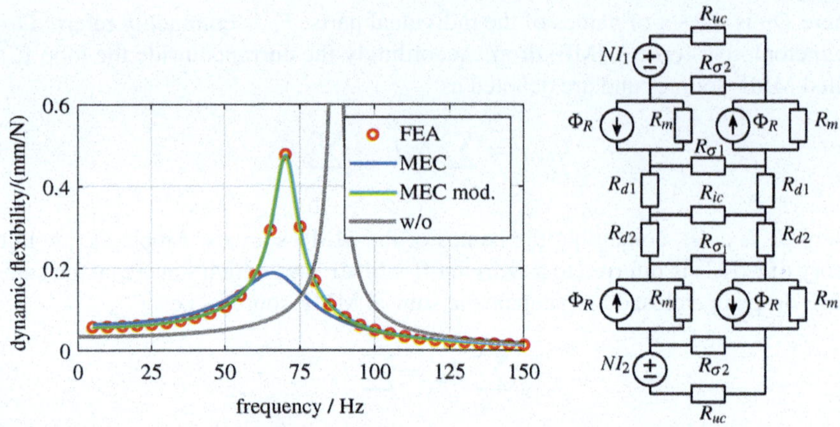

(a) (b)

Fig. 11 **a** Dynamic flexibility of the symmetric model shown in Fig. 10a. Red circles: time dependent FEA—Blue line: MEC network shown in Fig. 10b—Green line: modified MEC with leakage paths (cf. Fig. 11b)—Gray line: system without magnets. **b** MEC network with additional elements to model leakage

where N is the number of turns of the coils, R is the ohmic resistance of the coils and $\mathbf{\Phi} = [\Phi_1 \ \Phi_2]^T$ is a column matrix containing the magnetic fluxes linked with the coils. The dynamic flexibility of the nonlinear set of equations is calculated using the Harmonic Balance Method (HBM).[2] During the calculation the derivatives of the fluxes are evaluated numerically. The results are depicted in Fig. 11a.

For the validation process a finite element analysis (FEA) using the commercial program COMSOL Multiphysics has been carried out. The mesh for the analysis is depicted in Fig. 12a. In Fig. 12b the calculated flux is shown for a stationary analysis. A dynamic FEA is carried out using the moving mesh formulation of the software coupled with an ordinary differential equation for the mechanical subsystem. The dynamic flexibility, calculated with the FEA, is also shown in Fig. 11a. A significant difference to the result of the proposed MEC system exists. This can be explained by leakage effects, i.e. not all the magnetic flux follows the predefined path. Therefore the MEC model has been modified as illustrated in Fig. 11b with additional leakage paths. The dynamic flexibility of the enhanced model is also depicted in Fig. 11a. The FEA and the MEC analysis with the expanded model are in good agreement and thus, it is confirmed, that leakage is the main cause for the differences in the models. Consequently, the guidance of the magnetic flux through a structure has

---

[2] The parameters of the MEC are calculated by $\Phi_R = AB_R$, $R_{uc} = \ell_{uc}/(\mu_{fe}A)$, $R_{ic} = \ell_{ic}/(\mu_{fe}A)$, $R_m = \ell_m/(\mu_0 A)$, $R_{d1} = (d_0 - x)/(\mu_0 A)$, $R_{d2} = (d_0 + x)/(\mu_0 A)$, $R_{\sigma 1} = \ell_{\sigma 1}/(\mu_0 \ell_{\sigma 1} b)$, $R_{\sigma 2} = \ell_{\sigma 2}/(\mu_0 \ell_{\sigma 2} b)$. The values of the parameters used for the analysis are $m = 0.1\,\text{kg}$, $k = 3 \times 10^4\,\text{Nm}^{-1}$, $\hat{F} = 1\,\text{N}$ (force amplitude of excitation), $R = 0.015\,\Omega$, $d_0 = 3\,\text{mm}$ (nominal air gap length), $A = 100\,\text{mm}^2$, $N = 35$, $B_R = 1.2\,\text{T}$, $\ell_{uc} = 50\,\text{mm}$, $\ell_{ic} = 20\,\text{mm}$, $\ell_m = 2\,\text{mm}$, $\ell_{\sigma 1} = 4\,\text{mm}$, $\ell_{\sigma 2} = 13\,\text{mm}$, $b = 10\,\text{mm}$ (depth of iron core), $\mu_0 = 4\pi \times 10^{-7}\,\text{Hm}^1$, $\mu_{fe} = 5000\mu_0$.

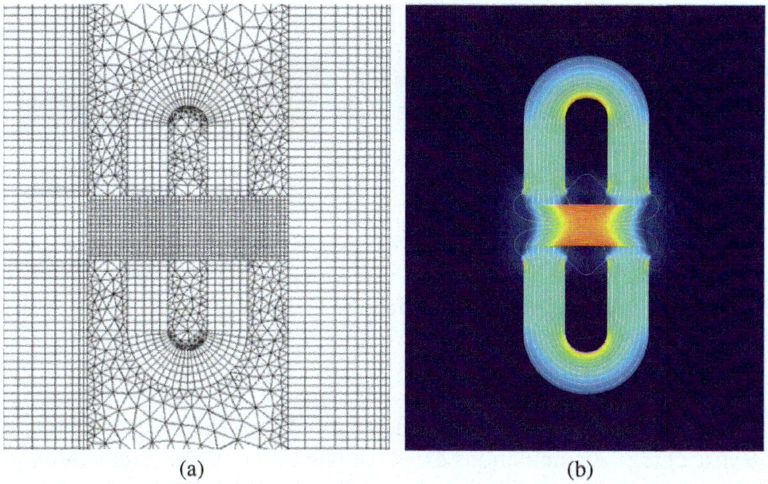

Fig. 12 **a** Mesh of the FEA model. **b** Simulation result from static FEA. Color Gradient: magnetic flux density—White lines: magnetic field lines

to be designed very carefully. Furthermore, flux leakage does effect the damping efficiency dramatically and therefore must be minimized in order to efficiently calm structural vibrations.

3 Analysis of Models Based on Eddy Currents

Another inductive damping device may be derived from the matrix in Fig. 1 by producing the modulation of the flux by moving a magnet in the vicinity of a conductive material. As a source of the magnetic flux, a permanent magnet is chosen and the transport is unguided. The induction is distributed over the conductive material. One representation of this set of realizations of the functionalities is shown in Fig. 13a and has been analyzed e.g. by Bae et al. [1]. Since the analytic calculation of eddy currents is rather complex and only applicable for simple geometries, a mixed formulation will be derived. Still, a short summary of the basic procedure as used in e.g. [1, 5, 12] is given.

3.1 Analytic Description of Eddy Currents

According to Ohm's law, the eddy current density $\mathbf{J}$ is given by

$$\mathbf{J} = \sigma \mathbf{E}, \tag{41}$$

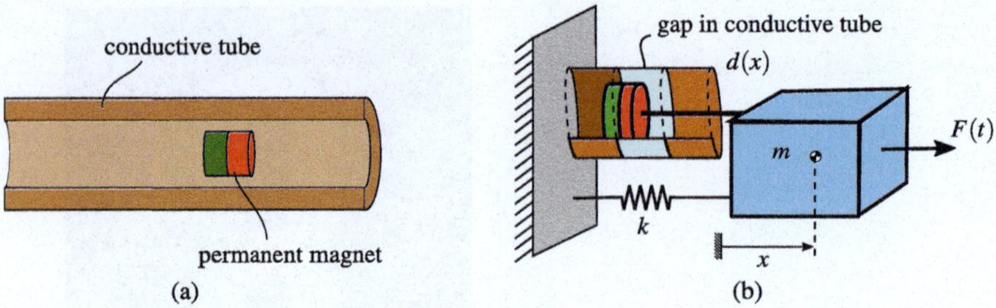

Fig. 13 **a** Model of eddy current damper as proposed by Bae et al. [1]. **b** Model of a single DoF oscillator featuring position-dependent inductive damping

where σ is the electric conductivity of the material and $\mathbf{E}$ is the electromotive force. If no electric charge accumulations exist, the electromotive force in a homogeneous conducting rigid object, moving translationally at the velocity $\mathbf{v}$ in a constant magnetic field $\mathbf{B}$, is given by

$$\mathbf{E} = \mathbf{v} \times \mathbf{B}. \tag{42}$$

The electromagnetic force on the object due to eddy currents can be calculated by

$$\mathbf{F} = \int_V \mathbf{J} \times \mathbf{B}\, \mathrm{d}V, \tag{43}$$

which is known as the Lorentz force equation [11]. Neglecting the magnetic field induced by the eddy currents, the magnetic field is a prescribed quantity and can be calculated with Biot-Savart's law. Inserting Eqs. (41) and (42) into Eq. (43) the force due to eddy currents yields

$$\mathbf{F} = \int_V \sigma\,(\mathbf{v} \times \mathbf{B}) \times \mathbf{B}\, \mathrm{d}V. \tag{44}$$

From this, the part of the force acting against the movement of the object and thus, as a damping force can be found as

$$F_d = v \int_V \sigma B_\perp^2\, \mathrm{d}V. \tag{45}$$

Herein v is the magnitude of the velocity and $B_\perp$ is the magnitude of the part of the magnetic flux density that is perpendicular to the velocity of the moving object. It can be concluded, that the damping force due to the eddy currents is linear in the velocity. This linearity in the velocity allows a numerical calculation of the damping force for a specific velocity. Afterwards a damping parameter can be calculated by dividing the damping force by this velocity. Further analysis may be carried out, using lumped models with the evaluated damping parameter. Note, that as the magnetic field of the induced eddy currents is neglected in this derivation, the method is only suitable, for (rather) low velocities.

3.2 Nonlinear Eddy Current Damping Element

The analysis in the previous subsection revealed that the damping force of a magnet moving in a conductive tube is proportional to the velocity, thus it behaves identical to linear viscous damping. In this section, the model of a permanent magnet moving in a conductive tube, as proposed by Bae et al. [1], is upgraded with geometric discontinuities for position-dependent damping behavior. Therefore a gap is introduced in the conductive tube as shown in the system in Fig. 13b.

To calculate the position-dependent damping parameter, a FEA model of the magnet in the conductive tube with a gap has been set up. In a time dependent study the magnet was moved with constant velocity v_0 through the conductive tube and the damping force F_d was calculated at each position. Afterwards, the resulting damping force F_d has been divided by the velocity v_0 to obtain the damping parameter. Furthermore, it has been normalized to a maximum value of one and stretched, so that the maximum damping value is reached at $\xi = \pm 1$. The resulting normalized position dependent damping parameter is depicted in Fig. 14a. For the FEA again the moving mesh formulation has been used to adapt the mesh during the simulation. Figure 14b shows the FEA model at a specific time step.

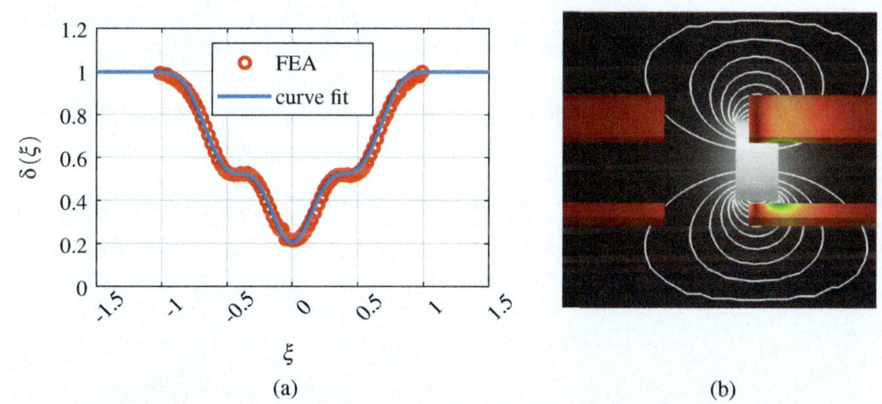

Fig. 14 **a** Normalized damping coefficient of position dependent eddy current damping. Red circles: result of FEA—Blue line: fitted curve used for dynamic analysis. **b** Model of the eddy current damper with geometric discontinuities and position-dependent inductive damping. White lines: magnetic field lines—Gray gradient (in air): magnitude of magnetic flux density—Colored gradient (on conductive tube): magnitude of eddy current density

For the analysis of the system depicted in Fig. 13b, the normalized damping parameter has been fitted with the curve shown in Fig. 14a. The equations of motion for the system are given by

$$m\ddot{x} + d(x)\dot{x} + kx = \hat{F}\sin(\Omega t), \tag{46}$$

where m is the mass of the moving object, $d(x)$ is the position dependent damping coefficient (not normalized), k is the stiffness of the spring, $\hat{F}$ is the amplitude of an external harmonic force and Ω is its frequency. The position of the moving object is described by x. To minimize the number of parameters, the equation of motion is converted into dimensionless form. Based on the characteristics $\omega_0 = \sqrt{k/m}$ and thus $\tau = \omega_0 t$, the non-dimensional parameters $\eta = \Omega/\omega_0$ and $\hat{f} = \hat{F}/(k\ell)$ are introduced. Moreover, the scaled coordinate $\xi = \frac{x}{\ell}$ will be used, where ℓ is the reference length used for the stretch of the damping parameter. Introducing the re-scaled parameters and coordinates the damping term may be transformed as

$$d(x) = d_0\delta\left(\frac{x}{\ell}\right) = d_0\delta(\xi) \quad \text{and} \quad \frac{d_0}{m} = 2D\omega_0 \tag{47}$$

where D is the damping factor and $\delta(\xi)$ is the normalized damping parameter, as depicted in Fig. 14a. The dimensionless equation of motion is given by

$$\xi'' + 2D\delta(\xi)\xi' + \xi = \hat{f}\sin(\eta\tau). \tag{48}$$

To solve the nonlinear differential equations, again a simple shooting method is used. For the analysis, the maximum value of the damping parameter has been set to $D = 1$. Figure 15a shows the frequency response of the nonlinear damped single degree of freedom device. The graph of the dynamic flexibility is depicted in Fig. 15b. It shows that for higher excitation levels the resonance peak becomes lower, and thus the position dependent damping allows for an amplitude dependent damping

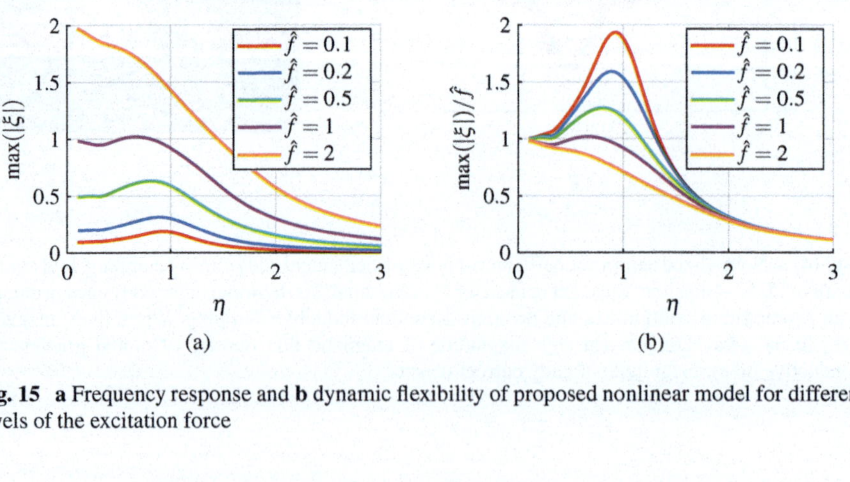

Fig. 15 **a** Frequency response and **b** dynamic flexibility of proposed nonlinear model for different levels of the excitation force

behavior. While small oscillations remain mainly unaffected, large oscillations are efficiently suppressed. This behavior might be favorable in situations, where for a better efficiency of a system a low damping ratio is necessary, but still large vibrations must be prevented.

3.3 Nonlinear Magnetically Damped Tuned Mass Damper

The proposed nonlinear damping device could as well be used as a magnetically damped tuned mass damper (TMD). A basic model of this is shown in Fig. 16. The equations of motion are given by

$$\begin{bmatrix} M+m & m \\ m & m \end{bmatrix}\begin{bmatrix} \ddot{x} \\ \ddot{z} \end{bmatrix}+\begin{bmatrix} 0 & 0 \\ 0 & d_0\delta(z/\ell) \end{bmatrix}\begin{bmatrix} \dot{x} \\ \dot{z} \end{bmatrix}+\begin{bmatrix} k_0 & 0 \\ 0 & k \end{bmatrix}\begin{bmatrix} x \\ z \end{bmatrix}=\begin{bmatrix} \hat{F}\sin(\Omega t) \\ 0 \end{bmatrix}, \tag{49}$$

where x is the coordinate describing the position of the primary mass M and k_0 is the stiffness of the spring connecting it with the environment. z is the coordinate describing the position of the TMD with the mass m and k is the stiffness of the spring connecting it with the primary mass. d_0 is the damping coefficient and $\delta(z/\ell)$ is the normalized position dependent damping parameter with the reference length ℓ as discussed in the previous subsection. To minimize the number of parameters, the scaled quantities

$$\omega_0 = \sqrt{\frac{k_0}{M}}, \quad \omega_T = \sqrt{\frac{k}{m}}, \quad D_T = \frac{d_0}{2m\omega_T}, \quad \mu = \frac{m}{M}, \quad \nu = \frac{\omega_T}{\omega_0}, \tag{50}$$

$$\hat{f} = \frac{\hat{F}}{k_0}\ell, \quad \tau = \omega_0 t, \quad \eta = \frac{\Omega}{\omega_0}, \quad \xi = \frac{x}{\ell}, \quad \zeta = \frac{z}{\ell} \tag{51}$$

are introduced. The dimensionless equations of motion read

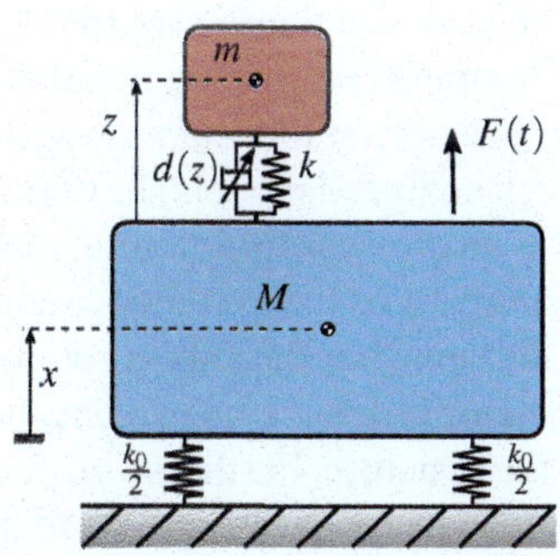

Fig. 16 Single degree of freedom oscillator with a magnetically damped tuned mass damper

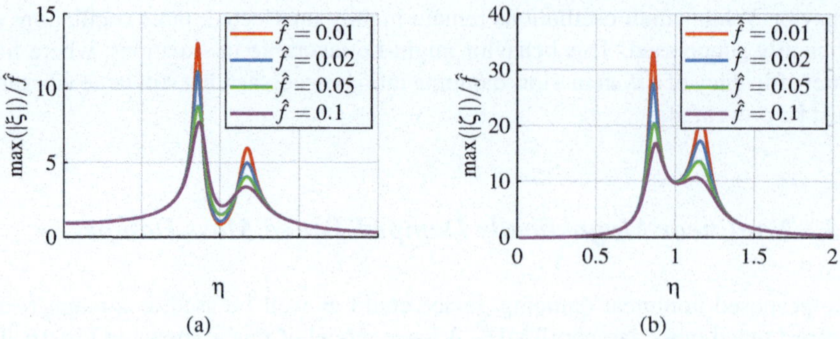

Fig. 17 Dynamic flexibility of the system with nonlinear damped TMD for different excitation levels. **a** Primary mass. **b** TMD

$$\begin{bmatrix} 1+\mu & \mu \\ \mu & \mu \end{bmatrix} \begin{bmatrix} \xi'' \\ \zeta'' \end{bmatrix} + \begin{bmatrix} 0 & 0 \\ 0 & 2\,\mu\nu D_T \delta(\zeta) \end{bmatrix} \begin{bmatrix} \xi' \\ \zeta' \end{bmatrix} + \begin{bmatrix} 1 & 0 \\ 0 & \mu\nu^2 \end{bmatrix} \begin{bmatrix} \xi \\ \zeta \end{bmatrix} = \begin{bmatrix} \hat{f}\sin(\eta\tau) \\ 0 \end{bmatrix}. \quad (52)$$

The dynamic flexibility charts for different excitation levels for the TMD and for the primary mass are shown in Fig. 17. The values of the parameters used for the analysis are $\mu = 0.1$, $\nu = 1$ and $D_T = 0.2$.

Due to the presence of damping in the TMD, the resonance amplitudes are limited. However, near the designed operating point (here: $\eta \approx 1$) the system behaves similar to a weakly damped TMD which may show very effective vibration compensation. Thus, this nonlinear damper might combine the benefits of weakly damped TMDs with the operational safety of optimally damped TMDs, which have smaller resonance amplitudes, than weakly damped TMDs.

4 Conclusion

In a first step, basic functional elements of inductive damping devices were identified and classified into a matrix. Using this schematic, a systematic derivation of possible damping designs may be obtained by re-combining several options.

Based on this matrix, two basic designs were derived and analyzed in more detail. All analyzed models show the possibility to efficiently reduce the vibration amplitude of an oscillating structure, modeled as a single degree of freedom oscillator. From the analysis of the systems, modulating the magnetic flux due to mechanical movement and guiding the flux through the structure, it was concluded, that saturation has a major influence on systems based on electromagnets, if only small air gaps occur in the structure. Furthermore, it was shown, that flux leakage paths must be implemented in an analysis, as they strongly decrease the damping performance.

The analysis of the considered eddy current damping elements showed, that neglecting the field of the eddy currents, the resulting damping force is proportional to the velocity. As the analytic calculation of eddy currents is only favorable for simple geometries, a coupled numeric-analytic analysis was presented, where the damping coefficient is calculated using FEA and the result is integrated into a lumped parameter mechanical model. Using this procedure, the damping parameter of a position dependent eddy current damper was evaluated and the dynamic behavior of the system was analyzed. It was shown, that the proposed model is capable of reducing predominantly large oscillations. Furthermore, the position-dependent damping element was used in a TMD. The system with the TMD behaved similar to a weakly damped TMD for small oscillations, but limited resonance amplitudes effectively.

References

1. Bae, J.-S., Hwang, J.-H., Park, J.-S., Kwag, D.-G.: Modeling and experiments on eddy current damping caused by a permanent magnet in a conductive tube. J. Mech. Sci. Technol. **23**(11), 3024–3035 (2009)
2. Bae, J.-S., Hwang, J.-H., Roh, J.-H., Yi, M.-S.: Development of an electromagnetic shock absorber. Int. J. Appl. Electromag. Mech. **49**(1), 157–167 (2015)
3. Bae, J.-S., Hwang, J.-H., Kwag, D.-G., Park, J., Inman, D.J.: Vibration suppression of a large beam structure using tuned mass damper and eddy current damping. Shock Vibr. **2014** (2014)
4. Behrens, S., Fleming, A.J., Reza Moheimani, S.O.: Electromagnetic shunt damping. In: Proceedings IEEE/ASME International Conference on Advanced Inteligent Mechatronics, pp. 1145–1150 (2003)
5. Laborenz, J., Siewert, C., Panning, L., Wallaschek, J., Gerber, C., Masserey, P.-A.: Eddy current damping: a concept study for steam turbine blading. J. Eng. Gas Turb. Power – Trans. ASME **132**(5), 052505-1–052505-7 (2010)
6. Laborenz, J., Krack, M., Panning, L., Wallaschek, J., Denk, M., Masserey, P.-A.: Eddy current damper for turbine blading: electromagnetic finite element analysis and measurement results. J. Eng. Gas Turb. Power – Trans. ASME **134**(4), 042504-1–042504-8 (2012)
7. Lian, J., Zhao, Y., Lian, C., Wang, H., Dong, X., Jiang, Q., Zhou, H., Jiang, J.: Application of an Eddy current-tuned mass damper to vibration mitigation of offshore wind turbines. Energies **11**(12) (2018)
8. Prtybylowicz, P.M., Szmidt, T.: Electromagnetic damping of a mechanical harmonic oscillator with the effect of magnetic hysteresis. J. Theor. Appl. Mech. **47**(2), 259–273 (2009)
9. Prtybylowicz, P.M., Szmidt, T.: Nonlinear response of a harmonically driven oscillator in magnetic field. Arch. Control Sci. **20**(1), 19–30 (2010)
10. Rosenboom, M., Hetzler, H.: Damping based on electromagnetic induction: a comparison of different minimal models. In: Proceedings in Applied Mathematics and Mechanics, vol. 20, no. 1 (2021)
11. Rothwell, E.J., Cloud, M.J.: Electromagnetics. CRC Press LLC, Boca Raton (2001)
12. Sodano, H.A., Bae, J.-S., Inman, D.J., Belvin, W.K.: Concept and model of eddy current damper for vibrations supprsion of a beam. J. Sound Vibr. **288**(4–5), 1177–1196 (2005)
13. Sodano, H.A., Bae, J.-S., Inman, D.J., Belvin, W.K.: Improved concept and model of Eddy current damper. J. Vibr. Acous. - Trans. ASME **128**(3), 294–302 (2006)
14. Sodano, H.A., Inman, D.J.: Modeling of a new active eddy current vibration control system. J. Dyna. Syst. Measur. Control – Trans. ASME **130**(2), 021009-1–021009-11 (2008)
15. Sudhoff, S.D.: Power Magnetic Devices. Wiley, Hoboken (2014)
16. Woodson, H.H., Melcher, J.R.: Electromechanical Dynamics - Part 1 Discrete Systems. Wiley, Hoboken (1968)

The Role of Damping in Complex Structural Dynamics: Data-Driven Approaches

Merten Stender and Norbert Hoffmann

1 Introduction and Problem Statement

Structural dynamics and computational simulation approaches have reached high levels of maturity over past decades. High-performance computing allows for simulating large structures at high precision. At the same time, linear and weakly nonlinear system identification techniques are available for the parameterization of mathematical models. Uncertainty quantification allows for propagating measurement errors, manufacturing tolerances, and the-like modeling inaccuracies into the simulation results. Viewing the problem of damping design from a conceptual perspective, one could therefore say that the problem should be solved today. Within the context of accurate modeling and determinism, model based approaches should allow to evaluate stability boundaries and even non-linear solutions after branching. In fact, this idea and approach has been and is about to be followed in a large number of disciplines. In some fields of application, the model based approach has today lead to tools to analyze damping behavior and to design damping elements [1]. In other fields, the purely model based approaches do, however, still fall short of allowing for robust system design, and in some fields, like e.g. brake noise, many researchers would today say that physics-based approaches do still not work at all [2, 3]. The key ingredients to a successful modeling of structural dynamics are i) correct mathematical models and ii) correct parameterization of those. Challenges arise when either of those ingredients cannot be met at high precision. For example, frictional processes are inherently connected with damping. However, they are difficult to measure and model, therefore the simulation of complex friction-affected structures is still a

M. Stender (✉) · N. Hoffmann
Hamburg University of Technology, Dynamics Group, Am Schwarzenberg-Campus 1, 21073 Hamburg, Germany
e-mail: m.stender@tuhh.de

N. Hoffmann
e-mail: norbert.hoffmann@tuhh.de

P. Eberhard (ed.), *Calm, Smooth and Smart*, Lecture Notes in Applied and Computational Mechanics 102, https://doi.org/10.1007/978-3-031-36143-2_5

challenge. The two research projects at Hamburg University of Technology (TUHH) lead by Prof. N. Hoffmann therefore addressed a data-driven perspective on damping in structural dynamics.

1.1 Nonlinear Dynamics

The field of nonlinear dynamics has emerged at least about half a century ago in mathematics and the sciences, and now can be considered as an established domain. Science and engineering have traditionally developed along the lines of thinking in terms of equilibria, stationarity, and then from there on to more complicated non-stationary behavior. In many cases, design processes proceed by assuming stationary operation, with instationarity forming perturbations to steady operation [4]. This has worked well in many cases in the past, but the limitations of the approach have become apparent [5, 6]. Environmental loads are very complex and can be described in terms of means and deviations only comparatively poorly. Also operational strategies are usually non-stationary, what can well be seen in the field of autonomous driving, where cars are practically never subject to the same operational state [7]. In the context of 'Industry 4.0', global plant control strategies will cause even more dynamic loads and higher performance requirements which have to be taken into account [8].

Chaotic dynamics [9] are typically related with the extreme sensitivity of a fully deterministic process to a small change in initial conditions. Very minor perturbations will cause an exponential divergence of trajectories, thereby limiting the prediction horizon. Considering uncertainties, numerical rounding and measurement noise, the implications of the concept of chaos are crucial. Although the findings in chaotic dynamics and complexity have made their way to many domains, like meteorology for weather and climate analysis, or also to the life-sciences, the field did have an impact on the design of complex dynamical machines mostly with respect to weak nonlinearities. Many further reaching achievements of nonlinear dynamics, which have broken ground in the complex system sciences [10] still need to find their way into machine dynamics design. Classically, machines are designed for a linear and regular behavior, although many recent data analyzes show that machine dynamics and machine vibrations usually exhibit rather irregular and chaotic motion [11–13]. One may even speculate that controlled chaotic motion could extend component lifetimes due to the broader spectrum of loads compared to strictly repetitive patterns of regular motion. However, the deliberate design of machines for low-amplitude chaotic operation conditions has, astonishingly, not found much attention yet.

1.2 Complexity and Structural Dynamics

In the field of nonlinear structural dynamics, a large number of advanced validation methods have been developed in recent years, including control-based continuation into the nonlinear domain [14], nonlinear time-series analysis [15], recurrence plot analysis [16], to name just a few. Independently from this, a wide range of disciplines from science, technology, medicine and other disciplines have been subjected to analysis and validation tasks which share many characteristics known from nonlinear structural machine dynamics. Over the years, these communities have come to summarize their systems as 'complex systems', and rather recently it has shown that many of the problems that arise can well be studied in the context of networks and graph theory. In the present field of nonlinear structural dynamics, it seems that the system and dynamics analysis methods from the field of complexity and network theory have hardly been studied, analyzed or exploited by now. Purely data-driven techniques and machine learning are currently making their way into many application-related fields of structural dynamics.

1.3 Damping Paradox

In externally excited systems, the dynamic response depends crucially on the dissipative mechanisms involved, and in almost all cases introducing dissipative design elements, i.e. dampers, reduces the strength, amplitude, or level of the dynamic response. For linear systems the corresponding theory has been standing for a long time, and for weakly non-linear systems it can be extended. For systems that may show self-excited vibration, the situation can be much different. Here, certain system parameters may fall in ranges where instability results from additive damping [17]. Finite-amplitude responses in the form of regular or irregular response result, including limit-cycles, quasi-periodic solutions, chaos. Hence, when it comes to either stability boundaries, or to characteristics of weakly or strongly non-linear solutions, the influence of dissipation can be very complicated. Often times, additional damping elements plainly stabilize a system. Very early on, however, it has already been noticed that an increase in damping, e.g. through putting additional dampers into a system, stability boundaries may become worse than without damping. Systems may even be destabilized through adding or increasing damping. This characteristic has actually, over time, been uncovered in a very large number of disciplines: physics, engineering, the geosciences [18–20]. The mechanical or mathematical origin of this destabilizing character of damping is most often to be found in the combination of the effects that additional damping elements may have on the overall system structure and dynamics: local dampers may alter the structure of the flow of energy through the structure, they may change amplitude and phase behavior, and these effects, intrinsic to the system, may suffice to turn a stable into an unstable system.

1.4 Friction and Damping

Mechanical friction is a complex process that spans multiple time and length scales [21]. The relevance of various frictional processes and phenomena is obvious for brake systems, clutches, drill strings and various other mechanical machines. Being a multi-physic and multi-scale process, friction is inherently difficult to model for a given contact between two bodies. Over-simplified approaches like Coulomb friction exist to capture the overall logic of friction. Realistic frictional contacts are however much more complex. Ultimately, frictional behavior contradicts classical understanding of linearizations and nonlinearities: classical, nonlinear behavior is expected for large deflections and deviations from an equilibrium. In contrast, friction can be approximated to behave linearly for large deflections or relative sliding velocities. In the range of small sliding velocities, frictional behavior is however strongly nonlinear. As a result, linearization about a point of operation may not always be appropriate for friction-affected systems, is however the standard in finite element simulations of brake systems. Moreover, friction can act as both an energy sink and energy source. Frictional damping will dissipate energy and transform motion into heat or deformation (frictional wear). Mechanical joints were shown to contribute the majority of dissipation to mechanical structures [22] through friction damping. At the same time, friction can also feed energy back into the structure and thereby excite finite-amplitude vibrations, commonly denoted as self- or friction-excited vibrations (FIV). Velocity-dependent friction can result in negative damping and thereby destabilize a structure, which is typically the case for stick-slip self-excited vibrations. Mode-coupling (binary flutter) instability [23] can be another root cause for FIV which exists also for constant friction. Here, a parameter variation in the frictional contact will couple two structural modes, one of which is destabilized.

Friction must thus be understood as a central aspect of many mechanical machines, crucially affecting the dissipation and the resulting vibration behavior. Owing to the complexity of frictional processes, their sensitivity, and narrow stability margins, understanding damping in mechanical structures is closely related with friction and self-exited vibrations. An important recent insight in self-excited system is that many of the systems under study demonstrate irregular rather than regular dynamics when investigated in detail [6, 11, 12]. Most often the resulting solutions appear to be periodic only at first sight [24], while in fact closer data analysis demonstrates that the seemingly periodic solutions rather do form tori or strange attractors. In a sense, this finding should not be considered to be too surprising. In many self-excited systems actually dynamics on a wide band of scales takes place, and the usually implicitly assumed separation of scales works in an approximate manner at most, if at all. Nevertheless, a wide range of theory and tools for irregular dynamics, which were often developed in stronger idealized mathematical environments, could be used and applied - although this is hardly ever done.

1.5 *Structure of This Work*

This work aims at summarizing the research efforts conducted in two three-year periods of the Priority Programme SPP 1897 under the project lead of Prof. Norbert Hoffmann at Hamburg University of Technology. As such, we re-visit central findings that have been published in the course of the SPP 1897 in 18 peer-reviewed journal publications. The illustration of the work does not follow a strict chronological order, but aligns with central aspects of research on data-driven damping identification in complex and friction-affected systems. Particularly, Sect. 2 introduces several approaches to signal processing, Sect. 3 revisits system identification techniques, Sect. 4 presents physics-based modeling methods, and Sect. 5 focuses on data-driven models. Section 6 summarizes highlights and several future directions of research, before Sect. 7 gives a comprehensive conclusion.

2 Signal Processing

Signal processing is a research field concerned with the extraction of information from (measured) data, which typically takes the form of univariate and multivariate time series in the area of structural dynamics. As such, signal processing is closely related with system identification tasks. Central aspects and challenges in signal processing for structural dynamics lie in the sparsity of the data: not every degree-of-freedom can be measured and the sampling frequencies of sensors are limited. Furthermore, measurements are typically contaminated with noise. This section presents how nonlinear signal processing techniques can be leveraged to understand complex vibrations, thereby extending the classical toolset beyond linear approaches such as the Fourier transform. As techniques from nonlinear time series analysis and recurrence analysis are not too common in structural dynamics, the central concepts of those methods are briefly revisited first.

2.1 *Phase Space Embeddings*

Since the 1980s a large number of conceptual approaches have been proposed for studying characteristic features of nonlinear dynamical systems based on measured time series [15]. Importantly, these time series are sparse in the sense that only a minimal set of degrees-of-freedom of the physical system can be sensed. At the core of the approach is the idea to characterize nonlinear invariant measures in phase space. This is possible in a number of different ways. Usually, the system and its dynamics are characterized from either a geometric or a dynamic perspective. Important examples include fractal dimensions, Lyapunov exponents as measures for stability of dynamics with respect to infinitesimal perturbations, and various forms of entropy to evaluate the uncertainty about the future state of a chaotic trajectory. A typical task is to perform a precise system characterization from a single time series, or from a set of time series. A time series $x_i = x(t_i)$ can be understood a finite representation of the trajectory $x(t)$ of some dynamical system. The data is converted into state vectors in some appropriately reconstructed phase space. A common method from dynamical systems theory to define such a phase space is time-delay embedding [25, 26]. One constructs $\mathbf{q}_i = \left[x_i, x_{i+\tau}, \ldots, x_{i+(m-1)\tau}\right]^{\top}$ to obtain an m-dimensional time-delay embedding of $x(t)$ with delay τ and state vectors $\mathbf{q}(t)$ in the reconstructed phase space $\mathbb{R}^m$. For deterministic dynamical systems, the reconstructed phase space is topologically equivalent to the original space if $m > 2D_{\mathrm{F}}$, where D_{F} is the fractal dimension of the support of the invariant measure generated by the dynamics in the true (but often at most partially observed) state space. The false nearest-neighbors (FNN) method [27] is an approach to derive a reasonable guess of how to choose m. If a reasonably large embedding dimension is determined, all dynamically relevant properties of the system are appropriately represented. Similarly, a delay τ is appropriate when the statistical dependence between the components of the embedding vectors $\mathbf{q}$ approaches zero. This can be achieved by choosing τ corresponding to properties of the auto-correlation function of the time series data. Another well established strategy for determining τ is to use the first minimum of the time-delayed mutual information [28]. Once the embedding has been achieved properly, all invariant measures of the dynamical system, like Lyapunov exponents or fractal dimensions, can be extracted following the traditional methods from the theory of dynamical systems.

2.2 *Recurrence Analysis*

A shortcoming or limitation of the use and application of embedding and phase space reconstruction methods arises when the embedding dimensions become larger, which makes direct visualization of the resulting dynamics difficult. Additionally, phase space reconstruction is theoretically only possible for purely deterministic signals, hence excluding noise-contaminated signals and processes that involve stochastic parts. As a consequence, recurrence plots [16] have been introduced, on which subsequent analysis, often referred to as recurrence quantification analysis, can be performed. Recurrence of states, meaning that states get arbitrary close to earlier states after some time, is a fundamental property of deterministic dynamical systems and is typical for many nonlinear or systems. From the set of (original or reconstructed) state vectors $\mathbf{q}$ representing a discrete sampling of the underlying system's trajectory (e.g., the chaotic attractor of a dissipative structural system), recurrences can be very well visualized by recurrence plots (RP), originally introduced by Eckmann et al. [29]. The RP is a graphical representation of the corresponding recurrence matrix $\mathbf{R}(\epsilon)$, usually defined as $R_{i,j}(\epsilon) = \theta(\epsilon - \left|\mathbf{q}_i - \mathbf{q}_j\right|)$, where $|\cdot|$ may denote any norm. A RP enables us to investigate the recurrences of the m-dimensional phase space trajectory through a two-dimensional graphical representation $R_{i,j}$ in terms of black and white dots indicating recurrent and non-recurrent pairs of vectors. The algorithmic parameter ϵ is a threshold value which determines whether two state vectors are close or not. RPs of dynamical systems with various types of dynamics exhibit distinct structural properties, which can be characterized in terms of their small-, medium- as well as large-scale features [30]. The study of recurrences by means of RPs has become popular with the introduction of recurrence quantification analysis (RQA) [31]. RQA measures use the distribution of small-scale features in the RP, namely individual recurrence points as well as diagonal and vertical line structures. One should also remark that even dynamical invariants, like the K2 entropy, mutual information, or fractal dimensions (i.e., the information and correlation dimensions) can be efficiently estimated from RPs.

2.3 *Applications and Findings*

Within the SPP 1897, several sub-projects were performed for structural dynamics diagnosis: given a single time series measurement of vibration, very detailed analysis was performed to understand the characteristics of the vibrations, transitions, and dynamical invariants. Friction brakes, and particularly disc brakes, have been studied for long times in order to understand and mitigate friction-excited vibrations that may annoy customers of automotive manufacturers. A range of different acoustics can be emitted by the brake system during braking, ranging from low-frequency *moan* sounds (100–800 Hz) to high-frequency *squeal* sounds (1–12 kHz). These vibrations have long been understood as regular limit-cycle solutions with a strict periodicity, more detailed analysis have revealed more complicated and chaotic dynamics [11–13].

Within the SPP 1897 we studied a large amount of real-world brake system vibrations with a focus on the qualitative and quantitative characteristics of the underlying nonlinear dynamics in [32]. Particularly, a systematic study was performed on the sound pressure level emitted during brake system testing on a noise dynamometer according to the SAE-J2521 standardized protocol at AUDI AG Ingolstadt. The measurement time series were subjected to a sliding window processing, a fully automated embedding process, and recurrence quantifiers were derived from the resulting recurrence plot representations. For the first time, a very large set of brake system vibrations were characterized by nonlinear invariant measures. Figure 1 illustrates an exemplary measurement signal, the reconstructed phase space derived through time-delay embedding, and the resulting recurrence plots. In the non-squealing regime at the beginning of the braking, no structure is observed in the reconstructed trajectories. Some long-term periodicities are found in the recurrence plot, potentially being linked with the disk rotation. During the onset of high-amplitude vibrations, a circular structure forms in the state space, thereby indicating strong periodicity with super-imposed fluctuations. Diagonal line structures begin to grow in the recurrence plot. In the fully developed vibration case, the trajectories form a limit cycle-like

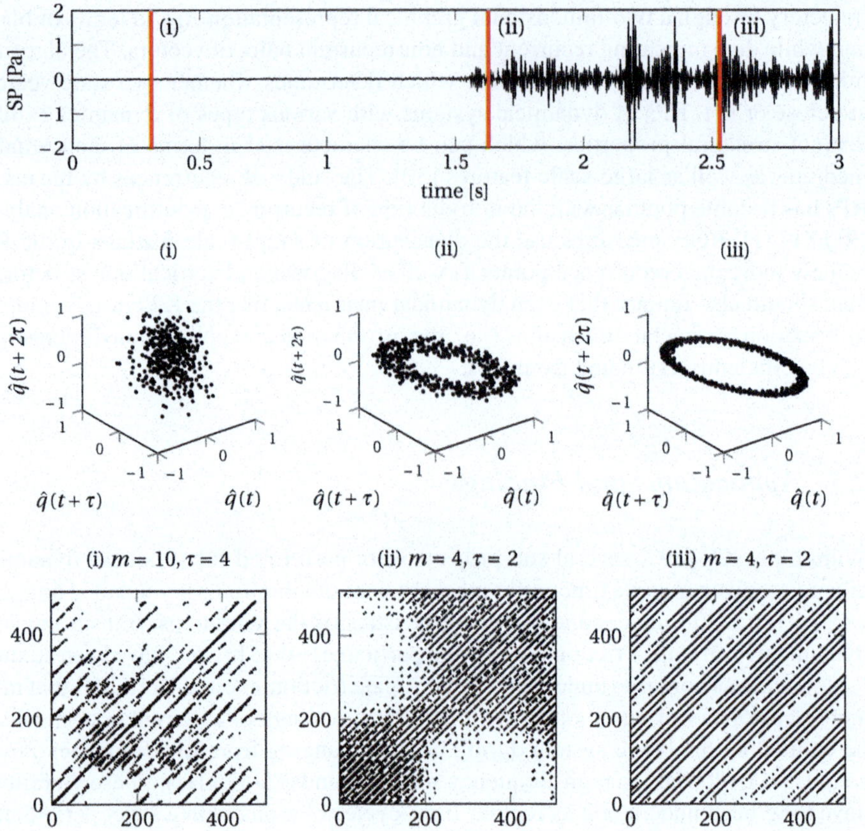

Fig. 1 Microphone measurement of the sound emissions of a single brake stop (top); three-dimensional representation of the reconstructed phase space using embedding parameters m and τ (middle); recurrence plots computed by the fixed-number of neighbors norm (bottom). Illustration adapted from [32]

signature, which is supported by long diagonal recurrence plot lines. Super-imposed modulations at a slower time scale can be read from the periodic interruptions in the recurrence plot. While all of these characteristics would be hardly distinguishable in time domain or in frequency domain, the state embedding and recurrence plot analysis allows for a very detailed analysis of the observed vibrations. The systematic analysis of the embedding parameters and the recurrence quantifiers for 156 min of braking, then separated into non-squealing and squealing epochs, reveals the main findings reported in [32]:

- The relevant dimensionality ($m = 8, \ldots, 12$) of steady sliding dynamics without sound emissions is higher than the dimensionality ($m = 3, \ldots, 6$) of high amplitude vibrations emitted during self-excited squeal events. The high-amplitude vibrations turn out more deterministic and only weakly chaotic, while quiet sliding can either be related with mostly stochastic or highly chaotic characteristics.

- Brake system vibrations exhibit clear signatures of temporal multi-scale dynamics, possibly introduced by different processes in the frictional contact, as well as external loading such as the disk rotation.
- Stationarity and instationarity of brake dynamics tend to depend on the perspective taken to analyze vibration time series. While the dynamics in time domain seem very transient and constantly changing, e.g. in amplitude, a recurrence matrix analysis reports rather constant invariant quantifiers. Even though the actual dynamics change, the qualitative properties and characteristics remain constant.

- **Effective damping and the distance of the instantaneous system state from the stability border** were able to be measured by recurrence quantifiers such as entropy and average line length, which has not been possible before. This result highlights the relevance of nonlinear signal processing techniques for understanding complex dynamics and damping in large multi-component structures.

A different work [33] studied the self-excited vibrations of two metallic bodies in dry frictional contact subjected to an imposed relative sliding velocity at La Sapienza Universita di Roma. The specimen and the complete experimental apparatus were supported by air bearings, thereby minimizing the influence of external vibrations or disturbances. Despite the simplicity of the set-up, rather complex vibrations arose from the sliding motion. Laser doppler velocimetry (LDV) measurements of the vibrations were analyzed using phase space embedding and successive recurrence analysis. Figure 2 illustrates an exemplary analysis, comparing a time-frequency analysis by short-time Fourier transform (STFT) to a rolling-window recurrence quantification analysis (RQA) over a duration of just 1.5 s of sliding. The STFT reveals many different strong periodicities as well as transitions between them. However, the analysis of the STFT would be limited to dominant frequencies and respective vibration amplitudes. The RQA however, is able to quantify the different epochs of the vibration signal, for example by the diagonal and vertical line lengths L, V. Some epochs appear very constant with respect to RQA, while transitions are clearly visible in the statistics of the vertical line lengths V. The interpretation of those metrics allows to draw conclusions about the degree of (ir)regularity, laminarity and also attractiveness, i.e. the actual damping at play. The labeled sections in Fig. 2 relate to different phases of vibration: purely harmonic oscillation (a), tori (b), tori with laminar fast-slow phases (c), multi-harmonic vibrations (d,e), chaotic vibrations (f) and low-deterministic or random-like vibrations (g).

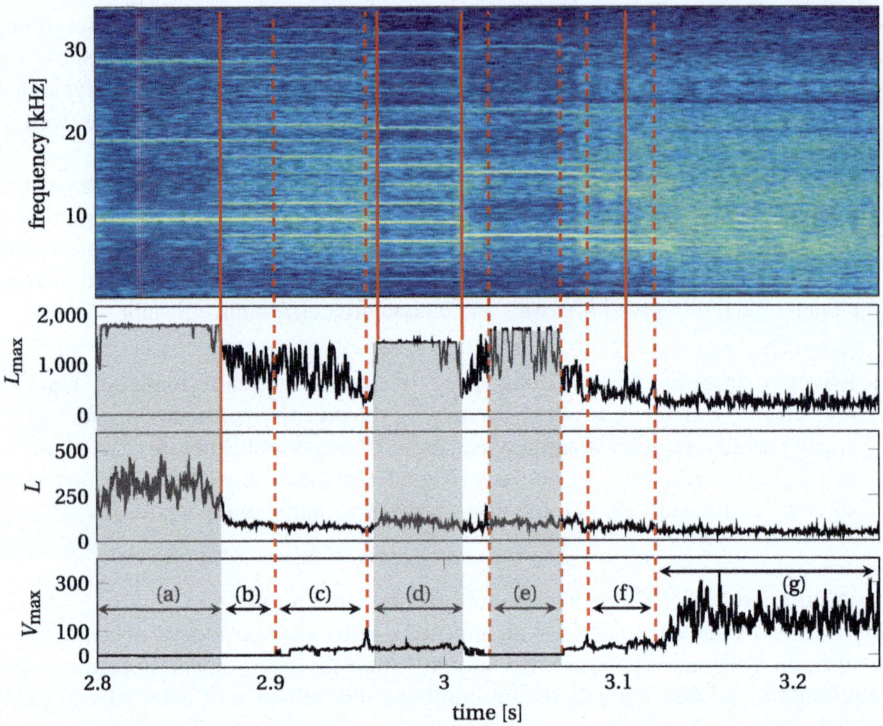

Fig. 2 Comparing a linear time-frequency analysis (top panel) with a nonlinear recurrence quantification analysis (bottom panel) for the motion of two friction-excited bodies, adapted from [33]

The research work conducted in the SPP project established nonlinear time series and recurrence plot quantification analysis as valuable tools for in-depth and nonlinear signal processing techniques. On the example of brake system applications and friction-excited dynamics, we were able to show that the resulting dynamics mostly form weakly chaotic attractors, that these dynamics can be stationary with respect to qualitative measures, and that damping and stability properties can be encoded in the instantaneous vibration signatures.

Besides the two works discussed before, the articles [34–36] were published in the course of the SPP on signal processing-related topics.

3 System Identification

Right after signal processing, the system identification will take place in a data-driven perspective on partially unknown systems like the ones that were treated in this project. System identification aims at deriving mathematical structures and descriptions from observations. These mathematical descriptions carry parameters that will allow the model to generalize to different situations, e.g. other initial conditions. Importantly, the identified parameters will be interpreted as stiffness or damping characteristics of a vibrating structure. Given complex dynamics that are strongly nonlinear, existing (linear) identification techniques for stiffness or damping are mostly not applicable.

Stender et al. [37] introduced a novel signal processing and damping estimation technique for self-excited vibrations at the example of brake squeal. Following the outlines given in the introduction, vibrations will be excited once the effective damping in the structure is not sufficient to dissipate the frictional energy fed into the system. In such scenario, vibration amplitudes will grow at an exponential rate $x(t) = C \cdot \exp(\lambda t)$, where λ is the largest positive eigenvalue of the linearized dynamical system. Therefore, λ directly encodes the effective damping at play, in this case negative damping. Many experimental studies on self-excited vibrations have related the finally obtained vibration amplitude with the damping properties of the linearized system: if large vibrations are observed, the corresponding real part of the eigenvalue is assumed to be large. However, such an understanding contradicts basic theoretical concepts of linear and nonlinear dynamics: amplitudes grow at exponential rate λ, but they will be limited by nonlinearities. Therefore, the final vibration amplitude is clearly related to nonlinear effects, which do not encode linear damping properties of the vibrating structure. The contribution of [37] is to provide a methodology to extract the growth rate λ from vibration measurements, and thereby allow for system identification of damping properties. Given a measurement time series, the point of instability and amplitude growth is found through amplitude-based criteria. Spectral filtering (using inverse Fourier transformation and wavelet bandwidth filtering) of the time series is performed in order to carve out the relevant frequency range of the final vibration. The instances of exponentially growing amplitudes is extracted, and an exponential fit to the amplitude envelope signal is performed. Thereby, the rate is estimated at which the amplitudes of a certain vibration frequency grow. In order to obtain more robust estimates of the grow rate, the data fitting range (time range describing the amplitude growth) is varied for more than 100 times. The resulting statistics give an indication about the estimation error and the uncertainty of the growth rate λ extracted from the experimentally measured time series. Finally, the growth rate can be used for model-updating of simulation models, typically large FE models. Figure 3 depicts the exponential fitting procedure to the filtered amplitude envelope function. Studying a complete test run of hundreds of brake stops, the relationship between growth rates and final vibration amplitudes was established. For the first time, it was shown that those two quantities are in fact de-correlated as theory would suggest.

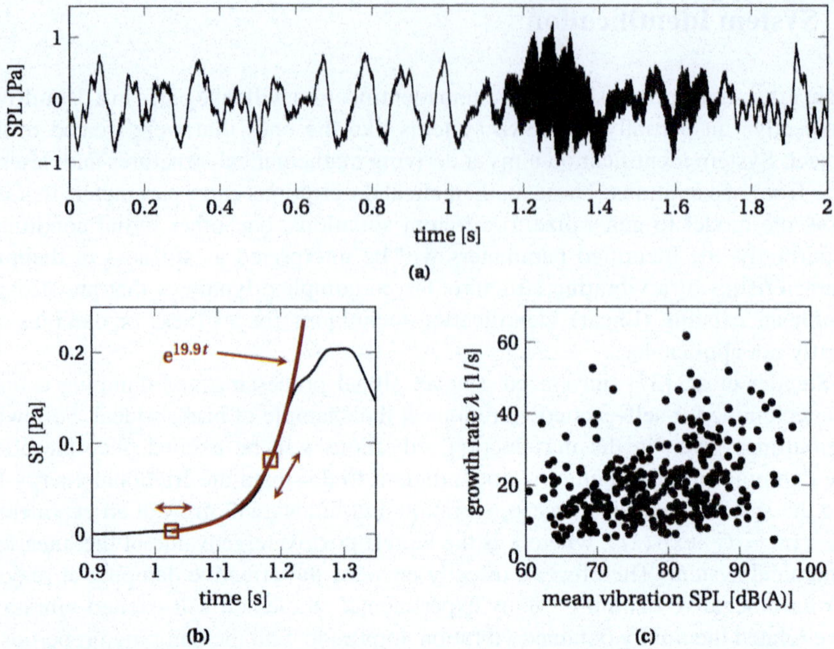

Fig. 3 Extraction of negative damping values from noise-contaminated vibration measurements of the sound pressure level (SPL): **a** sound emissions of a brake system exhibiting self-excited vibrations from 1.05 s onwards. **b** Identification of the exponentially growing amplitude envelope function after filtering and successive fitting of an exponential. **c** Relation between extracted growth rates λ and the final vibration amplitudes. Adapted from [37]

In a different work [38] in collaboration with University of Stuttgart, the effective damping properties of particle dampers were extracted from the forced response of a particle-damped beam structure. The dynamics of particle-damped structures are inherently multi-scale: compared to the overall structure's size and mass, the individual particles in the damper are small and light-weighted. The large number of particles interact with other particles and the damper's container walls, and hence with the structure to be damped. Using experimental data from a ring-down measurement, a time-scale separation was performed using the discrete wavelet transform. Thereby, the vibrational energies contained in different time scales can be extracted. On the slow scales, the bending modes of the beam structure dominate the response. On the very fast scales, the particle activity is reflected, which stems from collisions in the particle container. The effectiveness, i.e. the damping potential, of a particle damper can be estimated using computationally demanding discrete element simulations. Every particle collision and trajectory has to be computed in order to compute the cumulative dissipation force acting on the structure. It would therefore be beneficial to employ a different, much faster technique, to estimate the particle activity and damping potential from experimental studies.

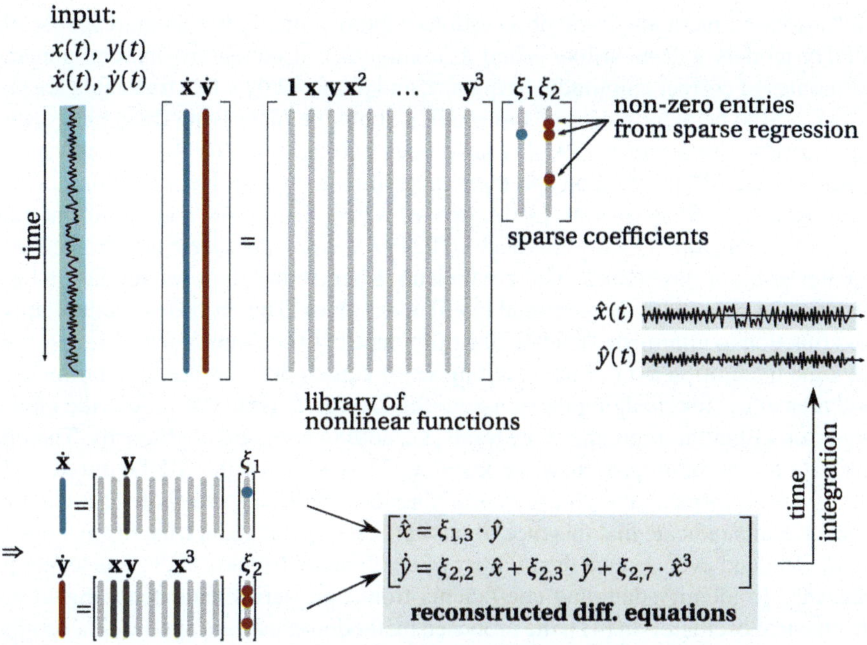

Fig. 4 Conceptual overview on SINDy methods: using a library of ansatz functions, we aim at solving the resulting system of equations $\dot{x} = f(x, t)$ while enforcing sparsity in the solution vector ξ, which coefficients will form the analytical differential equation that describes the data $x(t)$. Illustration adapted from [39]

In [38] it was found that the time scale separation works effectively, thereby separating the dynamics of the structure and the dynamics of the particles. A energy formulation at specific time scales allowed to extract equivalent particle activity, i.e. damping, metrics. The research showed very good agreement between those experimentally obtained activity metrics with the ones obtained from excessive simulations. The versatility of the proposed techniques allowed for measuring damping directly from time-series measurements at low costs, thereby allowing for fast experimental damper design studies.

Another contribution to data-driven damping identification followed the research lines of sparse identification of nonlinear dynamics (SINDy) proposed in [40]. Figure 4 illustrates the concept of finding analytical equations from time series using sparse regression. Overall, SINDy methods aim at deriving a set of analytical differential equations from given time-series data. These equations are descriptive for the dynamics observed and provided to SINDy, hence they can reflect certain properties of the underlying process that generated the observations. However, the mathematical models are not necessarily the same as the ones that would be obtained by classical physics-based methods: if the observations capture a purely harmonic response, the SINDy models will most-likely find a minimal set of equations that can generate harmonics of correct amplitude, frequency, and phase. Only when transients are provided to SINDy, more pieces of information about the underlying dynamics system are available. Two works [39, 41] have been published in the SPP for advancing aspects of the SINDy method and their application to system identification in structural dynamics. Stender et al. [39] proposed a post-processing routine to fine-tune the coefficients found by the classical SINDy procedure and sparse thresholded least squares solution procedure. The coefficients ξ are found by a sequential process, resulting in potentially sub-optimal coefficient values. The proposed method uses a loss function to minimize, typically the difference between input time series $x(t)$ and the time integration $\hat{x}(t)$ of the resulting set of equations. Constrained optimization will then allow for changing the non-zero SINDy coefficients ξ to reduce the modeling error without introducing more terms, i.e. adding non-zero coefficients. Thereby, the data fit can be improved while keeping the sparsity of the SINDy model. The study was performed on various types of classical oscillators and analysis configurations. Results indicate that this type of post-processing can reduce the fitting error by up to 30% and even reduce the number of coefficients further. Then, this technique was used to identify damping coefficients from time series data stemming from a mechanical oscillator. In [41], the proposed methodologies were applied to a geometrically nonlinearly vibrating and base-excited mechanical system. The experimental data exhibited some degree of noise contamination. The full state space was first reconstructed using embedding techniques. Then, SINDy and successive coefficient fine-tuning were applied to derive analytical mathematical models from a single vibration time series. Nonlinear and damping effects were identified successfully.

The true damping coefficients were identified at high accuracy without making any strict assumption about the library of SINDy ansatz functions. Hence, those methods represent a promising pathway towards a direct identification of dynamical models from data, particularly for finding uncertain parameters such as dissipative terms.

4 Physics-Based Modeling

Having identified mathematical models for vibrating structures, or having parameterized existing models, one will need to make predictions about the system dynamics under different excitations, parameter configurations, or initial conditions. A large set of numerical methods exist for these tasks, such as direct time integration, harmonic balance approaches, and continuation methods. Several activities in the SPP 1897 thus focused on the efficient determination of the nonlinear dynamics of mechanical structures.

One aspect of particular relevance is the phenomenon of multi-stability, even in small and weakly nonlinear dynamical systems. The complicated localization patterns and multi-stable states were studied in [42]. A chain comprised of multiple bi-stable frictional oscillators exhibits various states of vibration that co-exist at specific sliding velocities, thereby showing that in realistic friction brakes the uncertainty about emerging vibrations may be rooted in types of multi-stable dynamics. Closely related with multi-stability, the phenomenon of *isolated solution branches* poses major challenges to classical solvers for nonlinear dynamics, such as harmonic balance methods. Further, it was found that there is a possibility to harvest energy from forced systems that are susceptible to self-excited vibrations, even without requiring the system to exhibit large limit-cycle vibrations [43].

Isolated solution branches co-exist in certain parameter regions and do not link to bifurcation paths of equilibria or limit cycles. As path-following continuation methods will only find solutions that branch from a-priori known solutions, one will not be able to find isolated branches. An extension to the periodic motion concept was proposed [44, 45] in collaboration with Leibniz University of Hannover. This method allows for finding isolated solution branches using classical harmonic balance methods and nonlinear normal modes. At the core of the method, an artificial damping term is introduced into the differential equations, always setting the energy balance of the system to zero, i.e. obtaining an autonomous solution to the system. As the vibration amplitudes along the normal modes are increased, the absolute value of the artificial damping terms is tracked. Whenever this term vanishes, a true solution to the given system is found. That way, all solutions to a system can be found quickly, including isolated solution branches. Using this technique, small friction-excited systems were studied for all of their solutions at high energy levels. In [46] it was found that even minimal models for friction-excited vibrations in brake squeal can exhibit very complicated multi-stable dynamics. In certain weakly nonlinear parameter regimes up to five competing periodic solutions were found by the newly developed solution strategies. Moreover, we were able to show that the co-existing solutions even differ in their qualitative characteristics: hyper-chaotic solutions (having more than one positive Lyapunov exponent) were observed to exist in parallel to classical periodic limit cycle solutions. This finding has not been reported for friction-excited systems before, and may change the perspectives on vibrations observed experimentally: a system may not be in a regular or irregular configuration, it can in fact be in both configurations at the same time. Only the initial conditions and instantaneous perturbations will dictate if the system dynamics end up being periodic or strongly chaotic. In contrast to chaotic vibrations generated by the system itself, [35] studied the impact of a co-simulated chaotic friction model coupled to a classical minimal model for brake system vibrations. It was observed that the chaotic dynamics will propagate into the overall system dynamics, such as the ones of the brake disc, only to a certain degree. Synchronization effects and mass-proportional bandpass filtering will change the effective stability properties of the system, while the resulting system

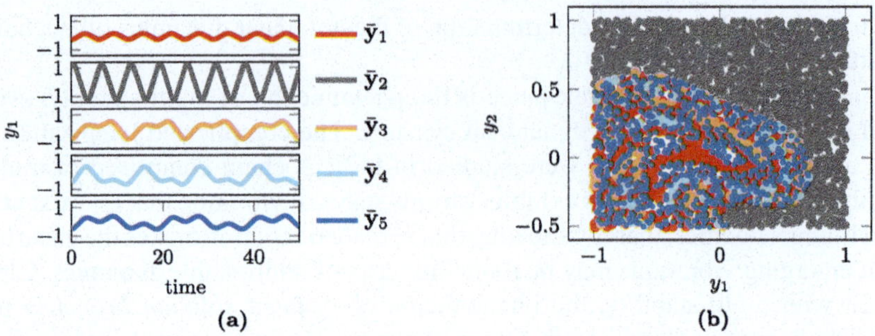

Fig. 5 Multi-stability in the Duffing oscillator $\ddot{y} - 0.08\dot{y} + y^3 = 0.2\cos(t)$: **a** five co-existing periodic solutions $\bar{\mathbf{y}}_i$ and **b** the corresponding state space, displacements y_1 and velocities y_2, and basins of attraction for each solution. Illustration adapted from [47]

dynamics were only weakly chaotic with different Lyapunov spectra than the friction model itself.

Our studies [34, 35, 46] highlight how complex dynamics can arise even in seemingly simple and small dynamical systems that were thought to be fully understood and behave regularly before. The interaction of multiple time scales, self-excitation and damping renders the dynamics particularly hard to predict using existing numerical techniques.

Understanding multi-stability as a central property of many nonlinear dynamical systems, classical concepts of stability loose some degree of relevance: most of classical stability concepts build on stability against local perturbations: will trajectories return to that solution if a *small* perturbation is imposed? The resulting stability statement is only valid locally in multi-stable systems, and the admissible size of those small perturbations is unknown. For practitioners those stability statements have only limited significance. Given that a system may show several very different dynamics that solely depend on the initial conditions, the shape and size of the respective basins of attraction are the relevant quantity, see Fig. 5. Given a range and distribution of initial conditions or external perturbations, the basin information will allow to derive the probability of arriving at a specific solution. The work in the SPP 1897 published an open-sourced toolbox [47] to derive the basin stability for multi-stable systems. In a fully automated manner, users can find all stable solutions of their dynamical systems, and obtain the probabilities of transitioning to those solutions. Thereby, classical stability concepts are extended towards a probabilistic estimate. Using the proposed methods and codes, the basin stability has been investigated for friction-excited systems [48] and fluttering airfoils [49].

5 Data-Driven Modeling

Albeit the maturity of mathematical modeling, signal processing and system identification techniques, there remain certain mechanical structures that cannot be modeled at acceptable precision today [3, 20]. One of which systems are full-scale realistic brake systems. The highly complicated friction interface, unsteady external loading conditions, a multitude of mechanical joints, and only partly-understood damping are the main challenges. Specifically, today one must accept that there is no technique that can robustly predict the onset of self-excited vibrations such as high-frequency brake squeal. Within the SPP project, a novel approach to brake squeal prediction using machine learning was researched and presented [50]. To overcome the limitations of physics-based modeling, deep artificial neural networks (NN) were trained to predict the squeal occurrence from a set of multivariate time series measurements of the loading conditions, such as rotational velocity, brake line pressure, and various temperatures. The sequential character of the input time series was explicitly treated using recurrent neurons that allow for taking time-dependent and history-dependent behavior into account. Training and validation data was acquired from dynamometer testing of four commercial disk brake systems, each test covering around 1500 brake stops. As the distribution of squealing and non-squealing brake stops were very uneven (non-squealing over-represented), typical classification metrics were not expressive for the accuracy of the trained models. Using the Mathews' Correlation Coefficient (MCC, ranging from (-1) to $(+1)$), scores of up to 0.78 were obtained, meaning that around 85% of all vibration events were predicted correctly. This result highlights the capabilities of deep neural networks for behavior prediction in structural dynamics. However, the prediction models performed differently on the data sets of all four braking systems. A global neural network for different brake systems did not achieve better behavior prediction performance than random. This result indicates the highly individual instability behavior of seemingly similar mechanical structures. Particularly, the excitation and damping mechanisms were found to be very specific to the individual mechanical structure, therefore requiring individual models for each structure.

Very recent work within the SPP 1897 was focusing on links between data-driven modeling and physics-based modeling. To this end, we proposed to use HAVOK (Hankel-alternative view of Koopman operators [51]) models, which derive analytical differential equations from data, similarly to the SINDy method. Classical descriptions of nonlinear dynamical systems $\dot{\mathbf{x}} = \mathbf{f}(\mathbf{x}, t)$ follow the discrete time update

$$\mathbf{x}_{k+1} = \mathbf{F}(\mathbf{x}_k) = \int_{k\Delta\tau}^{(k+1)\Delta\tau} \mathbf{A}\mathbf{x}(\tau) + \mathbf{f}_{\mathrm{nl}}(\tau)\mathrm{d}\tau, \quad \mathbf{x}_k = \mathbf{x}(k\Delta t) \tag{1}$$

where $\mathbf{F}$ is a nonlinear operator on the system states, $\mathbf{A}$ is the linear system matrix, and $\mathbf{f}_{\mathrm{nl}}$ are all nonlinear forces. A different perspective is taken in the Koopman theory, where instead of system states one will study measurement functions $g(\mathbf{x})$

that represent observables $\mathbf{s}_k = g(\mathbf{x}_k)$ of the system states $\mathbf{x}_k$. The Koopman operator $\mathcal{K}$ then advances the time step of the observables

$$\mathbf{s}_{k+1} = \mathcal{K}\mathbf{s}_k = \mathcal{K}g(\mathbf{x}_k) = g(\mathbf{x}_{k+1}). \tag{2}$$

In contrast to $\mathbf{F}$, the Koopman operator is a linear one (at the cost of being infinite-dimensional). The HAVOK methods now proposes to estimate a finite-dimensional operator $\bar{\mathcal{K}}$ from data. In this SPP project, several studies were undertaken to understand how the analytical equations derived from HAVOK approaches

$$\dot{\mathbf{v}}(t) = \mathbf{A}_{\mathrm{H}}\mathbf{v}(t) + \mathrm{B}v_r, \quad \mathbf{v} = \left[v_1, \ldots, v_{r-1}\right] \tag{3}$$

compare to the classical dynamical system representation of physics-based approaches. Here, $\mathbf{v}(t)$ represent eigen-time-delay coordinates derived from a Hankel embedding approach, i.e. the new coordinates or measurement functions derived from the input time series. The linear fit $\dot{\mathbf{v}} = \mathbf{A}_{\mathrm{H}}\mathbf{v}$ is good for the first $(r-1)$ components. The last r components represent a weak linear fit, and will be represented in the nonlinear forcing term $\mathbf{B}v_r$. Particularly, we investigated how the linear system representations $\mathbf{A}$ in Eq. (2) relate to $\mathbf{A}_{\mathrm{H}}$ in Eq. (3). Those results have not been published yet. It was found for the Lorenz system that the derived damping properties and the choice of initial conditions are clearly interrelated: the system representations derived from a time series that starts far off the attractor will carry more damping (i.e. negative parts of the eigenvalues) than a representation derived from a time series that starts in close vicinity to the attractor. While the volume contraction property of the chaotic attractor, i.e. a damping equivalent, does not change, the amount of transients taken into account for the HAVOK models does matter. Studies with linear, weakly nonlinear, and strongly nonlinear few-degree-of-freedom oscillators gave further insights into the models derived by HAVOK, i.e. analytical equations in the form of Eq. (3), from a single univariate time series: for weakly nonlinear dynamics the number of state variables $\mathbf{v}$ did match the true number of states $\mathbf{x}$, thereby finding the correct system dimensionality. Secondly, the eigenvalue spectrum of $\mathbf{A}_{\mathrm{H}}$ does match with the spectrum of the true eigenvalues of $\mathbf{A}$ in most situations. In this way, damping values can be estimated from the real parts of the eigenvalues. This pathway is a completely data-driven approach to deriving damping values from sparse time series data. As the nonlinearity increases, the system representation approaches a spectral decomposition of the observed dynamics: more states r are introduced in the HAVOK models, and the resulting eigenvalue spectra (imaginary parts) approach the most relevant frequency components of the signal. Concluding, the HAVOK approach represents a valid technique for deriving analytical equations from raw data, while at the same time bridging over to classical physics-based modeling and system representations. Further work in this field is required to arrive at a better understanding for how we can integrate data-driven models with physics-based models in a unified framework for model identification and behavior prediction.

6 Summary

The research activities within SPP 1897 were concerned with a data-driven perspective on damping in complex structural dynamics. The main findings of various research efforts can be summarized as follows:

- Many real-word mechanical structures operate at the edge of chaos. Multiple time and length scales are at play, and required adequate treatment, scale selection, and scale-bridging methodologies to capture the multitude of processes that interact.
- Nonlinear signal processing techniques are particularly useful for studying sparse experimental time series data. Assumptions about stationarity, determinism, or linearity tend to over-simplify the rich information contained in measured data.
- Transient data and processes carry a significantly higher degree of information that can be utilized for system identification. Particularly, damping estimates rely on transient trajectories evolving towards a steady-state response.
- Data-driven identification of damping properties is possible using sparse regression, Koopman-driven, and neural-network based approaches.
- Further research is required to link data-driven and physics-based approaches in a unified framework that is making use of both worlds' potentials.

More details of the investigations can be found in the respective publications achieved within the SPP 1897. Parts of the results were also published in the dissertation thesis of Stender [52] under open access.

7 Conclusion

The research projects within SPP 1897 led by Prof. Norbert Hoffmann focused on damping in structural dynamics, particularly in complex systems affected by friction and self-excited responses. A data-driven perspective was taken to approach damping effects and mechanisms in multi-scale, partially observable, and chaotic dynamics. The results of the research conducted within SPP 1897 were published in 18 peer-reviewed journal publications and has been presented at 19 conferences. Collaborations outside the SPP 1897 were undertaken with Politecnico di Bari [41, 42, 48, 49], La Sapienza Universita di Roma [33, 36], University of Technology Sydney [32, 35, 39, 50] among others. Collaborations within the SPP were undertaken with University of Stuttgart [38], Leibniz University Hannover [44–46], and Technische Universität Berlin [53].

Acknowledgements We like to thank the SPP 1897 project coordination at University of Stuttgart for running and organizing the programme. The discussions and gatherings within the Priority Programme were always highly interesting and very fruitful. Further, we like to thank the German Research Foundation (DFG) for funding the projects HO 3851/12 – 1 and HO 3852/12 – 2.

References

1. Nitsche, R., Gaul, L.: Lyapunov design of damping controllers. Arch. Appl. Mech. **72**(11), 865–874 (2003)
2. Ouyang, H., Nack, W., Yuan, Y., Chen, F.: Numerical analysis of automotive disc brake squeal: a review. Int. J. Veh. Noise Vib. **1**(3/4), 207 (2005)
3. Massi, F., Baillet, L., Giannini, O., Sestieri, A.: Brake squeal: linear and nonlinear numerical approaches. Mech. Syst. Signal Process. **21**(6), 2374–2393 (2007)
4. Charakopoulos, A.K., Karakasidis, T.E., Papanicolaou, P.N., Liakopoulos, A.: The application of complex network time series analysis in turbulent heated jets. Chaos (Woodbury, N.Y.) **24**(2), 024408 (2014)
5. Wiercigroch, M.: Chaotic vibration of a simple model of the machine tool-cutting process system. J. Vib. Acoust. **119**(3), 468 (1997)
6. Pilipchuk, V., Olejnik, P., Awrejcewicz, J.: Transient friction-induced vibrations in a 2-DOF model of brakes. J. Sound Vib. **344**, 297–312 (2015)
7. Gandia, R.M., Antonialli, F., Cavazza, B.H., Neto, A.M., Lima, D.A.d., Sugano, J.Y., Nicolai, I., Zambalde, A.L.: Autonomous vehicles: scientometric and bibliometric review. Trans. Rev. **39**(1), 9–28 (2019)
8. Xu, L.D., Xu, E.L., Li, L.: Industry 4.0: state of the art and future trends. Int. J. Prod. Res. **56**(8), 2941–2962 (2018)
9. Strogatz, S.H.: Nonlinear dynamics and chaos: with applications to physics, biology, chemistry, and engineering. Studies in Nonlinearity, 2nd edn. Perseus Books, Cambridge (2001)
10. Mitchell, M.: Complexity: A guided tour. Oxford University Press, Oxford and New York (2009)
11. Oberst, S., Lai, J.: Chaos in brake squeal noise. J. Sound Vib. **330**(5), 955–975 (2011)
12. Oberst, S., Lai, J.: Statistical analysis of brake squeal noise. J. Sound Vib. **330**(12), 2978–2994 (2011)
13. Wernitz, B., Hoffmann, N.: Recurrence analysis and phase space reconstruction of irregular vibration in friction brakes: signatures of chaos in steady sliding. J. Sound Vib. **331**(16), 3887–3896 (2012)
14. Renson, L., Gonzalez-Buelga, A., Barton, D.A., Neild, S.A.: Robust identification of backbone curves using control-based continuation. J. Sound Vib. **367**, 145–158 (2016)
15. Kantz, H., Schreiber, T.: Nonlinear Time Series Analysis. Cambridge University Press, Cambridge (2003)
16. Marwan, N., Romano, M.C., Thiel, M., Kurths, J.: Recurrence plots for the analysis of complex systems. Phys. Rep. **438**(5–6), 237–329 (2007)
17. Hoffmann, N., Gaul, L.: Effects of damping on mode-coupling instability in friction induced oscillations. ZAMM-J. Appl. Math. Mech./Zeitschrift für Angewandte Mathematik und Mechanik **83**(8), 524–534 (2003)
18. Kirillov, O.N., Verhulst, F.: Paradoxes of dissipation-induced destabilization or who opened Whitney's umbrella? ZAMM **90**(6), 462–488 (2010)
19. Kirillov, O.N.: Nonconservative Stability Problems of Modern Physics, vol. 14. Walter de Gruyter (2013)
20. Sinou, J.J., Jézéquel, L.: The influence of damping on the limit cycles for a self-exciting mechanism. J. Sound Vib. **304**(3–5), 875–893 (2007)
21. Vakis, A.I., Yastrebov, V.A., Scheibert, J., Nicola, L., Dini, D., Minfray, C., Almqvist, A., Paggi, M., Lee, S., Limbert, G., et al.: Modeling and simulation in tribology across scales: an overview. Tribol. Int. **125**, 169–199 (2018)
22. Tiedemann, M., Kruse, S., Hoffmann, N.: Dominant damping effects in friction brake noise, vibration and harshness: the relevance of joints. Proc. Inst. Mech. Eng. Part D: J. Autom. Eng. **229**(6), 728–734 (2015)
23. Hoffmann, N., Fischer, M., Allgaier, R., Gaul, L.: A minimal model for studying properties of the mode-coupling type instability in friction induced oscillations. Mech. Res. Commun. **29**(4), 197–205 (2002)

24. Akay, A.: Acoustics of friction. J. Acoust. Soc. Amer. **111**(4), 1525–1548 (2002)
25. Takens, F.: Detecting Strange Attractors in Turbulence: Dynamical Systems and Turbulence, Warwick 1980, pp. 366–381. Springer (1981)
26. Ruelle, D., Takens, F.: On the nature of turbulence. Les rencontres physiciens-mathématiciens de Strasbourg-RCP25 **12**, 1–44 (1971)
27. Rhodes, C., Morari, M.: The false nearest neighbors algorithm: an overview. Comput. & Chem. Eng. **21**, S1149–S1154 (1997)
28. Schreiber, T.: Measuring information transfer. Phys. Rev. Lett. **85**(2), 461–464 (2000)
29. Eckmann, J.P., Kamphorst, S.O., Ruelle, D.: Recurrence plots of dynamical systems. Europhys. Lett. (EPL) **4**(9), 973–977 (1987)
30. Marwan, N., Kurths, J.: Line structures in recurrence plots. Phys. Lett. A **336**(4–5), 349–357 (2005)
31. Webber, C.L., Marwan, N. (eds.): Recurrence Quantification Analysis. Understanding Complex Systems. Springer International Publishing, Cham (2015)
32. Stender, M., Oberst, S., Tiedemann, M., Hoffmann, N.: Complex machine dynamics: systematic recurrence quantification analysis of disk brake vibration data. Nonlinear Dyn. **267**(1), 105 (2019)
33. Stender, M., Di Bartolomeo, M., Massi, F., Hoffmann, N.: Revealing transitions in friction-excited vibrations by nonlinear time-series analysis. Nonlinear Dyn. **47**(7), 209 (2019)
34. Stender, M., Tiedemann, M., Hoffmann, N.: Characterization of complex states for friction-excited systems. PAMM **17**(1), 45–46 (2017)
35. Stender, M., Tiedemann, M., Hoffmann, N., Oberst, S.: Impact of an irregular friction formulation on dynamics of a minimal model for brake squeal. Mech. Syst. Signal Process. **107**, 439–451 (2018)
36. Di Bartolomeo, M., Lazzari, A., Stender, M., Berthier, Y., Saulot, A., Massi, F.: Experimental observation of thermally-driven frictional instabilities on C/C materials. Tribol. Int. 106724 (2020)
37. Stender, M., Tiedemann, M., Hoffmann, L., Hoffmann, N.: Determining growth rates of instabilities from time-series vibration data: methods and applications for brake squeal. Mech. Syst. Signal Process. **129**, 250–264 (2019)
38. Gnanasambandham, C., Stender, M., Hoffmann, N., Eberhard, P.: Multi-scale dynamics of particle dampers using wavelets: extracting particle activity metrics from ring down experiments. J. Sound Vib. **454**, 1–13 (2019)
39. Stender, M., Oberst, S., Hoffmann, N.: Recovery of differential equations from impulse response time series data for model identification and feature extraction. Vibration **2**(1), 25–46 (2019)
40. Brunton, S.L., Proctor, J.L., Kutz, J.N.: Discovering governing equations from data by sparse identification of nonlinear dynamical systems. Proc. Natl. Acad. Sci. U.S.A. **113**(15), 3932–3937 (2016)
41. Didonna, M., Stender, M., Papangelo, A., Fontanela, F., Ciavarella, M., Hoffmann, N.: Reconstruction of governing equations from vibration measurements for geometrically nonlinear systems. Lubricants **7**(8), 64 (2019)
42. Papangelo, A., Hoffmann, N., Grolet, A., Stender, M., Ciavarella, M.: Multiple spatially localized dynamical states in friction-excited oscillator chains. J. Sound Vib. **417**, 56–64 (2018)
43. Stender, M., Tiedemann, M., Hoffmann, N.: Energy harvesting below the onset of flutter. J. Sound Vib. **458**, 17–21 (2019)
44. Jahn, M., Stender, M., Tatzko, S., Hoffmann, N., Grolet, A., Wallaschek, J.: The extended periodic motion concept for fast limit cycle detection of self-excited systems. Comput. & Struct. 106–139 (2019)
45. Tatzko, S., Stender, M., Jahn, M., Hoffmann, N.: Limit cycle computation of self-excited dynamic systems using nonlinear modes. PAMM **20**(1) (2021)
46. Stender, M., Jahn, M., Hoffmann, N., Wallaschek, J.: Hyperchaos co-existing with periodic orbits in a frictional oscillator. J. Sound Vib. **472**, 115–203 (2020)

47. Stender, M., Hoffmann, N.: bSTAB: an open-source software for computing the basin stability of multi-stable dynamical systems. Nonlinear Dynamics (2021)
48. Stender, M., Hoffmann, N., Papangelo, A.: The basin stability of bi-stable friction-excited oscillators. Lubricants **8**(12), 105 (2020)
49. Nitti, A., Stender, M., Hoffmann, N., Papangelo, A.: Spatially localized vibrations in a rotor subjected to flutter. Nonlinear Dynamics (2021)
50. Stender, M., Tiedemann, M., Spieler, D., Schoepflin, D., Hoffmann, N., Oberst, S.: Deep learning for brake squeal: brake noise detection, characterization and prediction. Mech. Syst. Signal Process. **149**, 107181 (2021)
51. Brunton, S.L., Brunton, B.W., Proctor, J.L., Kaiser, E., Kutz, J.N.: Chaos as an intermittently forced linear system. Nat. Commun. **8**(1), 19 (2017)
52. Stender, M.: Data-driven techniques for the nonlinear dynamics of mechanical structures. Doctoral thesis, Technische Universität Hamburg (2020)
53. Stender, M., Schmid, D., Hoffmann, N.: Multiple scales in complex friction-induced disk brake vibrations. In: 15th Experimental Chaos and Complexity Conference, Madrid, vol. poster presentation (2018)

HyCEML – Hybrid CFRP Elastomer Metal Laminates Containing Elastomeric Interfaces for Deliberate Dissipation

Alexander Jackstadt, Vincent Sessner, Wilfried Liebig, Luise Kärger, and Kay Weidenmann

1 Introduction

In combining the advantages of metals and fiber-reinforced polymers (FRPs), fiber metal laminates (FMLs) are intended to surpass the properties of the monolithic materials [20]. However, as most lightweight materials and structures, FMLs tend to be prone to vibrations under real-world loading conditions. To overcome this, FMLs can be complemented by viscoelastic elastomer layers in order to achieve a desired level of damping [18]. Following the principles of constrained layer damping (CLD) [12, 14], a viscoelastic and highly compliant layer is laminated in between two stiff constraining layers. The large transverse shear deformations in the constrained viscoelastic middle layer induced under bending lead to the dissipation of vibration energy. By varying the laminate's parameters such as lay-up, layer thickness or the damping material itself [11], this intrinsic damping mechanism can be optimized to deliberately dissipate undesired vibration energy.

A. Jackstadt (✉) · L. Kärger
Karlsruhe Institute of Technology (KIT), Institute for Vehicle System Technology, Rintheimer Querallee 2, 76131 Karlsruhe, Germany
e-mail: alexander.jackstadt@kit.edu

L. Kärger
e-mail: luise.kaerger@kit.edu

V. Sessner · W. Liebig
Karlsruhe Institute of Technology (KIT), Institute of Applied Materials, Engelbert-Arnold-Straße 4, 76131 Karlsruhe, Germany
e-mail: wilfried.liebig@kit.edu

K. Weidenmann
Augsburg University, Institute for Materials Resource Management, Am Technologiezentrum 8, 86159 Augsburg, Germany
e-mail: kay.weidenmann@mrm.uni-augsburg.de

P. Eberhard (ed.), *Calm, Smooth and Smart*, Lecture Notes in Applied and Computational Mechanics 102, https://doi.org/10.1007/978-3-031-36143-2_6

1.1 Materials

Within this work, different laminates which can be classified as hybrid CFRP elastomer metal laminates (HyCEMLs) are investigated. The laminates comprise the constituents listed in Table 1. The metal layers consist of an aluminum wrought alloy EN AW-2024-T3 ALCLAD AMS-QQA-250/5 delivered as sheets of thickness $t_{Al} = 0.3$ mm. The carbon fiber-reinforced polymer (CFRP) layers are cured unidirectional prepreg sheets HexPly M77/38/UD150/CHS-12K-70 by Hexcel with a nominal cured ply thickness of $t_{CFRP} = 0.15$ mm. Two elastomer materials fulfill the role of the constrained damping layers within HyCEML. Both types of elastomer are provided by Gummiwerk KRAIBURG GmbH & Co. KG from their KRAIBON product range, namely SAA9579-52 and HAA9275-45. The first one will in the following be referred to as soft, whereas the latter is denoted as hard.

Within this work, multiple lamination schemes were manufactured and investigated, as shown in Fig. 1 and Table 2. All variations were manufactured in the same hot-mold process at a temperature of 150 °C at a pressure of 40 bar for 300 s. Further details on the manufacturing process and specimen preparation can be found in [16, 17, 19].

Table 1 Quasistatic linear elastic material parameters of the constituents used in HyCEML

		Aluminum	CFRP	Elastomer	
				Soft	Hard
Young's modulus in GPa	E_1 E_2 E_3	73.1	113.7 7.75 7.75	0.045	0.483
Shear modulus in GPa	G_{12} G_{31} G_{32}	28.0	3.76 3.76 2.75	0.015	0.15
Poisson's ratio	ν_{12} ν_{31} ν_{32}	0.34 0.31 0.40	0.34	0.48	0.45
Mass density in kg m^{-3}	ρ	2780	1496	1180	1250

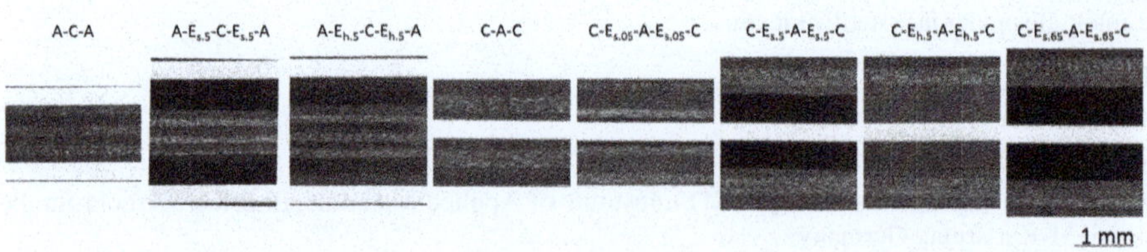

Fig. 1 Micro sections and naming conventions of FML and HyCEML stacking sequences considered within this study. The abbreviations for the different stacking sequences are listed in Table 2

Table 2 Overview of lamination schemes and naming conventions. Individual constituents, aluminum (A), CFRP (C) and elastomer (E) from Table 1, are abbreviated accordingly. Subscripts denote the individual layer thickness, whereas superscripts denote material specifications such as orientation or type of elastomer used. Symmetric or repeating parts of a laminate are denoted with $()_{sym}$ or $()_3$, if for example, 3 repetitions are present

Lay-up	Cured laminate thickness	Abbreviation
$[A_{0.3}/(C^{0°}_{0.15}/C^{90°}_{0.15})_3/A_{0.3}]$	1.5 mm	A-C-A
$[(C^{0°}_{0.15}/C^{90°}_{0.15})_{sym}/A_{0.3}/(C^{0°}_{0.15}/C^{90°}_{0.15})_{sym}]$	1.5 mm	C-A-C
$[A_{0.3}/E^{hard}_{0.5}/(C^{0°}_{0.15}/C^{90°}_{0.15})_3/E^{hard}_{0.5}/A_{0.3}]$	2.5 mm	A-$E^{h}_{.5}$-C-$E^{h}_{.5}$-A
$[A_{0.3}/E^{soft}_{0.5}/(C^{0°}_{0.15}/C^{90°}_{0.15})_3/E^{soft}_{0.5}/A_{0.3}]$	2.5 mm	A-$E^{s}_{.5}$-C-$E^{s}_{.5}$-A
$[(C^{0°}_{0.15}/C^{90°}_{0.15})_{sym}/E^{hard}_{0.5}/A_{0.3}/E^{hard}_{0.5}/(C^{0°}_{0.15}/C^{90°}_{0.15})_{sym}]$	2.5 mm	C-$E^{h}_{.5}$-A-$E^{h}_{.5}$-C
$[(C^{0°}_{0.15}/C^{90°}_{0.15})_{sym}/E^{soft}_{0.05}/A_{0.3}/E^{soft}_{0.05}/(C^{0°}_{0.15}/C^{90°}_{0.15})_{sym}]$	1.6 mm	C-$E^{s}_{.05}$-A-$E^{s}_{.05}$-C
$[(C^{0°}_{0.15}/C^{90°}_{0.15})_{sym}/E^{soft}_{0.5}/A_{0.3}/E^{soft}_{0.5}/(C^{0°}_{0.15}/C^{90°}_{0.15})_{sym}]$	2.5 mm	C-$E^{s}_{.5}$-A-$E^{s}_{.5}$-C
$[(C^{0°}_{0.15}/C^{90°}_{0.15})_{sym}/E^{soft}_{0.65}/A_{0.3}/E^{soft}_{0.65}/(C^{0°}_{0.15}/C^{90°}_{0.15})_{sym}]$	2.8 mm	C-$E^{s}_{.65}$-A-$E^{s}_{.65}$-C

2 Understanding the Damping Behavior of HyCEML

While the first scientific publications on CLD such as [8, 12, 14] date back more than half a century, there is very limited published work on the damping properties of hybrid CLD laminates in the context of lightweight structures. In particular, the influences of lamination schemes, the damping material's properties and the influence of different loading and boundary conditions have not been addressed. Therefore, the following experimental and analytical methods have been developed in order to accurately investigate the damping behavior of hybrid CLD laminates.

2.1 Experimental

In order to understand the damping behavior of HyCEML, a two-step experimental approach is adopted. First, the individual constituents expected to significantly contribute to damping are characterized. Furthermore, the different laminates are characterized as well.

2.1.1 Methods

Dynamic mechanical analysis (DMA) is used to determine the materials' viscoelastic, thus frequency-dependent, mechanical properties. For this purpose, specimens are subjected to a sinusoidal excitation in terms of a constant strain amplitude at multiple discrete frequencies. This is conducted over a wide range of temperatures so that the determined mechanical behavior can be extrapolated to a wider range of frequencies and strain rates respectively, using the principle of time temperature superposition applicable to polymers. The procedure is outlined in [6, 15]. The resulting complex Young's modulus

$$E^*(f) = E'(f) + \mathrm{i}E''(f) = E'(1 + \mathrm{i}\tan(\delta)) \tag{1}$$

is then available as a function of frequency at a chosen reference temperature T_{ref}. The real part E' is known as the storage modulus and quantifies the elastic contribution, whereas the loss modulus E'' describes the viscous part of the material behavior. The phase lag is described by the phase angle δ, and the loss factor $\tan(\delta)$ is a measure of the material damping.

2.1.2 Results

The master curves, determined in DMA under tension for the two elastomers used in this study, are shown in Fig. 2 in terms of storage modulus E', loss modulus E'' and loss factor $\tan(\delta)$. As the loss factor is a measure for the dissipation of the induced strain energy, the materials reach their highest damping in the peak of the loss factor $\tan(\delta)$. This occurs in the vicinity of the material's glass transition region, which

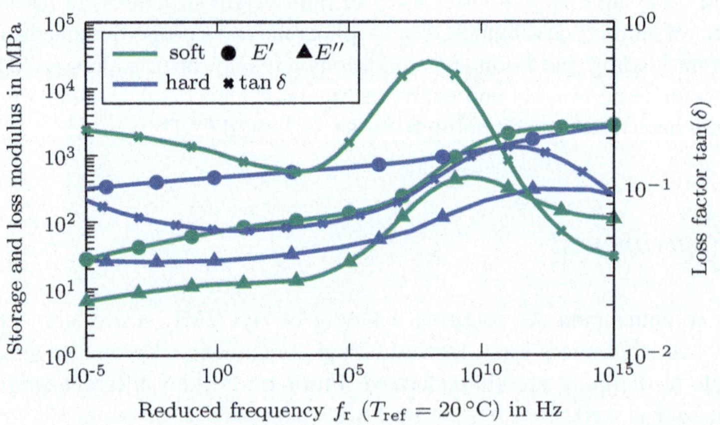

Fig. 2 Experimentally determined master curves of the two elastomers used in this work. Green shows the soft compound, whereas the blue curves belong to the hard compound [15]

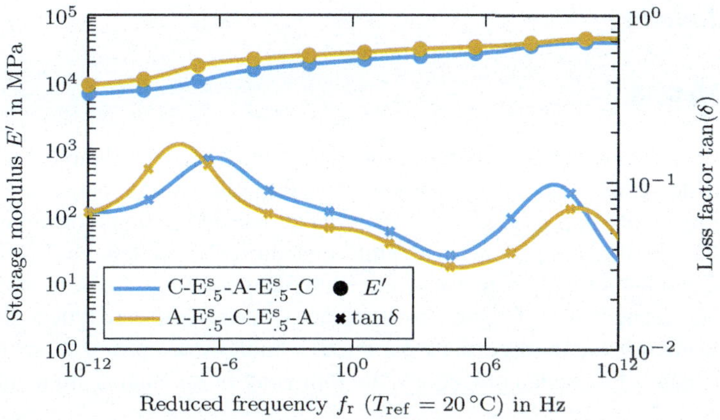

Fig. 3 Experimentally determined master curves of laminates C-E$^{s}_{.5}$-A-E$^{s}_{.5}$-C and A-E$^{s}_{.5}$-C-E$^{s}_{.5}$-A [15]

is found at a lower frequency for the soft elastomer. Furthermore, the loss factor is significantly higher in case of the soft elastomer. For high excitation frequencies, both elastomers show a similar stiffness. However, the soft elastomer shows a significantly more pronounced decrease of stiffness with decreasing excitation frequency compared to the hard elastomer.

To assess the damping capabilities of different HyCEML configurations, bending DMA was conducted. The results in terms of storage bending modulus and the corresponding loss factors are shown in Fig. 3 for two different laminates. In Fig. 3, the bending storage modulus shows that in the given frequency range and a reference temperature of $T_{ref} = 20$ °C, laminate A-E$^{s}_{.5}$-C-E$^{s}_{.5}$-A is slightly stiffer, with this effect being more pronounced for lower frequencies. The stronger decline in stiffness of laminate C-E$^{s}_{.5}$-A-E$^{s}_{.5}$-C towards lower frequencies is explained by the fact, that this laminate contains a higher fraction of CFRP than A-E$^{s}_{.5}$-C-E$^{s}_{.5}$-A. Due to its polymeric matrix, the CFRP also shows a frequency-dependent material behavior, as analyzed in [10], and thus a softer behavior towards lower frequencies. Considering the loss factor in Fig. 3, both curves show two maxima each. Whereas the peaks at high frequencies correspond to the glass transition region of the soft elastomer layers, the peaks at low frequencies result from the glass transition region of the CFRP. Generally, laminate C-E$^{s}_{.5}$-A-E$^{s}_{.5}$-C shows higher damping due to the elastomer layers lying closer to the plate's center. Thus, higher shear deformations are induced. An exception to this behavior is found below the CFRP's glass transition region, where laminate A-E$^{s}_{.5}$-C-E$^{s}_{.5}$-A shows higher damping as the CFRP layers constitute the plate's center and thus undergo higher shear deformations than those in laminate C-E$^{s}_{.5}$-A-E$^{s}_{.5}$-C.

2.2 Modeling

2.2.1 Method

While the experimental procedures outlined in Sect. 2.1 offer valuable insights into the laminates considered, the investigation of CLD in a wider range of variations concerning lamination schemes and material properties is time-consuming if not impossible. Furthermore, numerical finite element (FE) models tend to be rather large and computationally expensive, since full 3D models have to be used due to the strong heterogeneity in HyCEML's constituents. Consequently, analytical procedures can be of great value in large-scale parametric studies and optimizations of hybrid CLD laminates. For such a procedure, the following requirements are identified:

- Consideration of frequency-dependent material properties such as stiffness and material damping
- Minimal amount of degrees of freedom
- Prescribed kinematics suitable for a wide range of laminates

In order to fulfill these requirements, a plate theory based on a variable kinematics approach, namely the Generalized Unified Formulation (GUF) [1, 2] has been developed by the authors [4–6]. For each layer k within the laminate, an axiomatic approach for the displacement and out-of-plane stress components

$$\begin{aligned}
u_x^k(x,y,z) &= U_{x,\alpha_{ux}}^k F_{\alpha_{ux}}(z)\Phi_{u_x}(x,y)\\
u_y^k(x,y,z) &= U_{y,\alpha_{uy}}^k F_{\alpha_{uy}}(z)\Phi_{u_y}(x,y)\\
u_z^k(x,y,z) &= U_{z,\alpha_{uz}}^k F_{\alpha_{uz}}(z)\Phi_{u_z}(x,y)\\
\sigma_{xz}^k(x,y,z) &= S_{xz,\alpha_{\sigma xz}}^k F_{\alpha_{\sigma xz}}(z)\Phi_{\sigma_{xz}}(x,y)\\
\sigma_{yz}^k(x,y,z) &= S_{yz,\alpha_{\sigma yz}}^k F_{\alpha_{\sigma yz}}(z)\Phi_{\sigma_{yz}}(x,y)\\
\sigma_{zz}^k(x,y,z) &= S_{zz,\alpha_{\sigma zz}}^k F_{\alpha_{\sigma zz}}(z)\Phi_{\sigma_{zz}}(x,y)
\end{aligned} \tag{2}$$

is applied. In Eq. (2), the indices α are summation indices according to Einstein's summation convention. For each displacement and out-of-plane stress component, the order of expansion is chosen individually. Each index α then runs from 1 to $N+1$ indicating that a specific layer displacement or out-of-plane stress component is modeled with $N+1$ so far unknown variables $U_{i,\alpha_{u_i}}^k$ and $S_{iz,\alpha_{\sigma_{iz}}}^k$ respectively. The functions used to expand displacement and out-of-plane stress components $F_{\alpha_{u_i}}(z)$ and $F_{\alpha_{\sigma_{iz}}}(z)$ are combinations of Legendre polynomials as outlined in [2, 5]. The in-plane dependencies of the solution are summarized in the functions $\Phi(x,y)$.

2.2.2 Results

The analytical procedure has been verified against refined, thus computationally expensive, FE models [4]. In the following, some analytically obtained results for simply supported plates as shown in Fig. 4 are presented. The plates considered here have the dimensions $a = b = 0.4\,\text{m}$. Laminates C-$E^s_{.5}$-A-$E^s_{.5}$-C and A-$E^s_{.5}$-C-$E^s_{.5}$-A are considered, and the solutions are obtained using an adapted Navier method as outlined in [4]. Figures 5 and 6 show the plate's response to a harmonic force excitation in terms of displacement amplitude $||\hat{u}_z^*||$ and phase angle δ over the excitation frequency. From Fig. 5, it can be seen, that laminate A-$E^s_{.5}$-C-$E^s_{.5}$-A appears slightly stiffer based on the location of amplitude peaks compared to C-$E^s_{.5}$-A-$E^s_{.5}$-C. Both laminates show an increased level of damping for higher frequencies, as observable in Fig. 6 from the phase angle.

Figures 7 and 8 show the plate's natural frequencies and modal damping ratios. While this confirms, that laminate A-$E^s_{.5}$-C-$E^s_{.5}$-A is the stiffer one, Fig. 8 shows that its modal damping ratios increase in a more consistent way. Conversely, laminate C-$E^s_{.5}$-A-$E^s_{.5}$-C shows particularly high damping ratios of modes that comprise only one half-wave in x- or y-direction, but significantly more half-waves in the other direction.

To summarize, an analytical framework was developed for the rapid analysis of CLD laminates. The number of degrees of freedom is magnitudes lower than that of an equivalent finite element method (FEM) model. As it also considers frequency-dependent material behavior, it is a valuable tool in the analysis and design of hybrid

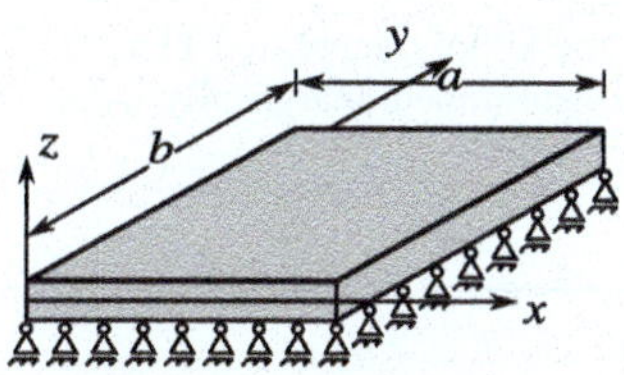

Fig. 4 Simply supported plate with global coordinate system

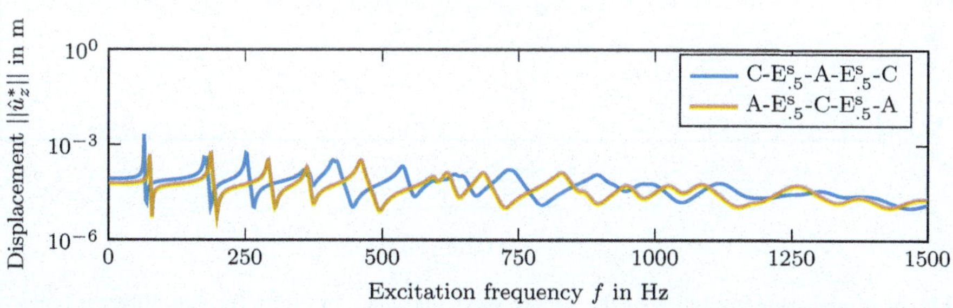

Fig. 5 Frequency response of the simply supported plate in Fig. 4 in terms of displacement amplitude $||\hat{u}_z^*||$ over excitation frequency f for two different types of HyCEML. The curves are determined analytically using the approach introduced in Sect. 2.2.1

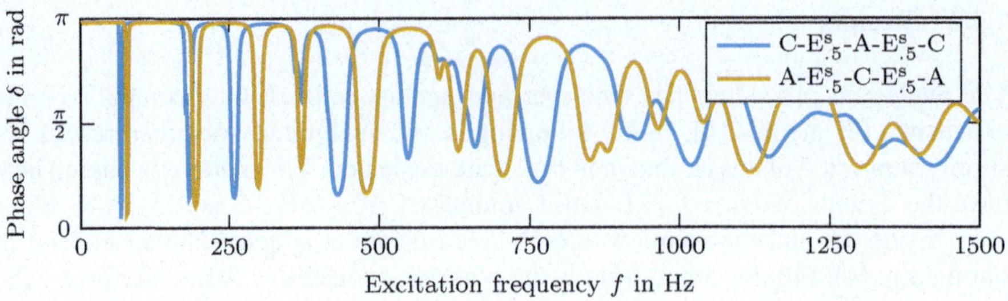

Fig. 6 Frequency response of the simply supported plate in Fig. 4 in terms of phase angle δ over excitation frequency f for two different types of HyCEML. The curves are determined analytically using the approach introduced in Sect. 2.2.1

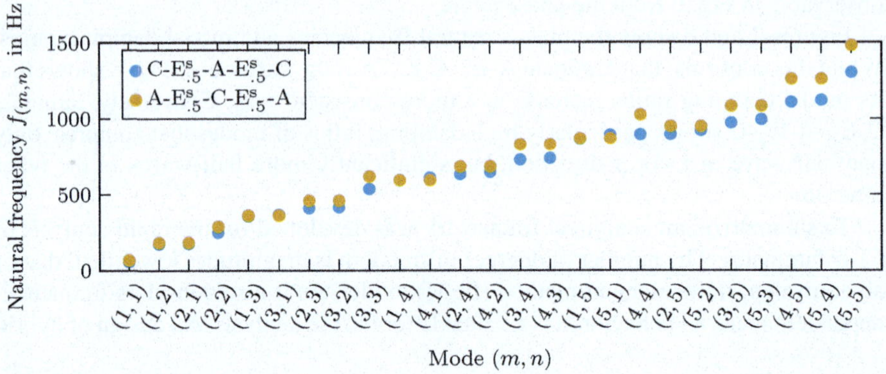

Fig. 7 Natural frequencies of the simply supported plate in Fig. 4 for two different types of HyCEML. The frequencies are determined analytically using the approach introduced in Sect. 2.2.1

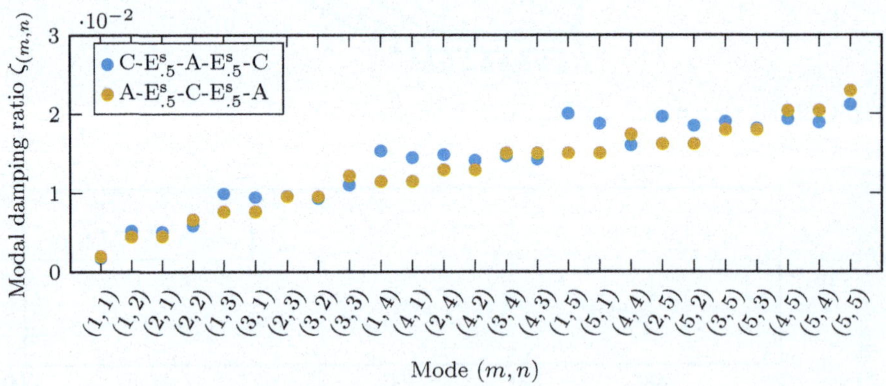

Fig. 8 Modal damping ratios of the simply supported plate in Fig. 4 for two different types of HyCEML. The damping ratios are determined analytically using the approach introduced in Sect. 2.2.1

laminates due to the multitude of design parameters which have an influence on the expected damping behavior.

3 The Mullins Effect in HyCEML

The damping layers in CLD applications are usually comprised of filled rubbers. In filled rubbers, however, the presence of a cyclic softening is to be expected. This cyclic softening effect has been the subject of numerous studies and is known as the Mullins effect after the author of some directional works on this effect. Although the Mullins effect is a large strain phenomenon, it should be considered in the modeling of CLD applications, as the strains observed in the damping layers exceed the global deformation of the laminate by far as shown in [5]. Furthermore, deformations during manufacturing or assembly and possible static loads during operation can trigger the Mullins effect and thus affect the mechanical behavior of the damping layer.

3.1 Experimental

As the Mullins effect is a large-strain phenomenon, it is only investigated for the soft elastomer, because the hard elastomer does not undergo large strains before failure. The experimental characterization of the Mullins effect is achieved using uniaxial cyclic tensile tests at constant strain rates. The specimen is loaded until it reaches a certain strain level and then unloaded until it is free of stress. An example of such a test result is shown in Fig. 9 for nominal strain levels of 10, 20, 40 and 80%. The curve clearly shows the typical softening behavior upon reloading up to the point of maximum strain. When this point is exceeded, the material's response follows the undamaged hyperelastic envelope. Furthermore, upon unloading to the stress-free state, significant residual strain is present. The softening in terms of a

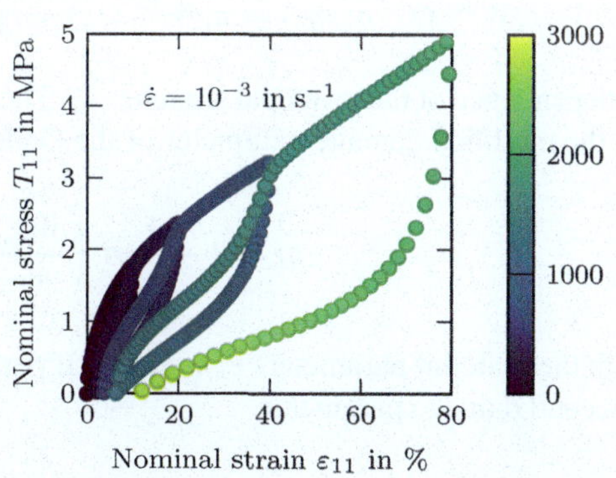

Fig. 9 Cyclic uniaxial tension test on the soft elastomer SAA9579-52 inducing the Mullins effect

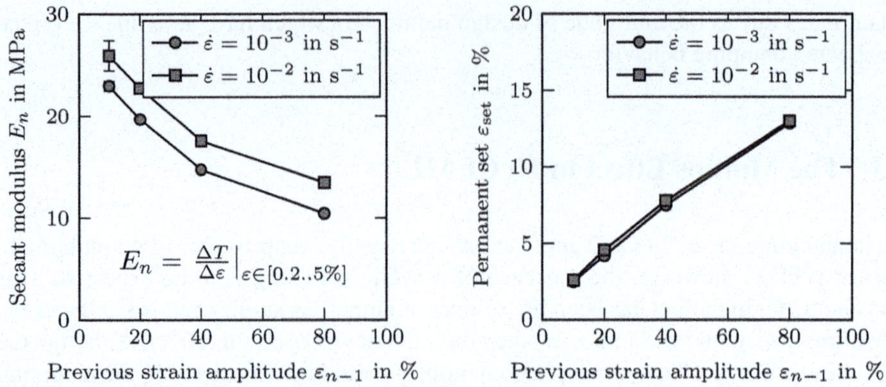

Fig. 10 Secant modulus (left) and permanent set (right) following cyclic loading to different strain levels. Results are shown for two different strain rates

secant modulus and the permanent set after each cycle are evaluated in more detail in Fig. 10. Although the overall secant modulus is higher in case of the higher strain rate $1 \times 10^{-2}\,\text{s}^{-1}$, the decrease is identical for both. The residual strain shows no difference between the two strain rates.

3.2 Modeling

In order to analyze the Mullins effect's role in CLD applications, nonlinear constitutive models are required, that can first and foremost depict the reduced stiffness of the material due to softening. As the observed permanent set can also be of interest, the proposed modeling approach also incorporates this. Following Dorfmann and Ogden [3], the deviatoric pseudo-elastic strain energy density function W^{dev} can be formulated as

$$W^{\text{dev}}(\lambda_i, \eta_1, \eta_2) = \eta_1 \tilde{W}^{\text{dev}}(\lambda_i) + (1 - \eta_2) N(\lambda_i) + \Phi(\eta_1, \eta_2) \tag{3}$$

in dependence of the principal stretches λ_i. The softening of the material is described by the modified damage parameter of the Ogden-Roxburgh (OR) model [13]

$$\eta_1 = 1 - \frac{1}{r}\text{erf}\left(\frac{W^{\text{dev}}_{\text{max}} - \tilde{W}^{\text{dev}}}{m + \beta W^{\text{dev}}_{\text{max}}}\right), \tag{4}$$

with the material parameters r, m and β. If permanent set is also taken into account, a second damage parameter

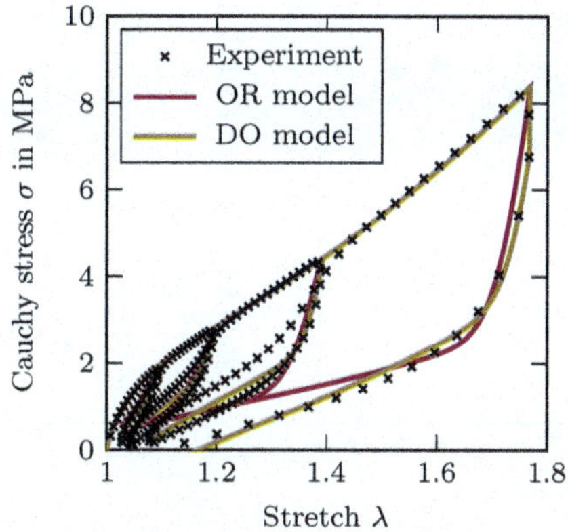

Fig. 11 Model predictions for the Ogden-Roxburgh (OR) and Dorfmann-Ogden (DO) models under uniaxial tension compared with experimental data

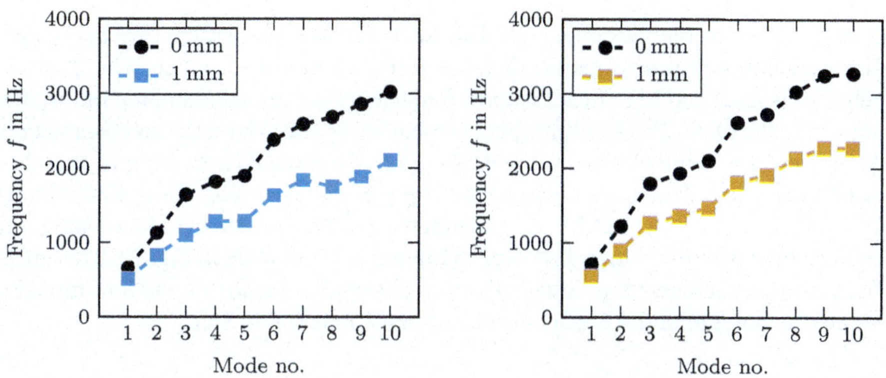

Fig. 12 Natural frequencies of laminates C-E$^{s}_{.5}$-A-E$^{s}_{.5}$-C (left) and A-E$^{s}_{.5}$-C-E$^{s}_{.5}$-A (right) with different levels of pre-straining

$$\eta_2 = \tanh\left(\left(\frac{\tilde{W}^{\text{dev}}}{W^{\text{dev}}_{\text{max}}}\right)^{\alpha\left(W^{\text{dev}}_{\text{max}}\right)}\right)\tanh(1)^{-1} \tag{5}$$

where α contains two more material parameters, is introduced, yielding the Dorfmann-Ogden (DO) model [3]. The parameters for both damage parameters are identified using a Differential Evolution algorithm. Details are omitted for brevity, but are found in [7]. Figure 11 shows the model predictions compared to the experimental findings from Fig. 9. Both models represent the experimental data well. When higher strain levels are reached, however, the DO model is more accurate in terms of representing the slope upon reloading as it takes into account the permanent set.

Figures 12 and 13 show the influence of the Mullins effect on the vibrational behavior of HyCEML based on a numerical case study. A simply supported plate

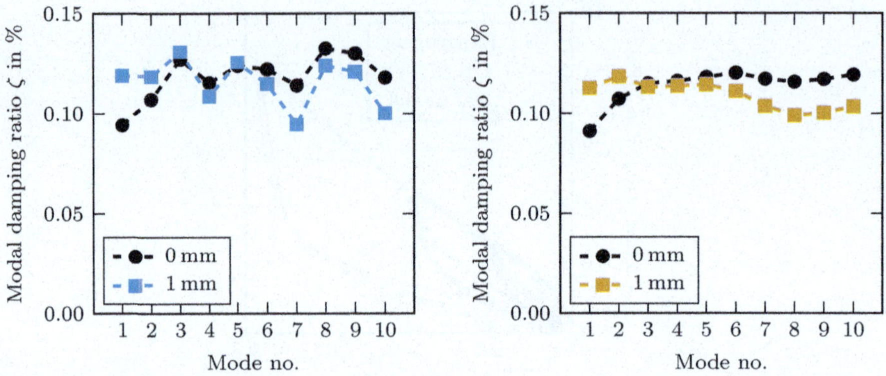

Fig. 13 Modal damping ratios of laminates C-$E^s_{.5}$-A-$E^s_{.5}$-C (left) and A-$E^s_{.5}$-C-$E^s_{.5}$-A (right) with different levels of pre-straining

corresponding to the specimens used in Sect. 4.1 was pre-strained by applying a displacement of 0 mm respectively 1 mm in the center of the plate. The dynamic behavior is analyzed in terms of natural frequencies and modal damping ratios after unloading the plate. The Mullins effect is modeled by the OR model and the material parameters are identified in [7]. From Fig. 12, a significant softening of the plate can be observed. The decrease in natural frequencies is similar for both laminates C-$E^s_{.5}$-A-$E^s_{.5}$-C and A-$E^s_{.5}$-C-$E^s_{.5}$-A. The influence of Mullins effect induced softening on the modal damping ratios, however, is limited as observable in Fig. 13. This effect has yet to be validated experimentally, as a possible influence on the loss modulus of the elastomer material is not considered in the material model.

4 Low-Velocity Impact Tolerance of HyCEML Regarding Its Damping Behavior

Historically, FMLs have been deployed, among other reasons, due to their high damage tolerance under impact loading [21]. While there is some research on how elastomer interlayers in FRP laminates can significantly improve their impact behavior, no studies, to the authors' knowledge, have been conducted on the influence of barely visible impact damage on the CLD mechanism.

Fig. 14 HyCEML plate in configuration A-$E^{s}_{.5}$-C-$E^{s}_{.5}$-A under low-velocity impact loading. The plate's in-plane dimensions are 150 mm and 100 mm, respectively. The specimen's upward movement in transverse direction is restricted by four clamps

4.1 Experimental

4.1.1 Methods

Drop Weight Impact Test

In order to inflict and assess the damage in HyCEML under low-velocity impact loading, drop weight experiments are conducted based on ASTM D7136. Thus, the boundary conditions of the specimen plates are considered simply supported. The impact energy is varied and takes values of 5 J, 10 J and 20 J. All the laminates listed in Table 2 are tested with each energy. Figure 14 shows the support fixture and a specimen plate during impact. The specimens have the dimensions $a = 150$ mm and $b = 100$ mm and are impacted in their center. The impactor has a hemispherical tip with the diameter $d = 20$ mm.

Modal Analysis

In order to assess the influence of various types of damage inflicted under low-velocity impact loading, as illustrated above, experimental modal analyses are performed. For reference, undamaged specimens are also analyzed. The specimen plates are excited using an automated modal hammer and the response is detected with a laser scanning vibrometer. Free-free boundary conditions are aimed for by placing the specimens on a soft polymeric foam material during the analysis. The natural frequencies and modal damping ratios are then determined from the frequency response function for each plate.

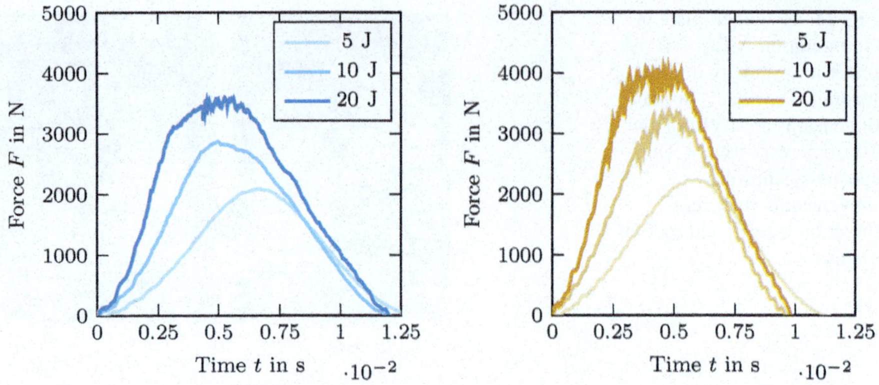

Fig. 15 Force time response of laminates C-$E^{s}_{.5}$-A-$E^{s}_{.5}$-C (left) and A-$E^{s}_{.5}$-C-$E^{s}_{.5}$-A (right) under low-velocity impact loading with different impact energies

4.1.2 Results

Drop Weight Impact Test

The results from the impact test illustrated in Sect. 4.1 are shown in the following. Figure 15 shows the force F detected by the impactor over time t for laminates C-$E^{s}_{.5}$-A-$E^{s}_{.5}$-C and A-$E^{s}_{.5}$-C-$E^{s}_{.5}$-A at different impact energy levels. For both laminates, a strain rate dependent stiffness is visible from the increasing slope with higher impact energies, which can be attributed to the viscoelastic behavior of the polymeric constituents in the laminates. Based on the smoothness of the curves for $E_{\text{impact}} = 5\,\text{J}$, no damage is detectable in both laminates. For $E = 10\,\text{J}$, however, a bend is visible in the response of laminate C-$E^{s}_{.5}$-A-$E^{s}_{.5}$-C indicating the onset of a damage mechanism. In the case of laminate A-$E^{s}_{.5}$-C-$E^{s}_{.5}$-A, oscillations start to occur from a force $F = 2500\,\text{N}$ onwards, suggesting continuous hardening of the aluminum layer due to plastic deformation. This effect occurs at higher energies and forces for laminate C-$E^{s}_{.5}$-A-$E^{s}_{.5}$-C due to the fact that, in this laminate, the aluminum layer is in the plate's center and thus undergoes less deformation under bending. The absence of pronounced bends in the curves of laminate A-$E^{s}_{.5}$-C-$E^{s}_{.5}$-A can be attributed to the absorption of energy by strong plastic deformation of the aluminum layers which in turn protects the CFRP layers from cracking as seen for laminate C-$E^{s}_{.5}$-A-$E^{s}_{.5}$-C. This mechanism is further discussed in [9].

The force over displacement curves for both laminates are shown in Fig. 16. The aforementioned onset of damage again is visible. Furthermore, the extent of plastic deformation can also be derived from the residual displacement at the end of the unloading phase when the impactor looses contact with the plate. This residual displacement is much more pronounced for laminate A-$E^{s}_{.5}$-C-$E^{s}_{.5}$-A, even for an impact energy of 5 J, confirming the aforementioned.

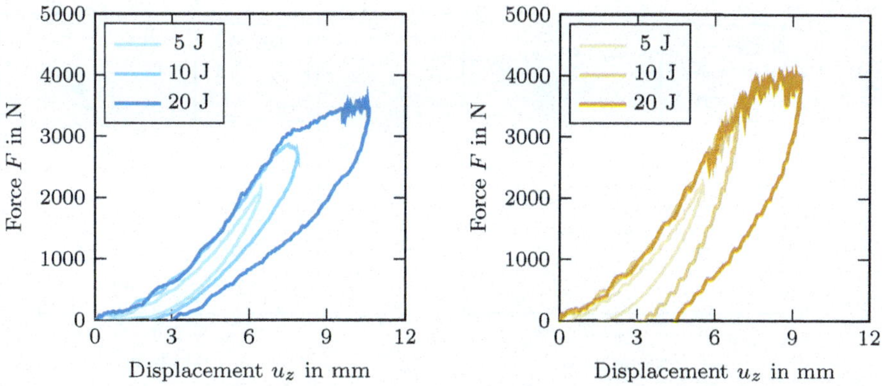

Fig. 16 Force displacement response of laminates C-$E^{s}_{.5}$-A-$E^{s}_{.5}$-C (left) and A-$E^{s}_{.5}$-C-$E^{s}_{.5}$-A (right) under low-velocity impact loading with different impact energies

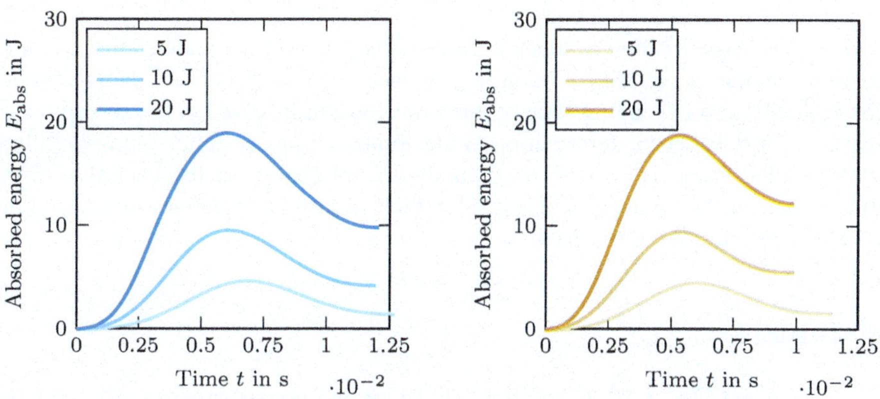

Fig. 17 Absorbed energy over time of laminates C-$E^{s}_{.5}$-A-$E^{s}_{.5}$-C (left) and A-$E^{s}_{.5}$-C-$E^{s}_{.5}$-A (right) under low-velocity impact loading with different impact energies

Figure 17 shows the energy transferred to the specimen during the impact event over time. The energy is calculated by integrating the force displacement curves from Fig. 16. From both curves with $E = 5\,\text{J}$ it can be seen, that for both laminates a similar amount of energy is absorbed by the specimen and thus not returned to the impactor. Since no severe damage could be detected for this energy level, it can be assumed that this amount of energy was dissipated during the impact event by viscoelastic deformation in the polymeric layers besides some general losses due to friction. For higher energies, laminate A-$E^{s}_{.5}$-C-$E^{s}_{.5}$-A generally shows a higher amount of absorbed energy after the impact event than laminate C-$E^{s}_{.5}$-A-$E^{s}_{.5}$-C. This indicates, that plastic deformation of the aluminum layers is the main contributor to the general absorption capabilities of the HyCEML material. This assumption has been validated using numerical models in [9].

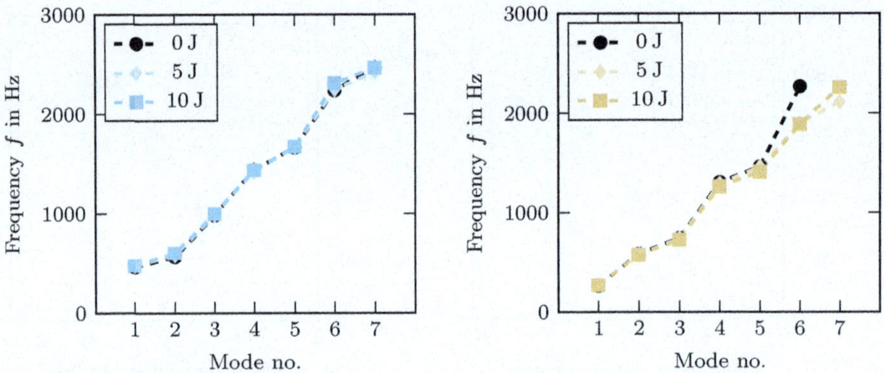

Fig. 18 Natural frequencies of laminates C-$E^s_{.5}$-A-$E^s_{.5}$-C (left) and A-$E^s_{.5}$-C-$E^s_{.5}$-A (right) impacted with different energies

From the experimental results presented above, it can be concluded that the lamination scheme of HyCEML strongly influences the occurrence of damage under low-velocity impact loading. The primary mechanism involved in the absorption of impact energy is plastic deformation of the aluminum layers. Since intra-ply failure of the CFRP layers is also observed, a significant change of the overall stiffness distribution in the laminate is expected, which in turn is assumed to influence the CLD capabilities of the material.

Modal Analysis

For laminates C-$E^s_{.5}$-A-$E^s_{.5}$-C and A-$E^s_{.5}$-C-$E^s_{.5}$-A and impact energies of 0 J, 5 J and 10 J, the natural frequencies determined in experimental modal analysis are shown in Fig. 18. In the case of laminate C-$E^s_{.5}$-A-$E^s_{.5}$-C, the deviation of the damaged plates compared to the undamaged one is negligible. In contrast, laminate A-$E^s_{.5}$-C-$E^s_{.5}$-A shows a visible influence of impact damage on the natural frequencies. For both impact energies, 5 J and 10 J, higher modes occur at lower frequencies, indicating a more damaged and thus more compliant material.

The modal damping ratio corresponding to the natural frequencies in Fig. 18 are shown in Fig. 19 for both laminates. It can be seen that the damping ratios of the damaged C-$E^s_{.5}$-A-$E^s_{.5}$-C plates differ from the undamaged ones. Both damaged plates seem to follow the same trend when deviating from the undamaged reference, as some modes show stronger damping while others are damped less. Generally, plates impacted with a higher energy also show higher deviations when compared to the undamaged plate. Laminate A-$E^s_{.5}$-C-$E^s_{.5}$-A behaves similar in this regard, however, the deviations from the damaged plates are higher than seen with laminate C-$E^s_{.5}$-A-$E^s_{.5}$-C. This effect is more pronounced for higher modes, as almost no deviations are visible for the first three modes.

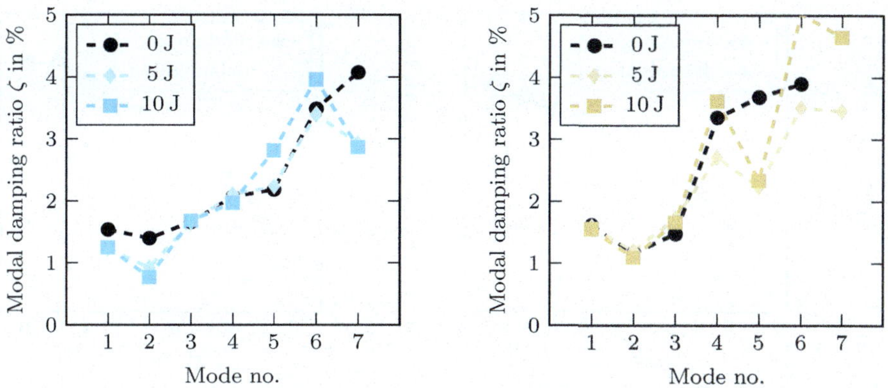

Fig. 19 Modal damping ratios of laminates C-$E^s_{.5}$-A-$E^s_{.5}$-C (left) and A-$E^s_{.5}$-C-$E^s_{.5}$-A (right) impacted with different energies

Following the results presented above, it can be concluded that the lamination scheme in HyCEML is critical to its damage tolerance with regard to the CLD damping capabilities. When comparing the laminates C-$E^s_{.5}$-A-$E^s_{.5}$-C and A-$E^s_{.5}$-C-$E^s_{.5}$-A, the more distinct plastic deformation in the latter suggest that this is the primary driver for a change in the damping capabilities. From laminate C-$E^s_{.5}$-A-$E^s_{.5}$-C it can be concluded that the global stiffness change due to damage is negligible as seen in the natural frequencies of undamaged and damaged plates. Consequently, the more pronounced change in modal damping cannot be attributed to the damage induced reduction of stiffness alone. More likely, the geometric change of the specimen results in an altered distribution of strain within the elastomeric damping layers. Furthermore, a change in the elastomer material's damping characteristics caused by the impact event itself is assumed to be a contributing factor.

4.2 Modeling

In order to further analyze the role of low-velocity impact damage on CLD, the resulting damage mechanisms are modeled numerically using the FEM solver Abaqus/Standard. As an example, the modeling of delaminations is illustrated in the following. Since the aim of this work lies in the behavior of damaged laminates under vibration, the initiation and evolution of damage is not modeled explicitly. Instead, delaminations are treated as defined regions with no interlaminar stiffness. Undamaged interfaces are assumed to be strictly bonded without any interface behavior. In order to prevent unphysical deformations and penetrations of layers adjacent to a delamination, a cohesive surface-to-surface contact is established between the top and bottom delaminated surfaces. Thus, an opening of the delamination is not

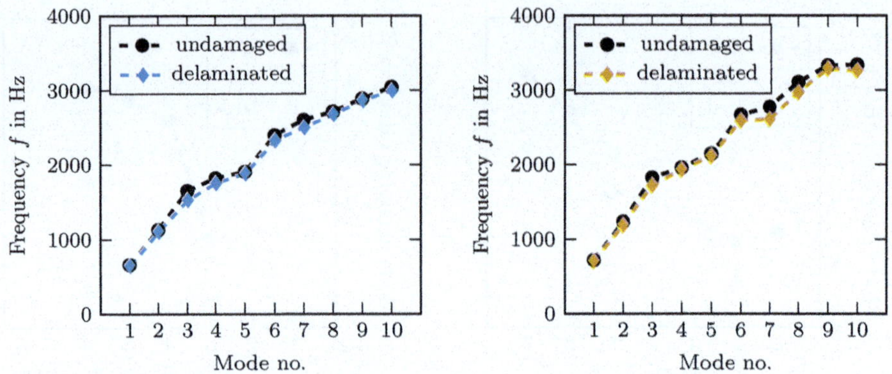

Fig. 20 Numerically determined natural frequencies of undamaged and delaminated simply supported laminates C-$E^s_{.5}$-A-$E^s_{.5}$-C (left) and A-$E^s_{.5}$-C-$E^s_{.5}$-A (right), showing marginal effect of delamination on the vibration behavior

prohibited. A complex eigenvalue extraction is performed by linear perturbation of the equilibrium equation of the system.

Based on micrographs of damaged specimens subjected to low-velocity impact as outlined in Sect. 4.1, delaminations between 0° and 90° CFRP layers have been found to predominantly occur under given loading. Figure 20 shows the natural frequencies of undamaged and delaminated simply supported specimens. The delaminated specimens are modeled to contain a circular delamination with a radius of 20 mm with its center aligned with the point of impact. For both laminates, the delamination lies in the bottom most interface between two CFRP layers with different orientations 0° and 90°. From Fig. 20 it is observed that the modeled delamination does not have an effect on the natural frequencies of the plate for both laminates. Figure 21, however, shows the corresponding modal damping ratios. For these values, a difference between the undamaged and delaminated laminates can be seen in case of laminate C-$E^s_{.5}$-A-$E^s_{.5}$-C. The deviation, however, is smaller than the ones seen experimentally in Fig. 19. No deviations between undamaged and delaminated plates is seen for laminate A-$E^s_{.5}$-C-$E^s_{.5}$-A. This can serve as an indicator, that the CLD mechanism in HyCEML is not as sensitive to inter-ply failure within the laminate following a low-velocity impact events as it is to plastic deformations. Consequently, the experimentally observed plastic deformation of the impacted laminates appears to be the main contributor to the observed change in the vibrational behavior, especially for laminates with outer aluminum layers.

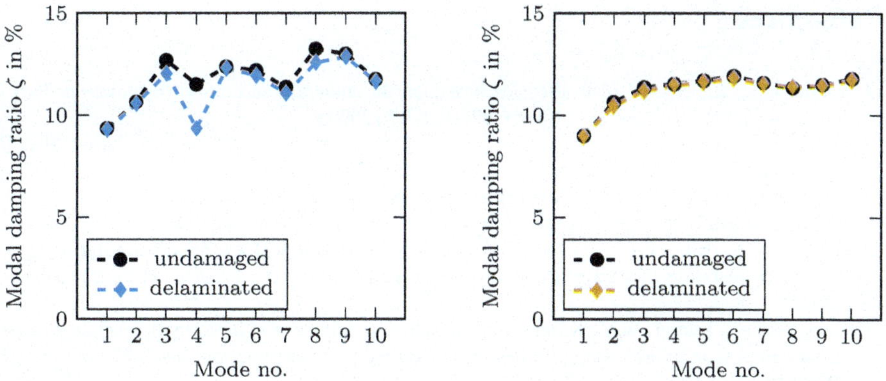

Fig. 21 Numerically determined modal damping ratios of undamaged and delaminated simply supported laminates C-$E^s_{.5}$-A-$E^s_{.5}$-C (left) and A-$E^s_{.5}$-C-$E^s_{.5}$-A (right)

5 Conclusion

Within this work, experimental methods for the characterization of the vibrational behavior of HyCEML have been developed and applied in order to analyze the influence of parameters like temperature, excitation frequency, and amplitude on the CLD mechanism of such laminates. Furthermore, the influence of material properties and stacking sequence within a laminate has been identified as a major contributor to the resulting damping behavior. In order to facilitate the optimization of such laminated structures, an efficient analytical modeling approach has been developed, which allows the identification of vibrational parameters for arbitrary materials and laminates. Subsequently, the CLD mechanism has been further investigated regarding its influenceability by a number of damage phenomena. On the one hand, the Mullins effect occurring in most filled elastomers has been characterized in the damping material used in this study. Constitutive models depicting this strain-dependent softening have been parameterized and found to accurately depict the material behavior. A numerical case study has shown, that pre-straining a plate by moderate deflections can greatly influence the laminates' natural frequencies. On the other hand, low-velocity impact has been considered and the resulting types of damage have been analyzed. Plastic deformation and delaminations both contribute to a change in the modal damping of HyCEML, with the former being the dominant effect.

Acknowledgements This work is funded by the Deutsche Forschungsgemeinschaft (DFG, German Research Foundation) SPP1897 "Calm, Smooth, Smart—Novel approaches for influencing vibrations by means of deliberately introduced dissipation", project "HyCEML—Hybrid CFRP/elastomer/metal laminates containing elastomeric interfaces for deliberate dissipation" (Karlsruhe Institute of Technology). The work is also part of the Young Investigator Group (YIG) "Green Mobility", generously funded by the Vector Stiftung. We express our utmost gratitude to Dr.-Ing. Dominik Schmid at TU Berlin for conducting the modal analysis presented in Sect. 4.1. Furthermore, the authors would like to acknowledge Gummiwerk KRAIBURG GmbH & Co. KG for providing the elastomer materials.

References

1. Demasi, L.: ∞6 Mixed plate theories based on the generalized unified formulation. Part I: governing equations. Compos. Struct. **87**(1), 1–11 (2009)
2. Demasi, L.: ∞6 Mixed plate theories based on the generalized unified formulation. Part II: layerwise theories. Compos. Struct. **87**(1), 12–22 (2009)
3. Dorfmann, A., Ogden, R.W.: A constitutive model for the Mullins effect with permanent set in particle-reinforced rubber. Int. J. Solids Struct. **41**(7), 1855–1878 (2004)
4. Jackstadt, A., Kärger, L.: Extension of an analytical solution of a unified formulation to the frequency response of composite plates with viscoelastic layers. PAMM **20**(1), e202000234 (2021)
5. Jackstadt, A., Liebig, W.V., Kärger, L.: Analytical modeling and investigation of constrained layer damping in hybrid laminates based on a unified plate formulation. Int. J. Mech. Sci. **216**, 106964 (2022)
6. Jackstadt, A., Liebig, W.V., Sessner, V., Weidenmann, K.A., Kärger, L.: Application of a mixed variational higher order plate theory towards understanding the deformation behavior of hybrid laminates. PAMM **19**(1), e201900048 (2019)
7. Jackstadt, A., Frölich, F., Weidenmann, K.A., Kärger, L.: Modeling the Mullins effect of rubbers used in constrained-layer damping applications. PAMM **21**(1), e202100098 (2021)
8. Kerwin, E.M.: Damping of flexural waves by a constrained viscoelastic layer. J. Acoust. Soc. Am. **31**(7), 952–962 (1959)
9. Li, Z., Zhang, J., Jackstadt, A., Kärger, L.: Low-velocity impact behavior of hybrid CFRP-elastomer-metal laminates in comparison with conventional fiber-metal laminates. Compos. Struct. **287**, 115340 (2022)
10. Liebig, W.V., Jackstadt, A., Sessner, V., Weidenmann, K.A., Kärger, L.: Frequency domain modelling of transversely isotropic viscoelastic fibre-reinforced plastics. Compos. Sci. Technol. **180**, 101–110 (2019)
11. Liebig, W.V., Sessner, V., Weidenmann, K.A., Kärger, L.: Numerical and experimental investigations of the damping behaviour of hybrid CFRP-elastomer-metal laminates. Compos. Struct. **202**, 1109–1113 (2018)
12. Oberst, H., Frankenfeld, K.: Über die Dämpfung der Biegeschwingungen dünner Bleche durch fest haftende Beläge. Acta Acust. Acust. **2**(6), 181–194 (1952)
13. Ogden, R.W., Roxburgh, D.G.: A pseudo-elastic model for the Mullins effect in filled rubber. Proc. R. Soc. A: Math., Phys. Eng. Sci. **455**(1988), 2861–2877 (1999)
14. Ross, D., Ungar, E.E., Kervin, E.M.: Damping of plate flexural vibrations by means of viscoelastic laminae. In: Ruzicka, J.E., (ed.) Structural Damping, pp. 49–87. Pergamon Press (1960)
15. Sessner, V.: Charakterisierung und Modellierung des Dämpfungsverhaltens von hybriden Faser-Metall-Elastomer-Laminaten. Karlsruher Institut für Technologie (KIT), Karlsruhe (2021)
16. Sessner, V., Liebig, W.V., Weidenmann, K.A.: Modal damping behavior of plane and 3D curved constrained layer damping CFRP-elastomer-metal laminates. Compos. Part C: Open Access **2**, 100037 (2020)
17. Sessner, V., Jackstadt, A., Liebig, W.V., Kärger, L., Weidenmann, K.A.: Damping characterization of hybrid carbon fiber elastomer metal laminates using experimental and numerical dynamic mechanical analysis. J. Compos. Sci. **3**(1), 3 (2019)
18. Sessner, V., Stoll, M., Feuvrier, A., Weidenmann, K.A.: Determination of the damping characteristics of fiber-metal-elastomer laminates using piezo-indicated-loss-factor experiments. In: Herrmann, A.S., (ed.), 21st Symposium on Composites, pp. 325–332. Trans Tech Publications Ltd. (2019)

19. Stoll, M., Sessner, V., Kramar, M., Technau, J., Weidenmann, K.A.: The effect of an elastomer interlayer thickness variation on the mechanical properties of Fiber-Metal-Laminates. Compos. Struct. **219**, 90–96 (2019)
20. Vlot, A., Gunnink, J.W.: Fibre Metal Laminates: An Introduction. Springer, New York (2001)
21. Wu, G., Yang, J.M.: The mechanical behavior of GLARE laminates for aircraft structures. JOM **57**(1), 72–79 (2005)

Shape Memory Alloy (SMA) Damping for Smart Miniature Systems

Kiran Jacob, Shahabeddin Ahmadi, Pejman Shayanfard, Frank Wendler, and Manfred Kohl

1 Introduction

In many applications, uncontrolled mechanical vibrations are a potential source of noise and damage. On the large scale, dampers are used in massive structures like buildings and bridges to reduce the impact of seismic loads, or in automobiles, to improve travel comfort and safety [1]. In the past couple of decades, there is a strong drive towards miniaturization with integration of many sensors/actuators at small footprint [2]. Vibration damping and control for small and micro-scale systems is challenging, as most damper concepts used for large structures cannot easily be downscaled. Here, the authors develop miniature dampers and vibration control using Shape Memory Alloy (SMA) foil based devices for miniature robotics application and smartphone camera stabilization against hand movements. By using SMA foils, design flexibility and miniaturization are achieved.

SMAs are functional materials capable of undergoing large reversible deformation through which a lot of energy is dissipated. This unique behaviour is attributed to the underlying phase transformation of the material between austenite and martensite phases. Unlike many other velocity dependent damper concepts, the energy dissipation in SMA is heavily dependent on the loading strain. SMAs exhibit two types of behaviours called pseudoelasticity or one-way shape memory effect depending on the operating temperature. The material shows pseudoelasticity when operated above austenite finish temperature (A_f) and one-way shape memory effect below marten-

K. Jacob · S. Ahmadi · M. Kohl (✉)
Institute of Microstructure Technology, Karlsruhe Institute of Technology,
Hermann-von-Helmholtz-Platz 1, 76344 Eggenstein-Leopoldshafen, Germany
e-mail: manfred.kohl@kit.edu

P. Shayanfard · F. Wendler
Friedrich-Alexander-Universität Erlangen-Nürnberg, Institute of Materials Simulation,
Dr.-Mack-Str. 77, 90762, Furth, Germany
e-mail: frank.wendler@fau.de

P. Eberhard (ed.), *Calm, Smooth and Smart*, Lecture Notes in Applied and Computational Mechanics 102, https://doi.org/10.1007/978-3-031-36143-2_7

site finish temperature (M_f). In pseudoelastic SMA, the strain recovery occurs upon unloading. In a mechanical loading cycle, large amount of energy is dissipated by material hysteresis which is used to develop passive dampers. In one-way SMA, the strain recovery occurs when the material is heated above A_f. This principle is used to develop active dampers and actuators for vibration control.

This article is structured as follows: in Sect. 2 the pseudoelastic material behaviour and passive damper performance under shock loading are illustrated. In Sect. 3, concepts for active damping based on the one-way SMA effect are introduced, and a related multi-axial constitutive model is sketched. In Sect. 4, a 2-DoF vibration stabilization platform using one-way SMA actuators is described.

2 Damping Using Pseudoelastic SMA Devices

A cold-rolled $Ni_{50.5}Ti_{49.1}Fe_{0.4}$ (at. %) foil of 30 μm thickness is used to investigate the passive damping behaviour of SMA. This material shows pseudoelastic behaviour at room temperature. The stress-induced transformation between austenite and martensite is responsible for the energy dissipation in pseudoelastic SMA. In this section, the behaviour of the material, damper devices and passive damper systems is described.

2.1 *Material Behaviour*

The mechanical behaviour of pseudoelastic SMA is investigated by displacement-controlled uniaxial tensile loading of a strip of SMA foil having dimensions (l, w, t): 10, 3, 0.03 mm. Figure 1a shows the tensile loading at various loading strains. Above a critical stress of about 500 MPa, the phase transformation to martensite is initiated resulting in the plateau region in stress-strain behaviour. A large strain of 0.04 is accommodated during the transformation to martensite. The material recovers to its initial state upon unloading. A large amount of loading energy is dissipated via hysteresis in this cyclic operation. The energy dissipation increases with increase in loading strain. Figure 1b illustrates the effect of pre-straining (ε_{pre}) on energy dissipation while retaining a constant loading strain of 0.01 with reference to the loading behaviour at zero pre-strain. The material has a non-linear behaviour with an initial elastic region until about 0.02 strain where the material remains in austenite and thereafter a plateau region where the transformation to martensite occurs. At pre-strain of 0.01 (ε_{pre1}) and loading strain of 0.01, not much energy dissipation is observed owing to predominant elastic behaviour. Adjusting the pre-strain close to the beginning of transformation plateau at 0.02 (ε_{pre2}), an almost repeatable energy dissipation is observed for loading cycles. Fixing the pre-strain in the elastic region enables the material to completely recover to its starting condition. Further increase in pre-strain to 0.03 (ε_{pre3}), high energy dissipation in the first loading cycle is attained

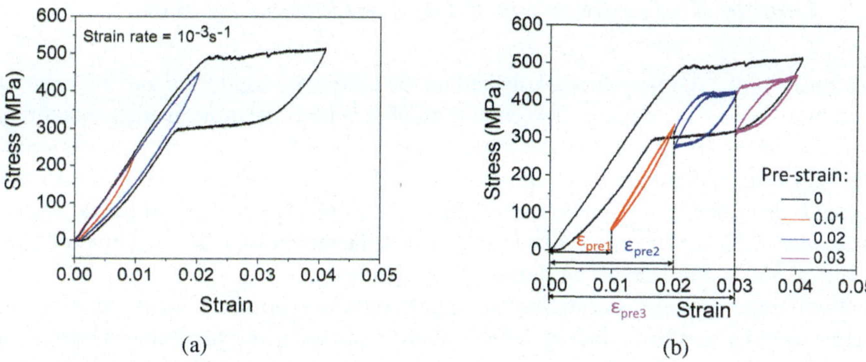

Fig. 1 Material hysteresis behaviour under uniaxial tensile loading for various loading strains (**a**) and at fixed loading strain of 0.01 (**b**)

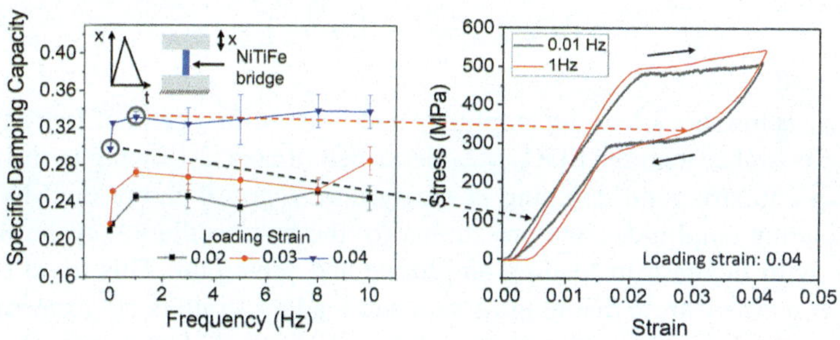

Fig. 2 SDC behaviour at various frequencies and loading strains showing increase in hysteresis until adiabatic loading is reached above 1 Hz

but subsequent cycles show reduced energy dissipation. Therefore, pre-strain and loading strain are crucial parameters for optimizing dissipation.

The energy dissipation of the material at various frequencies and loading strains is shown in Fig. 2. The energy dissipation arising from the material hysteresis is quantified using the parameter called Specific Damping Capacity ($SDC_{material}$), which is the ratio of energy dissipation through hysteresis to the loading energy. Until loading frequencies upto 1 Hz there is an increase in $SDC_{material}$ due to increase in hysteresis. This effect is associated with the self-heating of the material during the transformation to martensite. Owing to this temperature rise in the material, an increase in loading stress is required for transformation. Above 1 Hz, adiabatic heating is attained, resulting in a frequency independent $SDC_{material}$ behaviour.

2.2 Double Bridge Device and Passive Damper System

Designing of the damper device is based on the load to be supported and the external excitation. A passive damper device with double bridge structure is micromachined by laser cutting from SMA foil as shown at the top of Fig. 3 (drawing not to scale). The half-bridge dimensions of the device are $5 \times 0.3 \times 0.03$ mm^3 (L_0 w, t). The device is designed to operate in out-of-plane direction. The bridges of the device are strained to 0.05 for an out-of-plane loading displacement of 1.6 mm. The pre-strain of the device is adjusted along the in-plane direction.

The damper system is constructed from two double bridge devices sandwiching a mass of 18 g as shown in Fig. 3. The mass is pulled using an electromagnet and released by deactivating the electromagnet. The free oscillation of the mass is tracked using a laser displacement sensor. The SDC for the system is evaluated for the first oscillating cycle as

$$SDC_{system} = 1 - \left(\frac{x_1}{x_{max}}\right)^2, \tag{1}$$

where x_{max} is the shock loading amplitude and x_1 is the first peak of free oscillation. SDC_{system} is higher than the $SDC_{material}$ described in Sect. 2.1 owing to the presence of air drag and structural damping in the system. Figure 4 shows the SDC_{system} at various loading amplitudes and pre-strains for the first oscillation cycle. SDC_{system} increases with increase in loading amplitude and pre-strain. This trend continues until the combined strain due to pre-strain and loading strain is not exceeding complete martensite transformation strain of 0.05. Therefore, for a particular loading amplitude, pre-strain optimization enables improved energy dissipation. Maximum SDC_{system} of 70 % is reached utilizing maximal material hysteresis for energy dissipation, implying that only 30 % of the loading energy is remaining. After the first oscillation cycle, the mass undergoes several low amplitude oscillations before reaching equilibrium position. Owing to the lack of guidance for the mass, various modes of oscillations are observed in this settling behaviour.

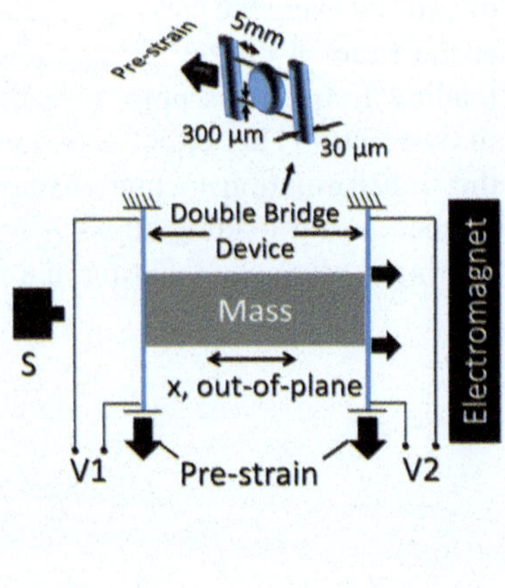

Fig. 3 The schematic of damper test setup [3] and an image of the double bridge device is shown at the top (drawing not to scale). Legends: S-Displacement sensor, V1,V2-Terminals for heating pulse (used while using one-way devices)

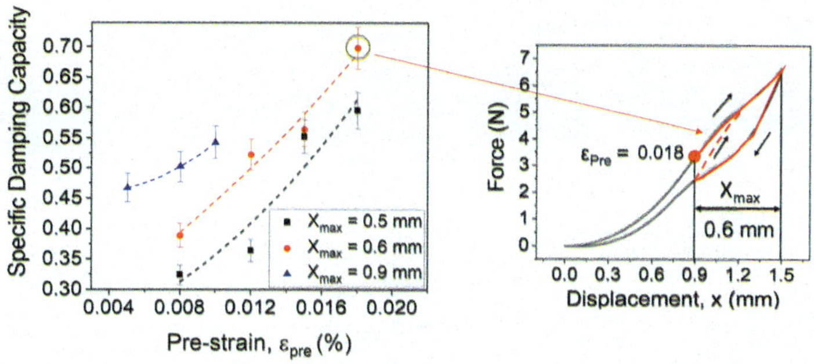

Fig. 4 SDC behaviour of pseudoelastic damper system at various pre-strains and shock loading amplitudes

2.3 Single Bridge SMA Device and Passive Damper System

A prospective application of the SMA damper is the shock absorption in miniature robotics, where vertical shock loads are of interest. As illustrated in Sect. 2.2, high energy dissipation is achieved under shock loading. The lack of mass guidance resulted in 3D mass movement after release from the electromagnet. In this section, the behaviour of SMA shock absorber under vertical shock loading with guided mass movement is illustrated.

A single bridge damper device design is used as the mass is balanced using the guidance. The bridge has a dimension of w= 1 mm and L_0= 6 mm, allowing for an out-of-plane displacement of 1.9 mm at a strain of 0.05 in the bridge.

The test setup for the shock absorber is illustrated in Fig. 5. A mass of 47 g is connected to the damper device and its movement is guided using a low friction linear bearing. The shock loading is provided using an electromagnet. The movement of the mass is tracked using a laser displacement sensor. Data acquisition and control system is used to acquire the measurement and to control the electromagnet.

Fig. 5 Schematic of test setup for vertical shock loading

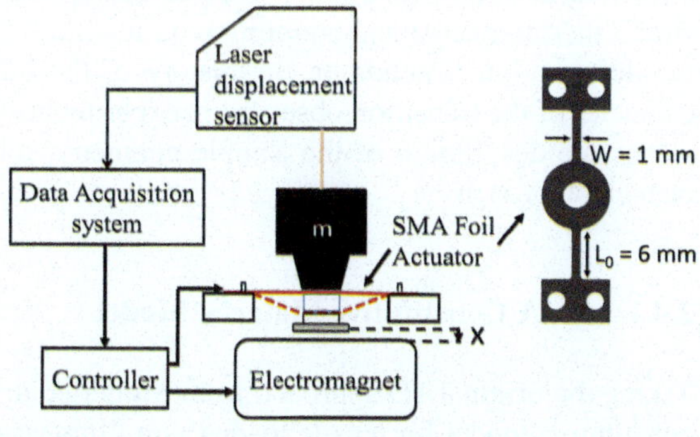

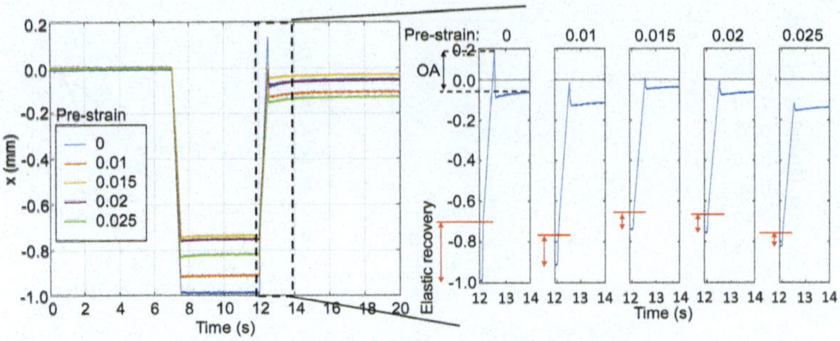

Fig. 6 The shock response of the system at various device pre-stains and the zoomed in settling behaviour showing an increase in overshoot amplitude with increase in elastic recovery

The shock response of this device under various pre-strains is shown in Fig. 6. The shock loading amplitude is selected to limit the total created strain on the SMA device to 0.045. In all loading experiments, the mass stabilizes immediately after the first overshoot demonstrating excellent shock damping performance. A closer look on the settling behaviour reveals the contributions of elastic recovery and material hysteresis. Elastic recovery, resulting in a sudden mass movement, occurs immediately after release as marked by the red lines. Apparently, the elastic recovery changes with the pre-strain and loading conditions. Accordingly, the overshoot amplitude (OA) follows the same trend as elastic recovery showing high OA at high elastic recovery. This study reveals the effectiveness of using SMA foil based passive damper as miniature shock absorber.

2.4 *Development of a FEM Model for Pseudoelastic Damping*

The mechanical response of SMAs is both nonlinear and history dependent due to a first order structural phase transition from austenite (A) to martensite (M), which is related to the release (A $\rightarrow$ M) or the uptake (M $\rightarrow$ A) of latent heat. The strong thermomechanical coupling requires transient, time-resolved simulations as the deformation depends on the history of the internal state variables. The local character of the transition observed in experiments necessitates a spatially resolved, mesoscale description of the sample geometry, for which the authors use a finite element approach.

2.4.1 SMA Constitutive Material Model

Taking the original 1D approach from Müller et al. [4] as a basis, a plane-stress 2D constitutive model for tensile loaded thin films was developed in [5]. The model is based on a thermodynamic description using a Gibbs free energy density

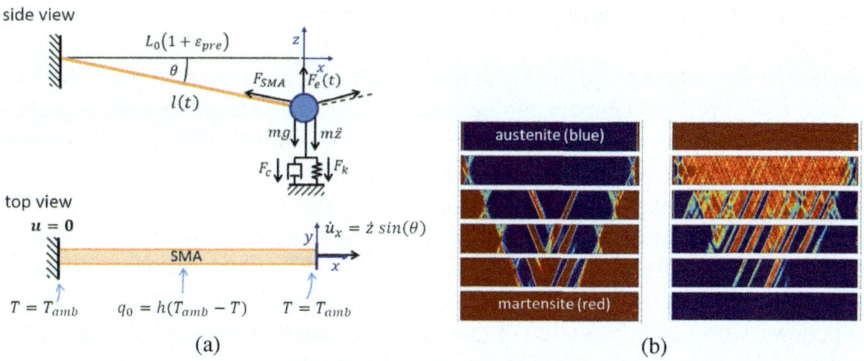

Fig. 7 (**a**) Force balance assumed in the ODE for mass displacement z (top) and 2D FE domain with boundary conditions (bottom) [6]. (**b**) Top → bottom: time evolution of martensite fraction in SMA bridge during loading (left) and unloading (right), at equidistant time steps of 10 s

$$g(\varepsilon, \sigma, T, x_\alpha, \nabla x_\alpha) = \Psi_{mech}(\sigma, \varepsilon) - \sigma\varepsilon + \Psi_{chem}(T) + \Psi_{PF}(x_\alpha, \nabla x_\alpha),$$

that depends on local values of strain ε, stress σ, temperature T, the fraction of phases x_α, and their gradients ∇x_α. The existence of tension and compression accommodated martensite variants ($\alpha = M_+, M_-$) and austenite ($\alpha = A$) is assumed. Twinned martensite is represented as 50–50 mixture of M_+ and M_-. The mechanical part of the free energy Ψ_{mech} forms a non-convex continuous three-well potential in linear strain space, whereas the chemical part Ψ_{chem} controls the energy difference between austenite and martensite phase. A phase-field energy term Ψ_{PF} for a coarse-grained treatment of A-M interface energies was incorporated [5, 7]. Rate equations for the phase fractions X_{M_-}, X_A and X_{M_+} are formulated as

$$\begin{aligned}
\dot{x}_{M_+} &= -x_{M_+} p^{x_{M_+}A} + x_A p^{Ax_{M_+}} - \frac{\omega}{W}\delta\Psi_{PF}/\delta x_{M_+} \\
\dot{x}_{M_-} &= -x_{M_-} p^{x_{M_-}A} + x_A p^{Ax_{M_-}} - \frac{\omega}{W}\delta\Psi_{PF}/\delta x_{M_-} \\
\dot{x}_A &= -\dot{x}_{M_+} - \dot{x}_{M_-},
\end{aligned}$$

where $\delta/\delta x_{M_\pm}$ denotes a variational derivative. By this approach the authors assume thermally activated kinetics describing transitions from compression-accommodated M_- over austenite A to tension-accommodated martensite M_+, where stress and temperature dependent transition rates $p^{\alpha\beta}(\sigma - \sigma^{\alpha\beta}, T)$ are used. The plateau stresses $\sigma^{\alpha\beta} = (\sigma_0^{\alpha\beta} + C^{\alpha\beta}(T - T_R))(1 + \xi)$ define the transformation criteria [5]. Here, a static noise field for the transformation stress level ξ is introduced to capture heterogeneous nucleation sites not observable on the device level. In this way, in simulations, local deformation pattern like Lüders-like bands are developing, where the properties of the noise distribution ξ are adjusted in comparison with DIC (digital image correlation) images from experiment.

2.4.2 FEM Setup for Simulating Bridge Vibrations

Due to the low thickness of the SMA films of a few tens of microns, bending deformation of the bridge dampers can be neglected, as the related stresses amount for only a few percent of those created by the tensile stretch. This was also verified by a good match between measured quasi-stationary force-displacement curves and calculated ones, where the stress-strain characteristics of the material was used. For transient simulations of pseudoelastic single bridges, the setup including mechanical and thermal boundary conditions is shown in Fig. 7a. The SMA domain is discretized with high spatial resolution in the FE model assuming 2D plane stress conditions (top view), whereas the equation of motion of the attached mass is given as the force balance (side view) of the proof mass, resulting in

$$m\ddot{z} - mg + 2F_{SMA}(t)sin(\theta) - F_e(t) = 0 \quad \text{with} \quad sin(\theta) = \frac{z}{\sqrt{z^2 + l_0^2(1+\varepsilon_{pre})^2}},$$

where l_0 denotes initial SMA length and ε_{pre} the applied pre-strain. This one-DoF ODE is solved simultaneously, where the coupling of mass position z and displacement, u, at the FE domain boundary is achieved by a boundary condition for the longitudinal displacement rate $\dot{u}_x = \dot{z}sin(\theta)$ assuming a stretched geometry with angle θ. For a quasi-stationary displacement of the mass in z-direction (Fig. 7a), the time evolution of the martensite fraction due to the occurring tensile load is shown in Fig. 7b.

The transient mass amplitude and martensite fractions are given in Fig. 8a for a single bridge device excited with a harmonic force of 0.1 N and 32 Hz. Simulations were conducted with the parameter set in Table 1 determined from characterization data of the quarternary alloy NiTiCoCu. Both curves show the evolution of a steady state after an initial settling phase. For operation of the devices close to the resonance

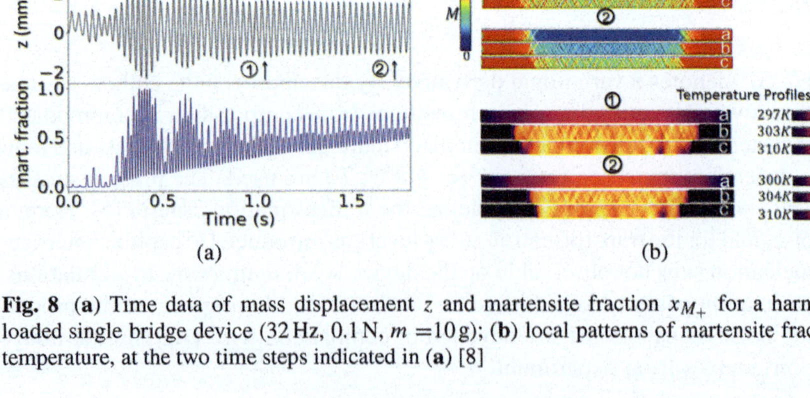

(a) (b)

Fig. 8 (**a**) Time data of mass displacement z and martensite fraction x_{M+} for a harmonically loaded single bridge device (32 Hz, 0.1 N, m =10 g); (**b**) local patterns of martensite fraction and temperature, at the two time steps indicated in (**a**) [8]

Table 1 Simulation parameters for a pseudoelastic NiTiCuCo damper device [7] and a one-way TiNi device [6]

Parameter	Symbol	Pseudoelastic SMA	One-way SMA
Proof mass	m	0.01 kg	0.018 kg
Length/width/thickness, SMA	$l/w/d$	10/0.35/0.02 mm	5/0.3/0.03 mm
Plateau stress A-M	σ^{AM}	242 MPa	495 MPa
Plateau stress M-A	σ^{MA}	134 MPa	100 MPa
Reference temp. for σ^{AM}/σ^{MA}	T_{ref}	294 K	349.7 K
Clausius-Clapeyron coeff. A-M	C^{AM}	10.4 MPa K^{-1}	5.8 MPa K^{-1}
Clausius-Clapeyron coeff. M-A	C^{MA}	14 MPa K^{-1}	6.2 MPa K^{-1}
Transformation strain	ε_T	0.0075	0.04
Elastic modulus, A	E_A	35.9 GPa	70.2 GPa
Elastic modulus, M	E_M	16.0 GPa	32.2 GPa
Latent heat	L	5600 J kg^{-1}	22130 J kg^{-1}
Thermal conductivity, A	k_A	18 Wm^{-1}K^{-1}	18 Wm^{-1}K^{-1}
Thermal conductivity, M	k_M	8.6 Wm^{-1}K^{-1}	8.6 Wm^{-1}K^{-1}
Heat transfer coefficient	h	30 Wm^{-2} K^{-1}	30 Wm^{-2} K^{-1}

of the bridge oscillator, generally a continuous accumulation of martensite is found, hence a steady decay of damping capacity follows. Representative local fine-banded patterns of phase fraction and temperature are given in Fig. 8b, which shows that martensite accumulation proceeds from the fixed boundaries of the SMA.

3 Damping Using One-Way SMA Devices

So far, the passive damping of SMA devices exploiting stress induced phase transformation is explained. By using SMA devices showing one-way shape memory effect, heating pulses can be utilized to stimulate phase transformation allowing for an external control of the energy dissipation. To investigate the active damping behaviour of SMA, cold-rolled NiTi foil of 30 μm thickness is used here. This material shows one-way effect at room temperature, retaining its strain upon loading owing to the reorientation of martensite. Strain recovery occurs by heating the material above austenite finish temperature to enable phase transformation to austenite. A large part of the loading energy is dissipated in this cyclic operation. In this section, the damping behaviour of one-way material based damper devices and damper systems is described [3, 9].

3.1 Material Behaviour

The phase transformation behaviour of a $Ti_{50.2}Ni_{49.8}$ (at. %) foil measured using Differential Scanning Calorimetry (DSC) is shown in Fig. 9a. The material transforms to austenite at temperatures above the austenite finish temperature A_f. When cooling down, the material transforms to martensite via an intermediate R-phase. However, macroscopically, the effect of R-phase is observed as a small plateau only in the first loading cycle near to a loading strain of 0.005 . Therefore the phase of the material below the R finish temperature (R_f) of 43.2 °C is referred to as the starting phase in this article.

A strip of one-way SMA foil having dimension 10×3×0.03 mm^3 is uniaxially loaded in displacement controlled mode to a loading strain of 0.04 at various frequencies in a thermal chamber, the corresponding SDC, estimated from the material hysteresis as explained in Sect. 2.1, is shown in Fig. 9b. The maximum SDC is achieved for loading at 25 °C. The SDC of the material decreases significantly with increase in temperature. At 60 °C, the material is in austenite and exhibits pseudoelastic behaviour demonstrating a lower SDC of about 45 %.

The damping capacity of the material at various loading strains and frequencies is shown in Fig. 10. The $SDC_{material}$ increases with increasing loading amplitude and attains a maximum of 86 % under quasi-stationary loading conditions. Also, the $SDC_{material}$ shows a reduction with frequency up to 1 Hz, which is caused by the increase in transformation stress arising from the self-heating of the material. The self-heating raises the material temperature above A_s resulting in enhanced elastic recovery. After reaching an adiabatic loading situation at 1 Hz, the $SDC_{material}$ shows a nearly frequency independent behaviour.

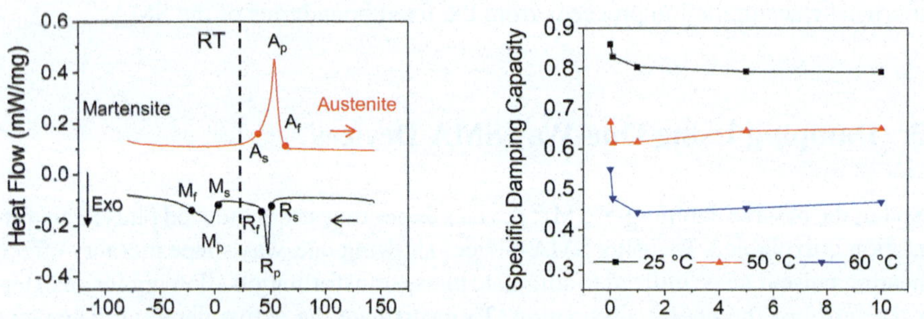

(a) (b)

Fig. 9 The phase transformation behaviour of one-way SMA (**a**) and its temperature dependent SDC behaviour at various ambient temperatures (**b**). Legends: Austenite–$A_{s,p,f}$: 47.2, 54.2, 59.1 °C, R-phase–$R_{s,p,f}$: 50.3, 47.9, 43.2 °C, Martensite–$M_{s,p,f}$: 2.0, –5.1, –28.5 °C (temperatures listed as start, peak, finish)

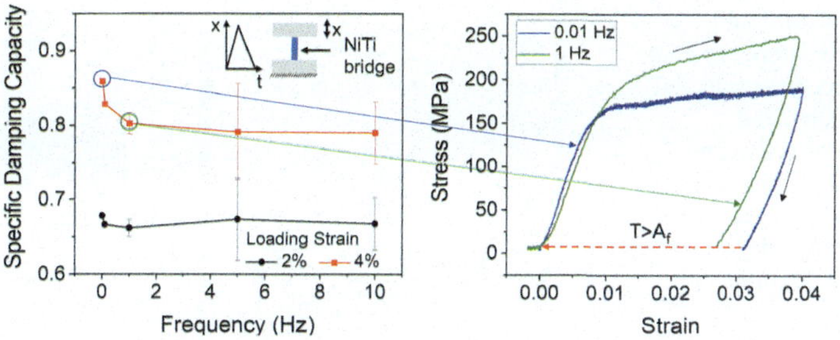

Fig. 10 SDC behaviour of one-way SMA strip under tensile testing showing reduction in damping capacity until reaching the adiabatic loading condition at 1 Hz

3.2 Double Bridge Device and Active Damper System

The double bridge device design explained in Sect. 2.2 is also used to develop one-way SMA devices to investigate their active damping performance. For active damping, the phase transformation to austenite is attained by electrical heating pulses. The pulse duration of 50 ms is chosen to enable the transformation within the first cycle of free oscillation, and leads to an inhomogeneous temperature distribution along the bridges.

The maximum temperature in the device (T_{max}) is shown in Fig. 11. The device heats-up from the initial room temperature to 80 °C when a heating pulse of 1 W is applied for 50 ms. Under this condition, more than 90 % of the bridge is at a temperature above A_f. The time to cool down to 90 % of the maximum bridge temperature lies between 1.2 and 1.3 s. This large cooling time limits the effectiveness of applying multiple heating pulses during free oscillation after shock loading.

The test setup explained in Sect. 2.2 is used to investigate the active shock damping performance. Terminals V1 and V2 in Fig. 3 are used to apply the heating pulses to enable phase transformation to austenite. Figure 12 shows the SDC_{system} at various pre-strains and shock loading amplitudes for this one-way damper system. The heating pulse is applied when the mass moves 0.3 mm after the release from the electromagnet. The SDC measurement for the first oscillation cycle shows an increase

Fig. 11 The maximum temperature on the double bridge device for various heating pulses of 50 ms duration, and related time to cool-down to 90 % of the raised temperature

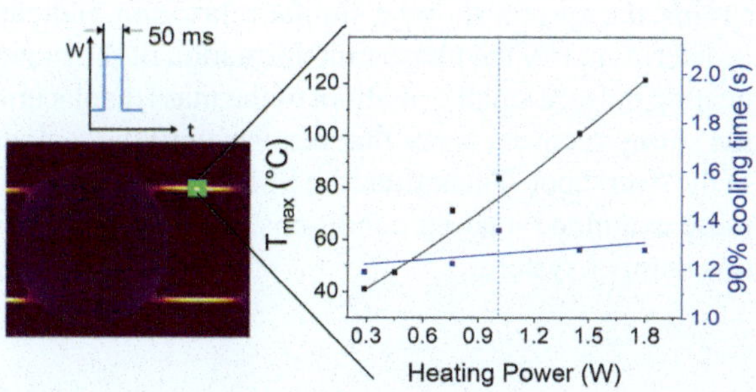

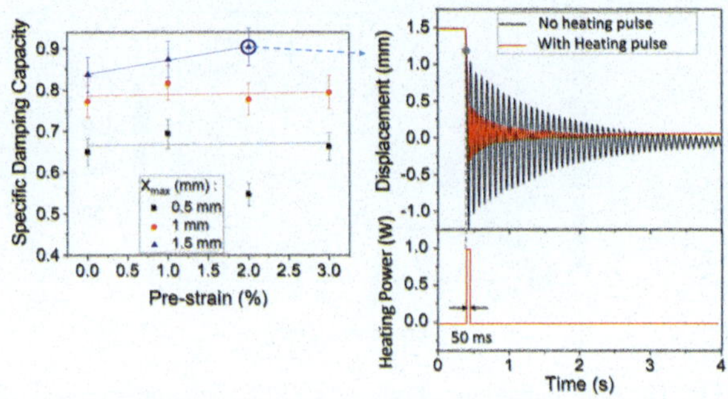

Fig. 12 SDC of double bridge active damper system depending on pre-strain and amplitude (left), and comparison of active and passive modes of operation (right) for a pre-strain of 2.0 % and a loading amplitude of 1.5 mm [9]

with the increase in loading amplitude and reaches a maximum of about 90 % resulting in a large reduction in oscillation amplitudes and settling time.

3.3 Single Bridge Device and Active Damper System

As for the pseudoelastic passive damper system in Sect. 2.3, the motion constrained to a single DoF is investigated for the one-way SMA device under vertical shock loading.

The thermal behaviour of the device after pulse and continuous heating is investigated to optimize the thermally-induced phase transition. Figure 13 shows the temperature distribution along a half-bridge. An improved homogeneity of the temperature distribution is achieved at a short heating pulse of 0.89 W for 100 ms indicating adiabatic conditions for the device below 100 ms. After the heating pulse, the cooling down time to reach the starting phase is about 0.75–1 s.

For the single DoF device, a short current pulse of 75 ms is applied on the SMA device at the instance of the electromagnet being turned OFF. A systematic study of the response for various mass and pre-strains is conducted (Fig. 14). The shock loading amplitude of the different pre-strains depicted in Fig. 14b is chosen to limit the total strain in the SMA to 0.045 to avoid overloading. The total strain is composed of pre-strain, loading strain of the device due to the supported mass, and strain introduced by the shock loading. Under test conditions of varying mass and pre-strains, the system shows a similar relaxation, indicating that this settling behaviour is determined by the phase transformation of the material alone. After the removal of heating pulse, a small overshoot of the mass displacement is visible. This arises due to the strong recovery force that persists until the temperature falls below R_f, resulting in the overshoot. Thereafter, the SMA device gradually relaxes to equilibrium. At all these examined loading conditions, oscillation-free fast settling is demonstrated by this damper system.

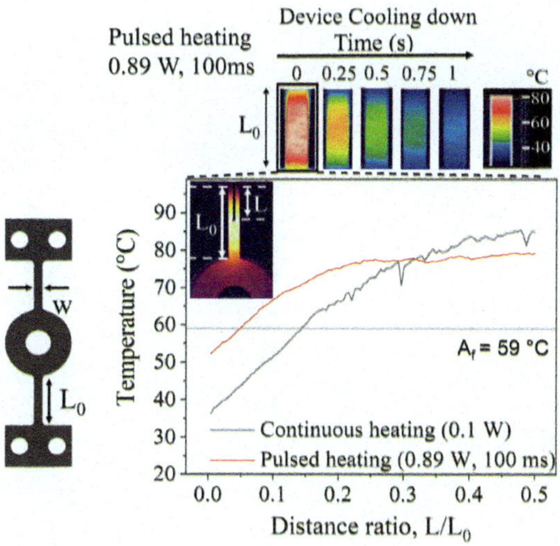

Fig. 13 Thermal distribution along the bridges with the application of continuous and pulsed heating [10]

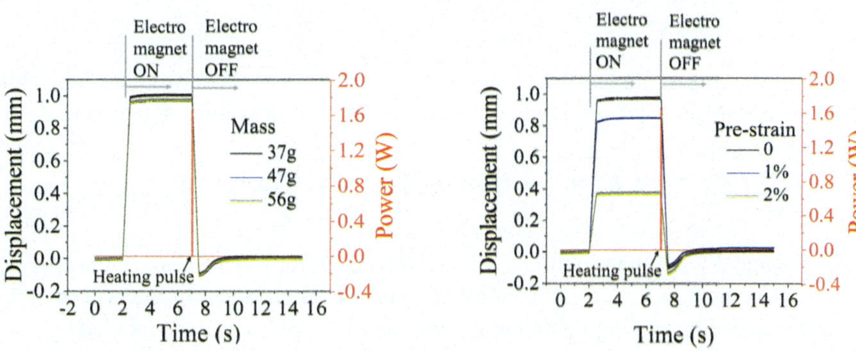

Fig. 14 The response of the vertical damper system to shock loading at various masses under zero pre-strain (**a**) and at various device pre-strains for a mass of 47 g (**b**)

3.4 Dynamical Modeling of One-Way Materials

For the linear bridge device made from the one-way material from Sect. 2.1, the FEM model developed in Sect. 2.4 was adapted, with the parameters given in Table 1. Contrary to the tensile pseudoelastic SMA, the transitions in the one-way material now include phase contributions x_α for all phases $\alpha \in \{M_+, M_-, A\}$ due to the heating to austenite and cooling down to a mixture of twinned martensite. Under normal operation, after cooling to room temperature the material will remain in the intermediate R-phase state, which is treated as a modified form of the common parent phase A

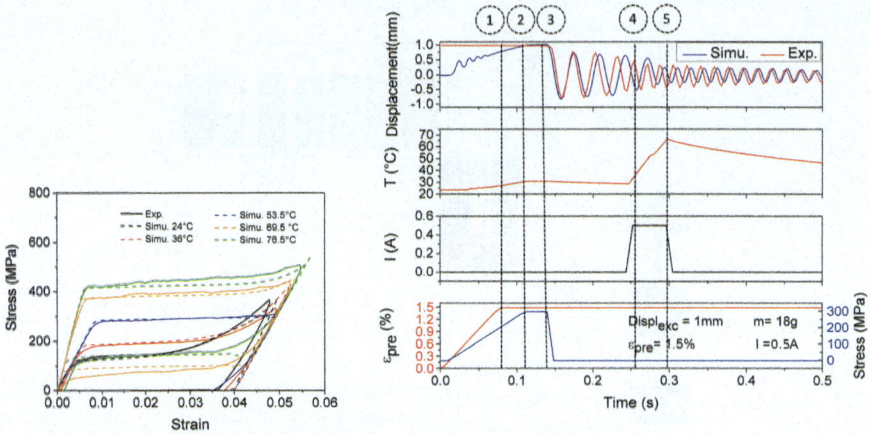

(a) (b)

Fig. 15 (**a**) Experiment (solid) and simulation (dashed): stress-strain response for one-way material; (**b**) Top panel: Simulation of proof mass displacement for a double bridge damper with active heat pulse (time steps 4–5), in comparison to experiment (red). Lower panels: temperature, heating current and pre-straining (ε_{pre}) to 1.5 % (steps 1–3)

(see Fig. 9a). This is accomplished by a transition path dependent latent heat (different for R→M, M→A or A→R transitions) and a temperature dependent elastic modulus

$$E_A(T) = E_R + \frac{1}{2}\left(1 + tanh\left(\frac{T - T'}{\Delta T'}\right)\right)(E_A - E_R),$$

where E_A and E_R are the moduli of A and R-phase (here, $E_M = E_R$), T' the approximate A-R transition temperature of 60 °C and $\Delta T'$ the transition interval of 5 °C. The pronounced hardening behavior apparent in the R-phase state is tackled by multiplying the plateau stresses with a hardening function $\sigma^{AM} = \sigma^{AM}(T)h(T, x_M)$, which depends on both temperature and martensite fraction. It is fitted to the tensile characterization data as $h(T, x_M) = 1 + \left(116.9 - 0.67T + 0.00955T^2\right) epx(\frac{x_M - 1}{0.15})$ and predicts well the tensile quasistatic loading in Fig. 15a. The simulation results for the mass displacement of a double bridge device under shock release is shown in Fig. 15b, where the initially ramped up load is held constant from time steps 2 to 3. In this experiment, a heating pulse (50 ms) is started delayed at point 4 in time to include the passive oscillation behavior. The increased amplitude reduction of the system is partly attributed to a stiffness increase, visible by a frequency increase from 29 to 59 Hz. From simulated stress and strain data, a specific damping capacity of 82 % is calculated for the first oscillation cycle after the pulse, which is nearly 50 % larger than for the same material in passive damping mode.

The dynamical simulations performed so far are based on a model originally developed for uniaxial loading. In case of film structure with higher structural complexity than simple bridges, a more complex strain path that depends on the loading history needs to be recorded. Furthermore, transitions between the low-temperature phase M and the metastable R-phase cannot always be excluded and simplified in the manner described above. To be able to numerically predict the associated response of three dimensional connected active/passive dampers, a phenomenological coupled multi-axial model for polycrystalline NiTi-based SMAs by Benešová et al. [12] and Sedlák et al. [13] is implemented in the ABAQUS finite element package within the user material subroutine UMAT for coupled temperature-deformation cycles. Based on the framework of continuum thermodynamics of irreversible processes, the model is capable of predicting dynamic loading including physical effects important in the multi-DoF SMA devices. Briefly, the model's features include: (i) a refined dissipation function coupling martensite transformation and reorientation processes, (ii) inclusion of the material responses associated with the transformation between austenite, martensite, and R-phase, (iii) the influence of tension-compression asymmetry; and the thermomechanical coupling, considering the strain-rate dependent latent heat generation/absorption during forward/reverse phase transformation where the localized martensite bands evolve. The following forms of free energy, dissipation function, and the heat equation are incorporated. The free energy is formulated as function of total strain $\varepsilon = e + \pi$ and inelastic strain π, martensite fraction ξ and temperature θ as

$$\begin{aligned}\Psi(\varepsilon(u), \pi, \xi, \theta) = \tfrac{1}{2} K\, tr(e)^2 + \mathscr{G}(\xi, \varepsilon) + \zeta(\pi, \xi) + \phi(\xi.\eta) + \delta_S(\pi, \xi)\\ + \tfrac{\nu}{2}|\nabla\pi|^2 + \tfrac{\nu}{2}|\nabla\xi|^2\end{aligned}$$

with ξ the martensite volume fraction, used to derive the driving force for transformation. The first two terms represent the elastic energy, $\zeta(\pi, \xi)$ a non-convex energy which, together with the last two gradient terms, is motivated by the localization of martensitic transformation, and $\phi(\xi, \eta)$ is the part of the chemical energy driving the thermomechanical coupling [12]. The dissipation function comprises the rate-independent part r_{RI} which depends on the transformation direction ($\dot{\xi}$), while the term r_{VI} denotes the viscous contribution [12]:

$$\begin{aligned}\mathscr{R}_{TOT} &= r_{RI}(\pi, \xi, \theta, \dot{\pi}, \dot{\xi}) + r_{VI}(\dot{\pi}, \dot{\xi}) \quad \text{with}\\ r_{RI}(\pi, \xi, \theta, \dot{\pi}, \dot{\xi}) &= \begin{cases} a_{reo}(\theta)|\dot{\pi}| + a_{AM}(\xi)\dot{\xi}, & \text{if } \dot{\xi} \geq 0\\ a_{reo}(\theta)(|\pi\dot{\xi}/\xi| + |\dot{\pi} - \pi\dot{\xi}/\xi|) + a_{MA}(\xi)\dot{\xi}, & \text{if } \dot{\xi} < 0\end{cases}\\ r_{VI}(\dot{\pi}, \dot{\xi}) &= \tfrac{\mu}{2}|\dot{\pi}|^2 + \tfrac{\mu}{2}|\dot{\xi}|^2\end{aligned}$$

By solving the heat equation

$$\dot{\omega} - div(\mathbb{K}(\xi, \theta)\nabla\theta) = r_{RI}(\pi, \xi, \theta, \dot{\pi}, \dot{\xi}) + 2r_{VI}(\dot{\pi}, \dot{\xi}) + \dot{\xi}\partial_\xi\phi(\xi.\eta)$$

the model can reproduce localized transformation at different strain rates related to heat transfer effects. This kind of approach is a prerequisite to the simulation of complex SMA structures in Sect. 4.2.

4 Two-DoF Vibration Stabilization Platform

The phase transformation between martensite and austenite for one-way SMA generates large recovery force allowing for active vibration control. The operating frequency of SMA actuator is limited by the cooling down time of the actuator. Therefore, a miniaturized SMA actuator is favourable for low frequency vibration control application below 100 Hz, owing to reduced power consumption for device heating and faster cooling. A prospective application is to compensate against the vibrations introduced by hand movement in smartphone camera where the typical operating frequency is below 10 Hz. In contrast to the vibration control using voice coil motor, less components are required for SMA based stabilizers facilitating miniaturization. The authors present a 2-DoF vibration stabilization platform for smartphone cameras to compensate for the vibrations arising from hand movements. The design criteria is to stabilize rotational movements of 1° along x and y axis based on the desirable specifications for Optical Image Stabilization (OIS) system [11].

4.1 Construction and Operation

A monolithic SMA device design is adopted for the stabilization platform to enable ease of assembly as shown in Fig. 16a. Using a NiTi foil of 15 μm thickness, four SMA actuators of bridge structure having a width of 65 μm are micromachined by laser cutting. A short current pulse of 0.2 A for 20 ms is used to raise the bridge temperature to 80 °C. After removal of heating pulse, the bridge requires 55 ms to reach the starting state (Fig. 16b), achieving a maximum operating frequency of 13 Hz.

The construction of the OIS system is illustrated in Fig. 17a and the assembled device (without top cover) is shown in Fig. 17b.

Figure 18 shows the operation of the OIS system. A short heating pulse of 20 ms is used to enable phase transformation to austenite. The antagonistic bridge pair SMA_1 and SMA_3 are activated sequentially to estimate the maximum tilting angle. The displacement is traced using laser displacement sensor on SMA_1. A maximum displacement of 65 μm is achieved, corresponding to a tilting angle of 1°.

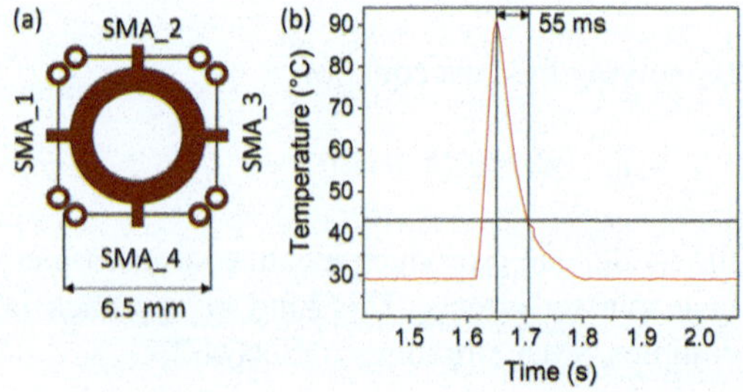

Fig. 16 (**a**) Monolithic SMA design for 2-DoF vibration stabilization and (**b**) its thermal response on a short heating pulse of 20 ms. The device cools down to the starting state in 55 ms

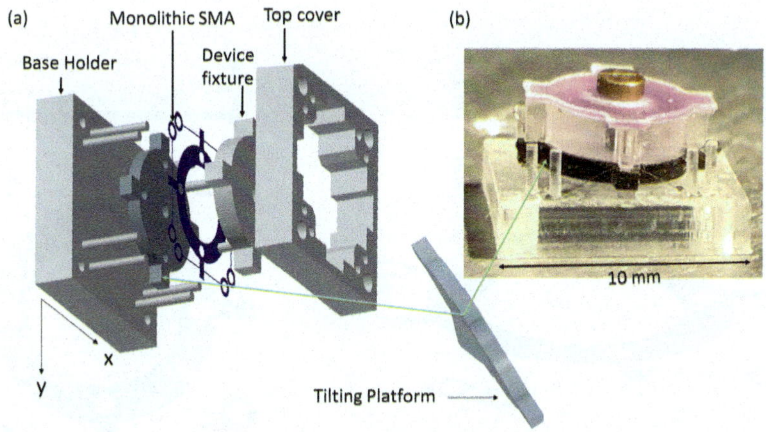

Fig. 17 The exploded view, illustrating the assembly of 2-DoF vibration stabilization platform (**a**) and the system after assembly (without top cover) (**b**)

Fig. 18 The stabilization angle of 1° is achieved in this system by sequential activation of antagonistic pairs SMA_1 and SMA_3

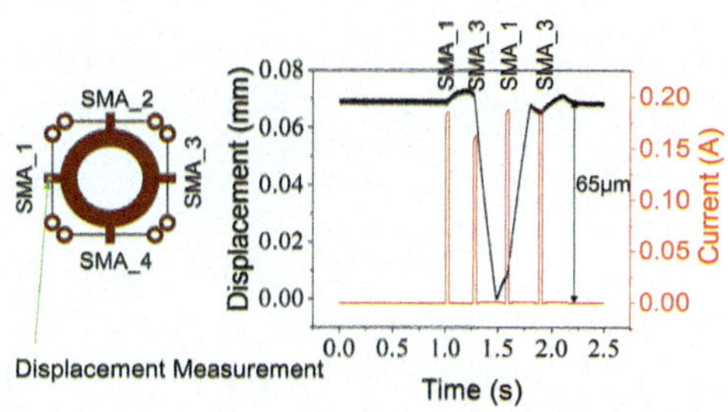

4.2 *Static Simulation of the Two-DoF Vibration Stabilization Platform*

The implemented multi-axial model is applied for an analysis of the stress-strain-phase transformation evolution upon single pulse heating. After the device has been pre-strained (Fig. 19a) the SMA_1 element is heated, where temperature changes homogeneously in the activated material and tilting to the observed angle follows. This leads to a coupled deformation of the opposed SMA_3 across the device. Figure 19b shows the total strain in the structure. The tilting, which occurs due to transformation-strain recovery in SMA_1 upon heating, results in additional progress of transformation in SMA_3. This leads to the actuation of SMA_1 under a variable bias triggered from SMA_3 (actuation under variable stress as in [14]). Figure 20a represents the related stress-temperature evolutions for SMA_1 and SMA_3 accompanied by the material's characteristic Clausius-Clapeyron transformation lines (blue lines for A→M, green lines for M→A phase transition), and in Fig. 20b the correlated stress-strain responses are shown simultaneously. The resulting coupling gives rise to the observed overshoot of the tilt actuator displacement during actuation in Fig. 18 (right).

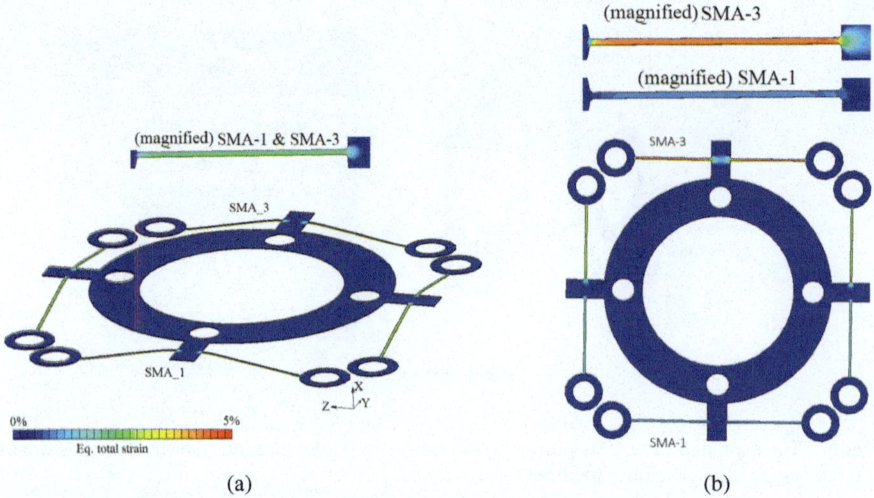

Fig. 19 Induced equivalent total strain after pre-straining of the device (**a**) and after actuation of bridge SMA_1, which causes tilting (**b**, top view)

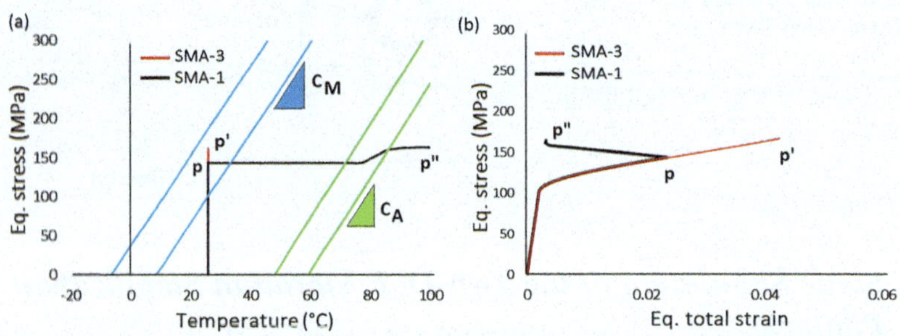

Fig. 20 Stress-temperature paths for two opposed bridges SMA_1 and SMA_3 from Fig. 19b, state after pre-straining for both SMAs: Point P, after heating: P' for SMA_3, P" for SMA_1 (**a**); The coupled stress-strain behavior in the SMAs (**b**). The blue and green lines represent the forward and reverse transformation stresses, their inclination is given by the Clausius-Clapeyron coefficients C_M and C_A

5 Conclusions

The energy dissipation of thin film/foil SMA materials mostly depends on loading strain, pre-strain and loading rate. Pre-strain and loading strain are crucial design parameters to maximize energy dissipation. In pseudoelastic SMA, pre-straining close to the beginning of transformation plateau results in achieving repeatable energy dissipation behaviour in all loading cycles. From a uniaxial tensile loading of the material, up to 34 % of loading energy is dissipated for the pseudoelastic SMA and 86 % for the one-way SMA material. In a damper system containing double bridge device, high SDCs of 70 and 90% are achieved using pseudoelastic and one-way SMA devices, respectively. Short current pulses below 100 ms enable more homogenized temperature distribution on one-way device achieving better phase transformation. For vertical shock loading with guided mass movement, the mass stabilizes without any oscillation after shock loading, showing large application potential as shock absorber in miniature robotics. For the dynamic simulation of bridge actuators, FEM models for 2D plane stress simulations of the bridge domains are combined with a thermally activated kinetic reaction model for phase transformations. For pseudoelastic and one-way SMAs, this approach well predicts the transient response, where localization of transformation and the accumulation of martensite phase in the SMA explain limitations of the damping capacity.

Miniaturized SMA actuators for 2-DoF tilt stabilization have demonstrated an operating frequency range up to 13 Hz. Ease of assembly is achieved using a monolithic device design. The more complex kinematics requires the use of multi-axial constitutive models with thermomechanical coupling to allow a precise prediction of the interference of strain and temperature across the distributed active SMA parts.

Acknowledgements The authors gratefully acknowledge funding by the German Science Foundation (DFG) within the priority program SPP1897. The authors appreciate the support of Nicholas Case in conducting single-bridge experiments.

References

1. Dixon, J.C.: The Shock Absorber Handbook. John Wiley & Sons (2008)
2. Wilson, S.A., Jourdain, R.P., Zhang, Q., Dorey, R.A., Bowen, C.R., Willander, M., Wahab, Q.U., Al-hilli, S.M., Nur, O., Quandt, E., Johansson, C.: New materials for micro-scale sensors and actuators: an engineering review. Mater. Sci. Eng. R: Rep. **56**(1–6), 1–129 (2007)
3. Jacob, K., Ahmadi, S., Wendler, F., Miyazaki, S., Gueltig, M., Kohl, M.: Shape memory foil-based active micro damping for portable applications. In: Transducers Conference, Berlin, Germany, vol. 23, pp. 590–593 (2019). (Jun 2019)
4. Müller, I., Seelecke, S.: Thermodynamic aspects of shape memory alloys. Math. Comput. Model. **34**, 1307–1355 (2001)
5. Wendler, F., Ossmer, H., Chluba, C., Quandt, E., Kohl, M.: Mesoscale simulation of elastocaloric cooling in SMA films. Acta Mater. **136**, 105–117 (2017)
6. Ahmadi, S., Jacob, K., Wendler, F., Padhy, A., Kohl, M.: Active damping of thin film shape memory alloy devices. Proc. Appl. Math. Mech. **20**, 202000310 (2020)
7. Ahmadi, S., Jacob, K., Wendler, F., Kohl, M.: Shape memory alloy film damping for smart miniature systems. Int. J. Smart Nano Mater. **9**(3), 199–215 (2018)
8. Ahmadi, S., Jacob, K., Wendler, F., Kohl, M.: Mesoscale simulation of shape memory alloy film damping. Proc. Appl. Math. Mech. **18**, 201800409 (2018)
9. Jacob, K., Sessner, V., Miyazaki, S., Kohl, M.: Active vibration damping and control using SMA foil-based devices. In: ACTUATOR 2021, pp. 200–203 (2021). ISBN 978-3-8007-5454-0
10. Jacob, K., Case, N., Miyazaki, S., Kohl, M.: Miniature shock absorber based on SMA foil actuation. In: ACTUATOR 2022 pp. 330–333 (2022). ISBN 978-3-8007-5894-4
11. La Rosa, F., Virzì, M.C., Bonaccorso, F., Branciforte, M.: STMicroelectronics. Optical image stabilization (2015). http://www.st.com/resource/en/white_paper/ois_white_paper.pdf. Accessed 14 Feb 2022

12. Benešová, B., Frost, M., Kadeávek, L., Roubíek, T., Sedlák, P.: An experimentally-fitted thermodynamical constitutive model for polycrystalline shape memory alloys. In: Discrete and Continuous Dynamical Systems Series S, vol. 14, pp. 3925–3952 (2021)
13. Sedlák, P., Frost, M., Benešová, B., Ben Zineb, T., Šittner, P.: Thermomechanical model for NiTi-based shape memory alloys including R-phase and material anisotropy under multi-axial loadings. Int. J. Plast. **39**, 132–151 (2012)
14. Shayanfard, P., Heller, L., Šandera, P., Šittner, P.: Numerical analysis of NiTi actuators with stress risers: the role of bias load and actuation temperature. Eng. Fract. Mech. **244**, 107551 (2021)

Lightweight Structures with Adaptive Dynamic Behavior Through Evanescent Morphing

Tom Ehrig, Christoph Hildebrand, Klaudiusz Holeczek, Niels Modler, and Pawel Kostka

1 Introduction

1.1 State of the Art

Improving the energy efficiency and achieving new performance levels of vehicles, machines and facilities is a continuous challenge for engineering and science. In the scope of structural design, a consistent implementation of lightweight design principles is widely recognized as an important tool to achieve this goal. Hybrid materials combining polymers, ceramics, lightweight alloys and composites as well as integral design with a reduced number of joints offer the possibility to develop light components with material type and distribution optimized for actual operating loads. An ultimately stiffness/strength-oriented design, however, often causes a problematic vibration susceptibility, especially of common thin-walled components and makes the application of additional damping measures necessary. Herein, some representative examples include:

T. Ehrig (✉) · C. Hildebrand · K. Holeczek · N. Modler · P. Kostka
Technische Universität Dresden, Institute of Lightweight Engineering and Polymer Technology, Holbeinstr. 3, 01307 Dresden, Germany
e-mail: tom.ehrig@tu-dresden.de

C. Hildebrand
e-mail: christoph.hildebrand@tu-dresden.de

K. Holeczek
e-mail: klaudiusz.holeczek@watttron.com

N. Modler
e-mail: niels.modler@tu-dresden.de

P. Kostka
e-mail: pawel.kostka@tu-dresden.de

P. Eberhard (ed.), *Calm, Smooth and Smart*, Lecture Notes in Applied and Computational Mechanics 102, https://doi.org/10.1007/978-3-031-36143-2_8

- special sandwich panel cores with layered damping materials [26] or filled with granular materials [17], including voids of optimized geometries [19],
- local dampers such as acoustic black holes [20], particle dampers [10, 21] and joints with viscoelastic damping [12],
- active magneto-/electro- or photorheological layers with the damping and stiffness controlled by magnetic, electric field or light, respectively [1, 9, 24],
- distributed active damping systems that use vibration sensors and actuators attached to complex-shaped structures, driven by real-time controllers [11, 18].

Common to active systems is a real-time calculation and generation of damping forces according to the instantaneous vibration position of the damped element. The necessary vibration measurement, signal routing, and a permanent power consumption combined with additional masses of the structure-integrated and external components often prove to be challenging and limit the application scope of active damping systems. Therefore, passive damping solutions are still used for a large part of technical applications. Especially free and constrained damping layers of viscoelastic materials are a widely used choice for flat panels, curved shell components [16] or structures with complex geometries [2]. A number of investigations aim at integration of such layers during the material manufacturing process [22] or at optimal distribution of damping segments [27].

In the case of widely used Constrained Layer Damping (CLD), a damping layer attached to the surface of a component is constrained by an additional top layer made of stiff material. Resulting shear deformations of the damping layer enable a particularly high damping power density of such treatment. However, the setup and material configuration of passive damping layers are typically designed for specific vibration conditions of the respective structures and do not allow any adaptation of their properties to varying excitation parameters.

1.2 Compressible Constrained Layer Damping

The presented research activities are pioneering work to develop a new kind of CLD, referred to as Compressible Constrained Layer Damping (CCLD) that implements a simple and almost massless mechanism for adaptation of damping properties. The conventional solid material of the viscoelastic layer is replaced here by a compressible one. The volume of the damping layer is considered as a structural cavity, which can be supplied with negative or positive pressure in order to control the damping layer thickness (Fig. 1) and thus the

- properties of the damping material - its densification or expansion affects the storage and loss shear modulus, G' and G'', respectively
- deformation kinematics of the layer - the shear deformation amplitude $\hat{\gamma}$ increases with the reduction of the thickness t_v.

Compared to typical active systems, the proposed actuating principle contains no explicit actuators. Instead, the damping layer itself or special structural cavities gen-

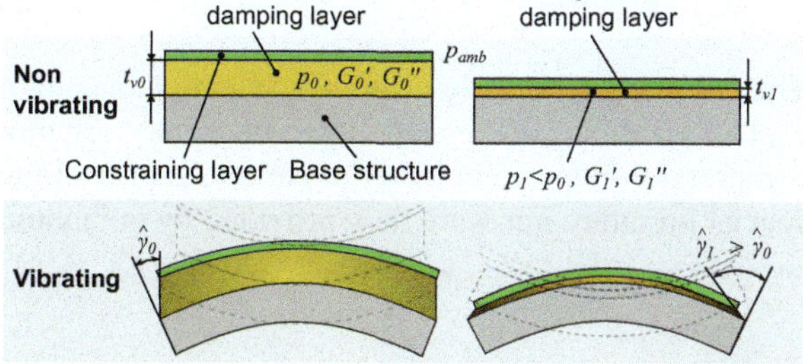

Fig. 1 The CCLD setup similar to the well-known CLD design, with the incompressible viscoelastic damping layer replaced by a compressible one. Actuating pressure $p_1 < p_0$ densifies the damping material and alters the shear deformation amplitude γ_1 due to variable damping layer thickness t_v

erate evanescent deformations when supplied with hydraulic fluid, compressed air or vacuum, which significantly reduces the implementation costs. In this way, the dynamic response of a CCLD-damped structure can be successively adapted on the basis of the current vibration conditions in order to achieve optimal dynamic response, e.g. in the sense of minimal vibration amplitude, minimal sound radiation, or maximal damping power etc. The term "evanescent deformation" used here refers primarily to the damping layer and it means such a geometrical change that it is irrelevant to the main function of the base structure. Additionally, curved base structures are subject to a certain degree of deformation and prestress caused by coupling effects due to CCLD actuation. These effects can also change the dynamic behavior and thus represent a part of the adaptation mechanism.

Considering the multiple phenomena and their interactions that occur in CCLD, the analysis of its adaptive damping properties as well as the elaboration of appropriate design guidelines require a complex work plan. The content of this chapter is structured accordingly as follows:

- Section 2 describes selection and characterization of compressible damping materials. Quasi-static and compression- and frequency-dependent viscoelastic properties of candidate materials are analyzed as a basis for understanding the damping layer behavior at various actuation conditions and vibration frequencies.
- Section 3 presents modeling and simulation approaches that use the acquired material data. First, an analytical model of simple beam structures is analyzed that allow a basic assessment of the CCLD potential as an adaptive damping treatment. Then, a parametric numerical model of a generic single-curved shell structure is described that explicitly includes both the complex deformation process during the CCLD actuation and the adaptive structural dynamic behavior.
- Further description focuses on experimental investigations on a lightweight single-curved shell structure with applied CCLD (Sect. 4). The experimental results are used at this stage for the validation of the developed numerical model.
- Finally, in Sect. 5 the application of the validated model for a simulation series is described. An assessment of the proposed damping solution is given based on a systematic variation of some CCLD key parameters.

2 Damping Materials

The following section describes the selection of suitable damping materials and their characterization. These steps were necessary for modeling purposes since neither compression-dependent material parameters for suitable materials were available in special literature nor were they provided by the manufacturers.

2.1 Material Selection

Even though a variety of viscoelastic materials are suitable for the application in a CCLD, the following material characteristics were identified as necessary for the realization of the CCLD principle:

- high damping to significantly influence the structural dynamic behavior,
- strong compression dependency of viscoelastic material properties for broad adjustment range of structural dynamic properties using evanescent morphing,
- high permeability for the actuation fluid and a technically feasible relation between compressive stress and compression, allowing an effective actuation,
- low density to suit the lightweight aspect.

The permeability for the actuation fluid significantly limits the material choice, making flexible open-cell foams a feasible material group. A product survey of commercial foams revealed the unavailability of sufficient quantitative specifications of material properties related to the above mentioned characteristics, hampering a direct selection of suitable materials. Especially the pressure-compression-shear damping/stiffness characteristics as a parameter of key importance could not be found for any compressible material even in specialized scholar publications. Therefore, a preselection of materials was conducted taking into account merely the available information about basic material features and the final selection was made based on the self conducted experimental material characterization.

Since the material density is an important lightweight indicator and it influences the damping as well as determines indirectly the actuation-relevant porosity of the foam, a broad range of densities was used as an initial material selection criterion. This led to the selection of twelve open cell polyurethane (PU) foams and a single melamine foam,[1] samples of which were procured and subjected to an in-house quality control and preliminary rheometric tests. Finally, three PU foams and the melamine foam with favorable viscoelastic properties and significantly different property profiles were used in further investigations.

Another material group that fulfills the criteria mentioned above are nonwovens. Since the production of nonwovens converts the fibers directly into fabrics and thus eliminates the yarn production process, nonwovens are a very cost-effective solution for many applications. Low density and remarkable energy absorption capacity are

[1] As a material with fundamentally different open cell morphology compared to PU foams.

Table 1 For the in-depth investigation selected foams and nonwovens

Sample	Type name	Manufacturer	Base material	Density kg/m^3
Foam A	Basotect® G+	BASF SE	Melamine resin	9.0
Foam B	RG1720	Schaumstoffe W. Wegerich GmbH	PU	16.3
Foam C	Visco 5030	Schaumstoffe W. Wegerich GmbH	PU	48.4
Foam D	Confor™ M CF40	Aearo Technologies LLC	PU	96.2
Nonwoven A	PP05	MK Filze GmbH	100 % PP fibers	120
Nonwoven B	PES 03-36	Filzfabrik Gustav Neumann GmbH	100 % PET fibers	360
Nonwoven C	Woll05	MK Filze GmbH	90 % wool fibers 10 % staple rayon	280
Nonwoven D	Nomex®	Filzfabrik Gustav Neumann GmbH	100 % meta-aramid fibers	160

the reasons why nonwovens have been used for years for acoustic or impact damping [15]. Although it seems likely that the superior energy absorption capacity could also be used for structural dynamic damping, there are only a few publications on the use of nonwovens or dry fiber assemblies for vibration damping. Nonwovens made of four main representative materials have been chosen (cf. Table 1):

- Polypropylene (PP) fibers as low density material, used e.g. for sound/heat insulation and soundproofing of loud-speakers or in vehicles,
- Polyethylene terephthalate (PET) fibers, as a low-cost, often recycled material,
- Wool fibers, as a renewable raw material, and
- Aramid fibers, which are known for their good damping properties e.g. in fiber composite materials and have a high temperature and chemical resistance.

2.2 *Material Characterization*

The CCLD-relevant properties were determined for the materials given in Table 1 based on procedures described in DIN EN ISO 845:2009-10 (density), DIN EN ISO 11357-2:2014-07 (glass transition temperature), and ASTM D 3576 (cell diameter of PU foams; due to different morphology, a similar procedure was derived for the melamine foam). In addition, microscope images were analyzed for all materials to evaluate the microstructure. For the nonwovens, also fiber diameter and fiber distribution were determined from these images.

The experiments related to the compression behavior and viscoelastic shear properties were conducted using an universal testing machine (Z2.5, Co. Zwick-Roell) respectively a rotatory rheometer (MCR 502-300, Co. Anton Paar) in parallel plate mode. Detailed information regarding the characterization of the foams can be found in [7], of the nonwovens in [8].

2.2.1 Quasi-static Material Properties

The quasi-static through thickness compression behavior, which determines the actuation characteristics of the CCLD, was investigated by uniaxial compression tests. Throughout this chapter, the compression level k is defined as:

$$k = \left(1 - \frac{t_{v1}}{t_{v0}}\right) \cdot 100\% \tag{1}$$

where t_{v1} and t_{v0} are the sample thicknesses in the compressed and uncompressed state, respectively. For an initial comparison of the materials, the nominal compressive stress as a function of the compression level is shown in Fig. 2. For the sake of clarity and due to the good repeat accuracy, only one loading cycle is presented. All investigated materials show a quantitatively similar, non-linear behavior, however, they can be compressed to different levels. For an adaptive damping system, this means that nonwovens reveal a smaller range of the compression level due to a pressure-driven actuation. This, in turn, leads to lower deformations of the structure, which is well compatible with the idea of evanescent morphing. The results shown later in this section, however, confirm that even such small deformations lead to significant changes in viscoelastic properties. For the tested foam materials, three typical regions [25] can be distinguished: cell wall bending, cell wall buckling and foam densification (shown exemplary for foam D in Fig. 3). It could be observed that all of the tested foams could be densified at least up to 88 % using atmospheric pressure.[2] In order to understand the deformation kinematics and to assess the compression reversibility at such high densifications, high-resolution in situ computed tomography (CT) enabling x-ray scanning in deformation controlled mode, was used (Fig. 3). It was found that even if the tested foams were compressed to one tenth of their initial thickness, they would return to their undamaged initial structure after relief.

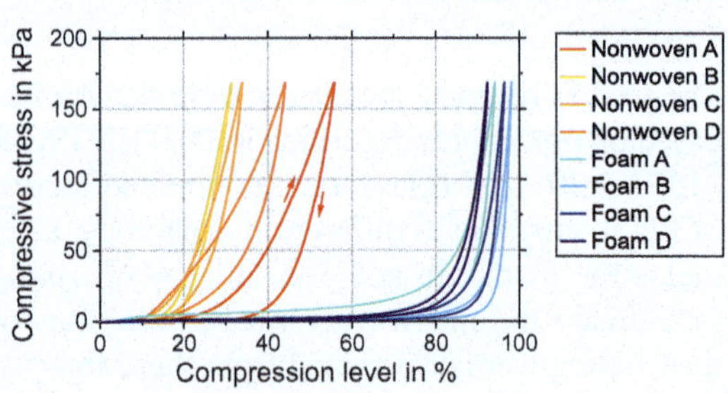

Fig. 2 Nominal compressive stress versus compression level k for loading and unloading at uniaxial compression. The flexible foams could be compressed to significantly higher levels at comparable compressive stress as nonwovens

[2] Which is a theoretical limit in the case of a vacuum actuation.

Fig. 3 Nominal compressive stress versus compression level k with in situ CT images of foam D at (I) 5 %, (II) 50 %, (III) 88 % compression and (IV) after relief of the compression force again at 5 % compression

2.2.2 Compression- and Frequency-Dependent Material Properties

In the scope of material characterization and following modeling of the structural dynamic behavior, a linear viscoelastic model of the foams as damping material was assumed at individual compression levels. This assumption was verified based on the analyses of raw rheometric data revealing harmonic, phase shifted stress response on applied harmonic strain. Therefore, the well established description of the viscoelastic properties using the complex shear modulus $G^* = G' + iG''$ was applied, where G' is the stiffness related storage modulus and G'' is the damping related loss modulus. The compression-dependent complex shear modulus—determining the adaptability of the CCLD treatment—was characterized in shear rheometric tests at different frequencies and temperatures for several compression levels with compressive stresses up to 100 kPa, anticipating a vacuum driven CCLD actuation.

In the case of foam materials, the time-temperature superposition (TTS) principle allowed a transformation of the test data in an augmented frequency range that is much broader than the testing machine's frequency span (here 0.1 ... 16 Hz). The spans of test temperatures and frequencies were selected in such a way that the measured data could be extrapolated for a fixed operating temperature of 23 °C into a new frequency range ending at 10^4 Hz, what required several thousand independent rheometric tests. Figuer 4 is provided to illustrate the frequency- and compression-dependent behavior of the storage shear modulus and the loss factor for foam D. The characterization of foams A-D is described in detail in [7].

Figure 5a shows an example of the characteristic hysteresis behavior for a foam and a nonwoven as determined with the rheometric investigations. The enclosed area of the loops is proportional to the energy dissipated per cycle. As can be seen, nonwovens have a significantly higher energy dissipation capacity than foams. The influence of the compression level is shown exemplary in Fig. 5b for nonwoven A. With increasing compression, all materials show an increase of the stiffness on the one hand and an increase of the dissipated energy on the other hand. Nonwovens, however, revealed an even stronger dependence of shear damping and stiffness upon compression. Further advantages of nonwovens are almost frequency-independent material behavior

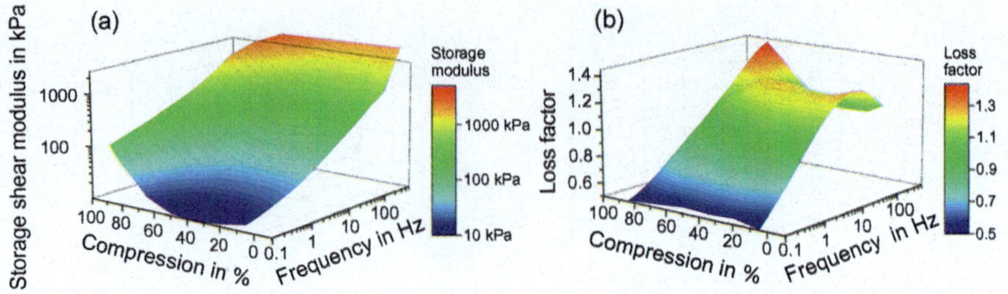

Fig. 4 **a** Storage shear modulus and **b** loss factor versus compression and frequency of foam D [4]

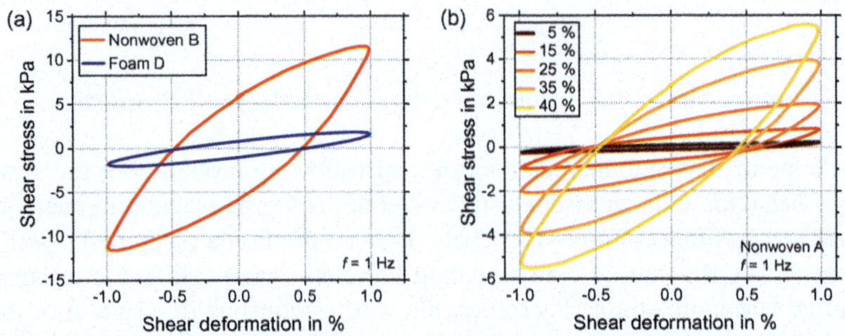

Fig. 5 **a** Comparison of the shear stress versus shear deformation for foam D and nonwoven B at comparable compressive stress; **b** variation of the compression level k from 5 to 40 % for nonwoven A, which showed the strongest dependency of shear damping and stiffness upon compression

at room temperature (except for nonwoven A made of PP) and significantly less temperature-dependent behavior.[3] However, nonwovens reveal a non-linear material behavior already at significantly smaller shear deformation amplitudes. This is due to the fundamentally different damping mechanism, which is mainly based on friction between the entangled fibers, and presents a challenge for modeling and design of a potential damping element. A more detailed discussion of the above mentioned effects can be found in [8].

Regarding the shear oscillation motion, an insignificant volume change during a vibration period was assumed, which should result in a low impact of fluid flow effects on damping properties. Thus, the influence of effects connected with fluid flow in the cell skeleton was neglected in the above presented material characterization.

2.2.3 Characterization at High Excitation Parameters

In addition to the above-described tests at low excitation parameters, effects occurring at high excitation amplitudes and frequencies were investigated. For these investiga-

[3] In fact, this is also the reason why TTS is not feasible with most nonwovens.

tions a test stand was developed and used in connection with a large electrodynamic shaker (V8-440, Co. Bruel&Kjœr) generating a broad range of vibration amplitudes and frequencies (up to 1000 Hz). After a virtual dimensioning and fundamental experimental tests, an add-on module for simple-shear tests, based on the inertial mass principle, was extensively tested and put into operation. With this test stand it is also possible to characterize the damping materials to the point of damage in order to determine the limits of the CCLD. A primary result of the conducted material characterization is a catalog of material data describing compression- and frequency-dependent properties of several foams and nonwovens. These results, revealing a complex picture, can be used as the necessary basis for the modeling and simulation of structures with CCLD in an extended range of excitation parameters.

3 Modeling of Lightweight Structures with CCLD Treatment

For a detailed analysis of the CCLD influence on the dynamic behavior of lightweight structures, models with different abstraction levels have been realized. First, an analytical model of bending beams and its application in exemplary parameter studies are described (Sect. 3.1). Then, further models were developed, e.g. with prestressed beams [13], which cannot be discussed here in detail due to limited space. Finally, the latest and most complex model–a fully parametric finite element (FE) model for single-curved shell structures–is presented in Sect. 3.2.

3.1 Analytical Model for a Simply Supported Beam with CCLD

For the analytical approach, a well established mathematical formulation of a simply supported beam with CLD [28] has been modified by adding the possibility of a compression-driven adaptation of the geometrical and material properties of the viscoelastic layer. This was accomplished by implementing the compression dependent properties of the viscoelastic layer: its thickness, mass density as well as storage and loss shear modulus.

The model was used in a series of simulations to study the CCLD potential regarding the damping capacity and its adaptation as well as to understand the physical relationships of parameters describing the system configuration. The length ($L = 550$ mm) and the width ($W = 55$ mm) of the base beam made of carbon fiber reinforced epoxy resin was kept constant while its thickness h_b was varied in such a way that its slenderness took the values of 10, 20 and 100. The thickness of the constraining layer, made of the same material as the base beam, was kept constant at 0.25 mm. The parameters of the viscoelastic layer were consecutively set according to material

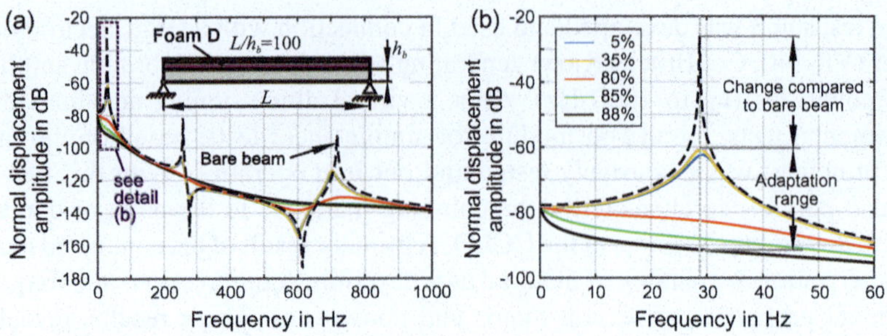

Fig. 6 Exemplary normal amplitudes at the beam's middle point *L*/2 calculated using the analytical model from [14]: **a** Comparison of the vibration amplitudes of the bare beam (dashed line) and beam with CCLD at different compression levels, **b** detailed view of the frequency range around the 1st eigenfrequency and graphical explanation of the key CCLD efficiency indicators

characterization results for the foams shown in Table 1. Since the shear deformation of the viscoelastic layer during the beam bending vibrations increases with the decrease of its thickness (cf. shear deformation amplitude γ_1 in Fig. 1), possibly thin layers were pursued in order to maximize the damping effect. Thus, the lower thickness limit in uncompressed state was set for each foam to exactly 10 mean foam cell diameters.

The normal displacement amplitude at the beam's middle point *L*/2 as a response to a unit, normal harmonic excitation force applied in the same point was calculated for every above mentioned slenderness value and foam material. The change of the dynamic behavior in comparison with the bare beam behavior and the adaptation range as a result of the CCLD actuation, were used as key effectiveness indicators. They are presented for an exemplary system configuration in Fig. 6. Results for other configurations are compactly presented in Fig. 7 for the bending modes 1, 3 and 5. A detailed description of the model used and a thorough discussion of the results can be found in [14]. The results of the simulations described up to this point confirm the principal correctness of the initial theoretical assumptions about the adaptive dynamic behavior of structures with CCLD treatment. Both significant vibration damping and the CCLD adaptability through effects occurring during compressive actuation of the damping layer were observed. The developed analytical modeling approach can be used for a quick preliminary system design. However, due to the complexity of real-world structures, more elaborate modeling is often required.

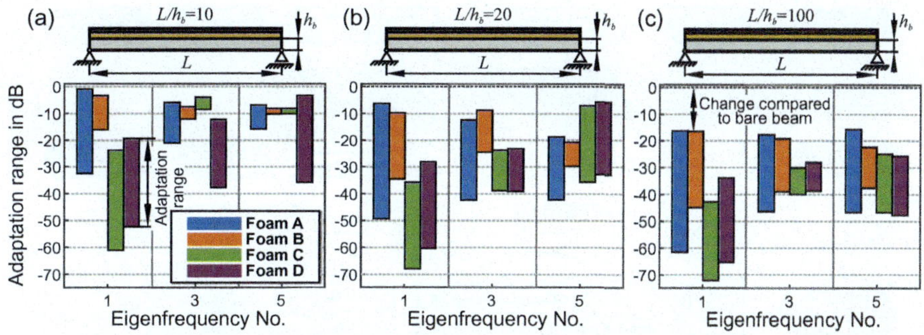

Fig. 7 Adaptation ranges of the CCLD-beam's dynamic behavior in comparison with the bare beam behavior for different foam materials and slenderness values: **a** 10, **b** 20 and **c** 100 [6]

3.2 *Numerical Model for Single-Curved Structures with CCLD*

A multi-step FE model was developed to estimate the potential of CCLD for single-curved structures. In the following, the setup of the model is described and a possible indicator to estimate the efficiency of the CCLD, the mean mobility, is discussed.

3.2.1 Setup of the Parameteric FE Model

The key component of the model is a two-step simulation procedure which was implemented in the commercial software COMSOL Multiphysics® 5.6 (Fig. 8). This simulation procedure is re-run with adapted input data for each set of selected CCLD design parameters. The modeled geometry was discretized using quadrilateral elements for the damping and constraining layer as well as tetrahedral elements for the base structure to reduce computation time. The different layers are modeled as union, therefore no relative motions of the layers to each other are possible. The model boundary conditions are defined as a pressure load on the damping layer and a fixed- and floating constraint on two opposing axial edges of the base structure.

In the first step, the deformation and the prestress, caused by the compression of the damping layer material by the CCLD-actuation, are computed using a stationary solver (no damping is taken into account). For a realistic simulation of the large material deformations, the hyperelastic Storaker model [23] was selected for the damping layer. For the fitting of the Storaker model, data sets from uniaxial compression tests mentioned in Sect. 2.2.1 were used separately for each considered foam material (cf. Table 1). A sample validation of the model was performed on a real CCLD setup assembled with foam D. The deformation behavior of the CCLD actuated in the pressure-controlled mode was measured with a 3D scanner and compared with the model outputs of this simulation step. Due to the good agreement, the described modeling approach was used in all further described simulations.

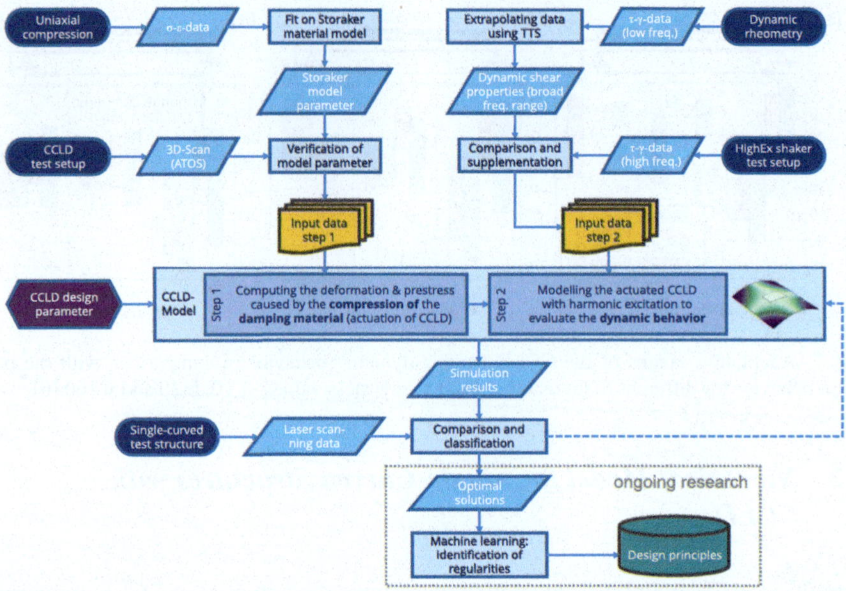

Fig. 8 Overview of the parametric FE model with input-output data

In the second step, the deformed and prestressed CCLD from the first step is taken over for a harmonic response analysis. Stationary vibration responses to a monoharmonic force excitation are simulated in a selected frequency range. To reduce simulation time, an adaptive control of the frequency step was implemented. In this simulation step, the damping layer material is switched to linear elastic material and parametrized with frequency-dependent values of shear modulus and loss factor (Sect. 2.2.2) taking into account the previously calculated compression level.

For the validation of the simulation results from this two-step simulation procedure, a single-curved test structure was manufactured and the simulation results were compared to the experimental data (Sect. 4).

3.2.2 Mobility as Indicator of CCLD Treatment Efficiency

To evaluate the impact of the proposed CCLD on the dynamic behavior of a structure, a variety of possible indicators can be defined. In this study, the mobility as a complex-valued frequency response function describing velocity per unit force was assessed. Therefore, the measured excitation force $F_{exc}(\omega)$ and response velocities $v(\omega)$ were used to calculate the mobility $M(\omega)$ as following:

$$M(\omega) = \frac{v(\omega)}{F_{exc}(\omega)} \tag{2}$$

The magnitudes of the mobilities at measurement points n were then averaged:

$$\bar{M}(\omega) = \sqrt{\frac{1}{N}\Sigma_{n=1}^{N} M_n(\omega)^2} \tag{3}$$

Here, $\bar{M}(\omega)$ is the mean mobility magnitude as a single convenient measure, which allows an assessment of the damping efficiency. N is the number of the velocity measurement points. It should be noted that the mobility is actually defined as the velocity that is perpendicular to the surface in relation to the excitation force. For the investigated curved structure, only the z components of the velocities[4] were taken into account to improve the comparability of the velocities measured by the laser scanning vibrometer and the simulation results.

4 Experimental Setup for Model Validation

The proof-of-principle for the CCLD treatment was performed in an earlier investigation phase on a flat panel structure and revealed very promising results [5]. In this section, the experiments on a single-curved shell structure are presented and the results are compared with the developed model.

4.1 Configuration of the Curved CCLD Sample Structure

The analyzed base structure is a cylindrical shell section with a constant radius r_b, thickness t_b, width w_b and section angle ϕ_b precisely milled from aluminum (EN AW-5083). This expensive manufacturing solution guarantees very high geometric precision and residual stress-free condition of the base structure, avoiding a number of random effects that could influence the observed vibration behavior. The CCLD treatment was set up as shown in Fig. 9. The CCLD patch was applied only partially and has a width w_v and section angle ϕ_v. The viscoelastic damping layer consists of the open-cell PU foam D with the thickness t_v depending on the compression level. An aluminum sheet with the thickness t_c was used as constraining layer. The additional mass due to the CCLD patch was only 0.12 kg (+4.36 %). The vacuum-sealing was assured by using a vacuum film, which was glued to the base structure and the constraining layer. To avoid introducing another viscoelastic layer and to exclude further phenomena such as curing of the adhesive in the open-cell pores, the individual layers were not glued together. The structure was held in position by the tightly applied vacuum film and later fixed by the actuation pressure, which enabled

[4] This refers to the components of the vibration velocities that are perpendicular to a plane defined by the corner points of the curved shell.

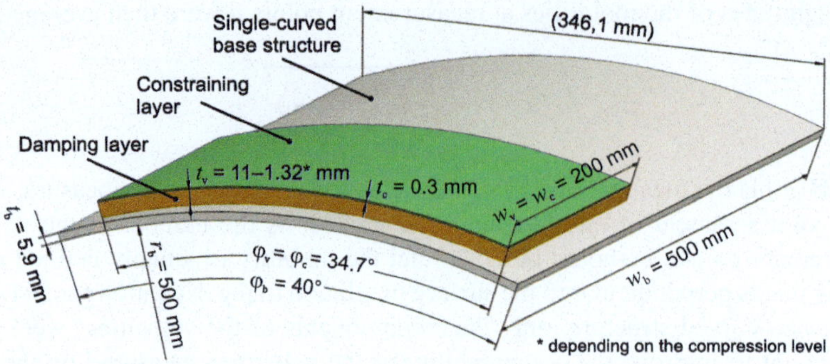

Fig. 9 Schematic drawing of the CCLD setup on a single-curved structure

Table 2 Applied actuation pressure and corresponding damping layer thickness ($t_{v0} = 11\,\text{mm}$)

Applied actuation pressure p_{act}	–1.1 kPa	–2.4 kPa	–5.4 kPa	–16.9 kPa	–31.7 kPa	–49.0 kPa
Damping layer thickness t_v	$0.95\,t_{v0}$	$0.65\,t_{v0}$	$0.35\,t_{v0}$	$0.2\,t_{v0}$	$0.15\,t_{v0}$	$0.12\,t_{v0}$
Compression level k	5 %	35 %	65 %	80 %	85 %	88 %

the coupling of the layers by frictional forces.[5] The applied actuation pressure was generated by a vacuum pump and adjusted with a valve to achieve the predetermined compression levels (Table 2).

4.2 Measurement Setup

The characterization of the dynamic behavior was carried out on the test structure hung up on an auxiliary frame, mounted on a vibration-insulated table (Fig. 10). An electrodynamic shaker (type 4810, Co. Bruel&Kjœr) was used to generate a sine sweep excitation force in the frequency range between 50 and 450 Hz. This force, applied near a corner of the base structure, was measured with an impedance head (type 8001, Co. Bruel&Kjœr). At the same time, the vibration response was recorded by means of a laser scanning vibrometer (type PSV-400, Co. Polytec) at 35 regularly distributed laser light reflecting spots, with a resolution of 48.9 mHz.

Several test series were carried out, first with the base structure without CCLD and then with the applied CCLD patch. The compression levels were set in increasing, decreasing and random order, according to values shown in Table 2. The repeat

[5] The simulation results (Sect. 4.4) show a remarkably good agreement especially for the lowest compression level ($k = 5\,\%$), which supports this assumption.

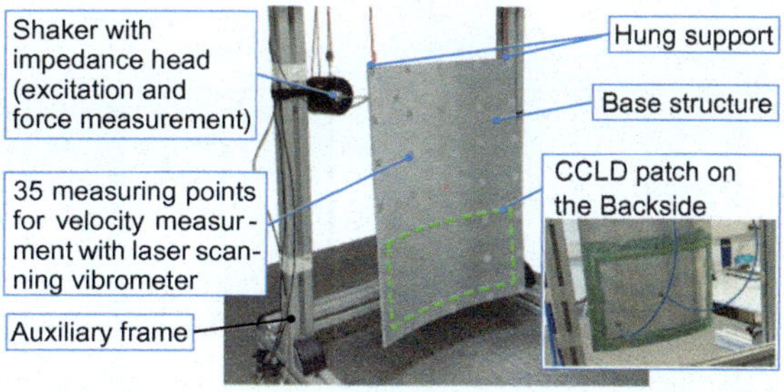

Fig. 10 Illustration of the measuring setup with the adaptive structure [4]

accuracy of all these measurements was excellent, so that only one measurement per compression level is used in the following discussion.

4.3 Discussion of the Measurement Results

4.3.1 Measured Mean Mobility at Different Compression Levels

The two simultaneous and compression-dependent phenomena: variability of material properties of the damping layer and its changing deformation kinematics (cf. Sect. 1.2) suggest a complex effect of the CCLD actuation on the dynamic behavior of the adaptive structure. Figure 11a shows the mean out-of-plane mobility (cf. Eq. 3) of the adaptive CCLD structure at different compression levels for the measured frequency range from 50 to 450 Hz, compared with the mean mobility of the bare base structure. Figure 11b and c provide a detailed illustration of the frequency range around the 2nd and 5th eigenfrequency (EF), respectively. Due to the additional mass of the CCLD patch, the EFs shift to a lower frequency range at low compression levels of the damping layer. Increasing shear stiffness at higher compression levels compensates this effect, and the EFs shift back to a higher range.[6] A significant reduction of the amplitudes compared to the base structure can be observed, but the relationship between compression level and reduction of the amplitude is rather complex. In the example shown, the amplitude of the 5th EFs decreases steadily with increasing compression (Fig. 11c), while the amplitude of the 2nd EF reaches its minimum amplitude at a compression level of 80 % (Fig. 11b).

4.3.2 Compression Level for Achieving the Minimal Mobility

The effect of CCLD in an adaptive mode of vibration damping can be illustrated by the example of the presented structure under a monoharmonic vibration excita-

[6] The reduction of the bending stiffness of the adaptive CCLD structure, caused by a decrease of the second moment of inertia with increasing compression, appears to play a minor role.

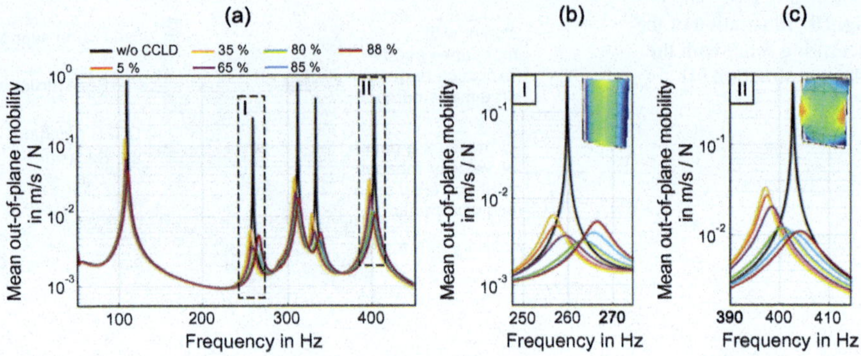

Fig. 11 **a** Mean out-of-plane mobility for different compression levels; **b**, **c** detailed view of the 2nd EF and the 5th EF (legend is valid for all three figures) [4]

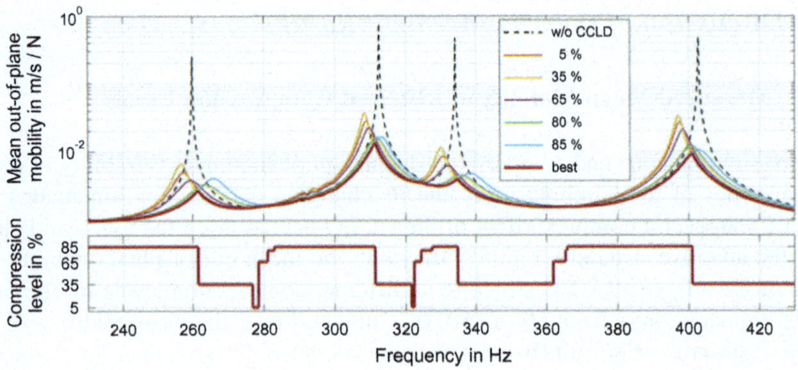

Fig. 12 Mean out-of-plane mobility for different compression levels and controlled state with the minimal achievable mobility (bold line) by adapting the compression level (step diagram)

tion. Figure 12 extends the mean mobility patterns by a step diagram showing the frequency-dependent compression level of the CCLD yielding the minimum mobility. This actuation profile, which includes all the levels analyzed, allows a significant calming of the sample structure (bold line) compared to the respective constant compression levels. In addition to minimum mobility, other criteria for CCLD control are also possible, e.g. maximum damping power, minimum sound radiation, etc. The CCLD control in technically common, more complex excitation scenarios remains a challenging task that will not be addressed here.

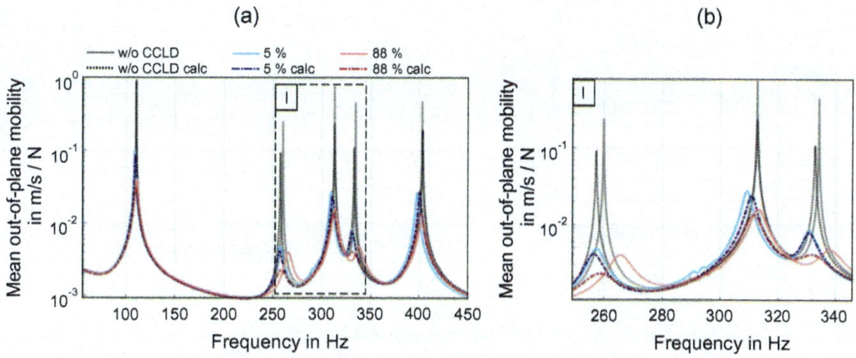

Fig. 13 Measured mean out-of-plane mobility (solid lines) compared to the simulation results (dotted lines) for **a** the measured frequency range and **b** detail of the 2nd to 4th EF [4]

4.4 Comparison of Experimental and Simulation Results

Achieving a good quantitative agreement between experimentally determined and simulated vibration properties in terms of natural frequencies and vibration amplitudes is often challenging in the case of more complex structures. In the present case, an additional difficulty results from the modeling of the CCLD actuation process that modifies many properties at material, geometric, and kinematic level.

In a first step of the validation procedure, the model and experiment outputs in terms of the mean mobility were compared for the bare structure. The actual validation step concerned a similar comparison at different CCLD actuation states. An example of experimental and simulated patterns for the two steps is shown in Fig. 13. The obtained results were evaluated as satisfactory and the model as valid for further simulation-based analysis of CCLD effect in other system configurations. A more detailed discussion on the validation procedure can be found in [4].

5 Model Application

The results shown in Sect. 4 demonstrate that CCLD can be an effective damping treatment. However, determining the optimal CCLD layout for a given use case appears to be a challenging task due to a large number of design parameters, including:

- Dynamic behavior the base structure,
- Number, distribution and geometry of CCLD patches (c.f. Fig. 9 as example),
- Material parameters, including (i) frequency-dependent, (ii) compression-dependent and (iii) temperature-dependent material behavior of the damping layer,
- Vibration excitation parameters.

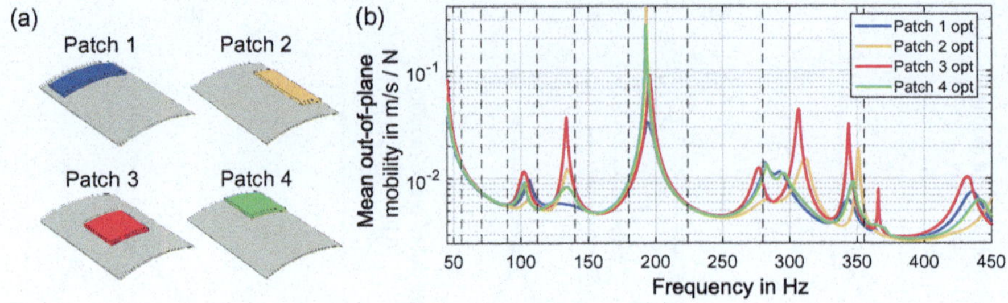

Fig. 14 **a** Schematic illustration of the four variants and **b** their mean out-of-plane mobility with adaptive control to achieve the minimal mobility

Furthermore, an optimal design depends on the respective objective function such as minimum vibration amplitude, minimum sound radiation, or maximum damping power, as well as on possible constraints including allowed space, position, and mass of the damping components. In the scientific approach followed here, a numerical experiment will be performed to comprehensively characterize the dynamic behavior under different parameter combinations. The results obtained will serve as the basis for a data-driven identification of CCLD design principles.

To illustrate the complex nature of the described task, an exemplary study will be presented. For a given base structure, only the position, width and section angle of the CCLD patch are varied. In all four variants shown in Fig. 14a, the CCLD patch has the same area and the additional mass in relation to the base structure is only 5 %. All other parameters remain constant in this study. Figure 14b illustrates the best possible condition i.e. the lowest mean out-of-plane mobility as described in Sect. 4.3.2. The responses for a monoharmonic excitation are obtained by successive control of the compression level for each of the four variants. It is clearly visible that the different patch configurations significantly influence the mobility, especially in the range of the EF's. Thus, with Patch 2, the EF at approx. 100 Hz can be damped effectively, whereas the EF at approx. 130 Hz can be damped significantly less. For Patch 1, on the other hand, the situation is exactly the opposite. The range between approx. 260 Hz and 380 Hz illustrates a similar effect particularly well. The vertical lines in Fig. 14b mark the 1/3 octave bands. A possible quantitative measure for the CCLD assessment can be the integrated mean out-of-plane mobility e.g. over 1/3 octave bands (Fig. 15). It is apparent that for the given CCLD parameters, the variant with Patch 1 is the best choice in the considered frequency range. However, it also becomes clear that, depending on the frequency range of interest, different variants are the best option. This small insight with only a few parameter sets already reveals a complex picture and should demonstrate the necessity of a large-scale parameter study.

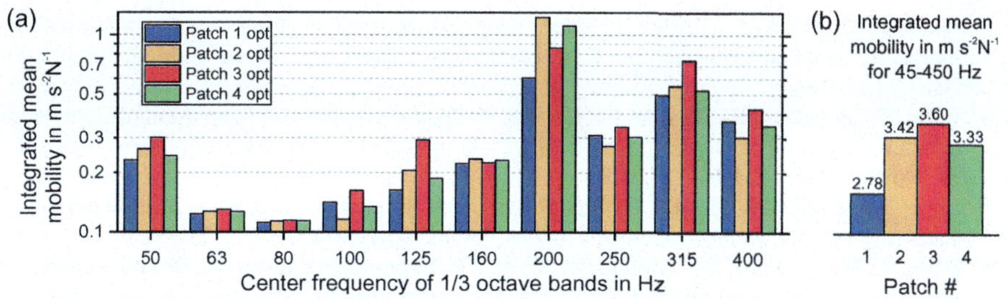

Fig. 15 **a** Comparison of the mean out-of-plane mobility integrated over the 1/3 octave bands for the four patch variants and **b** comparison of the mean out-of-plane mobility integrated over the entire frequency range (45–450 Hz)

6 Conclusion

The novel CCLD damping technique poses an efficient and nearly weight neutral solution to calm vibrating lightweight structures. In experiments carried out on a single-curved shell structure with a partial CCLD coverage and adaptive CCLD control it was possible to significantly reduce the maximum vibration amplitudes with only a small increase in the structure's mass. In order to choose suitable damping materials for the CCLD technique, different viscoelastic foams and nonwovens were investigated, characterized and necessary material properties were identified. In addition to the experimental investigations, a two-step simulation procedure was proposed and investigated. The therefore developed model uses the experimentally obtained material properties and is capable of predicting the structural dynamic behavior of a structure with applied and actuated CCLD. It shows a good agreement with the experimental results. This model provides the basis for a large-scale parameter study in the future. The hereby obtained results will then be used to develop guidelines for the design of CCLD patches and identify optimal CCLD design parameters like patch size, -position and -material. Further experiments with the test structures in different scenarios, such as in acoustic damping context [3] are planned.

References

1. Cho, M., Kim, J., Choi, S., Kim, G.: Non-contact tunable damping of a cantilever beam structure embedded with photo-rheological fluids. Smart Mater. Struct. **25**(2) (2016)
2. Dannemann, M., Täger, O., Modler, N.: Combined semi-analytical and numerical vibro-acoustic design approach for anisotropic fibre-reinforced composite structures. J. Sound Vib. **404**, 1–14 (2017)
3. Ehrig, T., Dannemann, M., Luft, R., Adams, C., Modler, N., Kostka, P.: Sound transmission loss of a sandwich plate with adjustable core layer thickness. Materials (Basel, Switzerland) **13**(18), 4160 (2020)
4. Ehrig, T., Hildebrand, C., Modler, N., Kostka, P.: Modelling and experimental verification of a curved lightweight structure with adaptive dynamic behaviour. In: Proceedings of the 28th International Congress of Sound and Vibration, ICSV 2022 (2022)

5. Ehrig, T., Holeczek, K., Kostka, P.: Experimental investigations of lightweight structures with fluidically actuated Compressible Constrained Layer Damping. Mater. Today Commun. **16**, 204–211 (2018)
6. Ehrig, T., Holeczek, K., Modler, N., Kostka, P.: Dynamic behaviour adaptation of lightweight structures by compressible constrained layer damping with embedded polymeric foams and nonwovens. Appl. Sci. **9**(17) (2019)
7. Ehrig, T., Modler, N., Kostka, P.: Compression and frequency dependence of the viscoelastic shear properties of flexible open-cell foams. Polym. Test. **70**, 151–161 (2018)
8. Ehrig, T., Müller-Pabel, M., Modler, N., Kostka, P.: Experimental investigations on compressed nonwovens as damping material for enhanced constrained layer damping. In: Proceedings of the 28th International Congress of Sound and Vibration, ICSV 2022 (2022)
9. Eshaghi, M., Sedaghati, R., Rakheja, S.: Dynamic characteristics and control of magnetorheological/electrorheological sandwich structures: a state-of-the-art review. J. Intell. Mater. Syst. Struct. **27**(15), 2003–2037 (2016)
10. Gnanasambandham, C., Fleissner, F., Eberhard, P.: Enhancing the dissipative properties of particle dampers using rigid obstacle-grids. J. Sound Vib. **484** (2020)
11. Haase, T., Unruh, O., Algermissen, S., Pohl, M.: Active control of counter-rotating open rotor interior noise in a Dornier 728 experimental aircraft. J. Sound Vib. **376**, 18–32 (2016)
12. Hammami, C., Balmes, E., Guskov, M.: Numerical design and test on an assembled structure of a bolted joint with viscoelastic damping. Mech. Syst. Signal Process. **70–71**, 714–724 (2016)
13. Holeczek, K., Koschichow, R., Schlieter, T., Ehrig, T., Kostka, P.: Numerical investigations of polymer-based fibre-reinforced structures with fluidically actuated Compressible Constrained Layer Damping. PAMM **18**(1) (2018)
14. Holeczek, K., Zhou, B., Kostka, P.: Evanescent morphing for tuning the dynamic behavior of composite lightweight structures: theoretical assessment. Mech. Adv. Mater. Struct. **10**(2), 1–107 (2019)
15. Karthik, T., Prabha-Karan, C., Rathinamoorthy, R.: Nonwovens: Process, structure, properties and applications. Woodhead Publishing India in textiles, WPI Publishing (2016)
16. Kliem, M., Høgsberg, J., Vanwalleghem, J., Filippatos, A., Hoschützky, S., Fotsing, E., Berggreen, C.: Damping analysis of cylindrical composite structures with enhanced viscoelastic properties. Appl. Compos. Mater. **26**(1), 85–113 (2019)
17. Koch, S., Duvigneau, F., Duczek, S., Woschke, E.: Vibration reduction in automotive applications based on the damping effect of granular material. In: Automotive Acoustics Conference 2017, vol. 102, pp. 43–57. Springer (2019)
18. Kostka, P., Holeczek, K., Hufenbach, W.: Structure-integrated active damping system: integral strain-based design strategy for the optimal placement of functional elements. Int. J. Compos. Mater. **3**(6B), 53–58 (2013)
19. Krause, D., Paetzold, K., Wartzack, S.: Additively manufactured components for structural applications in aircraft interior-two case studies. In: DFX 2016: Proceedings of the 27th Symposium Design for X, pp. 147–156 (2016)
20. Lee, J., Jeon, W.: Vibration damping using a spiral acoustic black hole. J. Acoust. Soc. Am. **141**(3), 1437 (2017)
21. Meyer, N., Schwartz, C., Morlock, M., Seifried, R.: Systematic design of particle dampers for horizontal vibrations with application to a lightweight manipulator. J. Sound Vib. **510** (2021)
22. Sessner, V., Liebig, W.V., Jackstadt, A., et al.: Wide scale characterization and modeling of the vibration and damping behavior of CFRP-elastomer-metal laminates-comparison and discussion of different test setups. Appl. Compos. Mater. **28**(5), 1715–1746 (2021)
23. Storåkers, B.: On material representation and constitutive branching in finite compressible elasticity. J. Mech. Phys. Solids **34**(2), 125–145 (1986)
24. Tan, A.S., Belkner, J., Stroschke, A., Sattel, T.: Damping adjustment utilizing digital electrorheological valves with parallely segmented electrodes. Smart Mater. Struct. **28**(7) (2019)
25. Weaire, D., Fortes, M.A.: Stress and strain in liquid and solid foams. Adv. Phys. **43**(6), 685–738 (1994)

26. Yang, J.S., Ma, L., Schmidt, R., Qi, G., Schröder, K.U., Xiong, J., Wu, L.: Z: hybrid lightweight composite pyramidal truss sandwich panels with high damping and stiffness efficiency. Compos. Struct. **148**, 85–96 (2016)
27. Zhang, D., Qi, T., Zheng, L.: A hierarchical optimization strategy for position and thickness optimization of constrained layer damping/plate to minimize sound radiation power. Adv. Mech. Eng. **10**(10) (2018)
28. Zheng, H., Cai, C., Tan, X.M.: Optimization of partial constrained layer damping treatment for vibrational energy minimization of vibrating beams. Comput. Struct. **82**(29), 2493–2507 (2004)

Acoustic Black Holes – Modelling, Shaping, Placement and Application

Steffen Hoffmann, Sebastian Rothe, and Sabine Christine Langer

1 Introduction

The development of new technical products is challenged by constantly growing requirements regarding, for example, energy consumption, environmental compatibility and handling. In order to meet these requirements, new design approaches and material combinations are necessary.

In most cases, the lightweight design-driven development, especially for mobility vehicles, leads to a worsening of the acoustic properties of the technical product in particular. With less insulating mass available, new measures are required to compensate the worsened acoustic characteristic. Acoustic Black Holes (ABHs) show immense potential to overcome the conflict between low mass and good acoustic properties, as they can be used as an efficient passive damping measure (by even reducing product weight).

This paper summarises main results on modelling, design and positioning studies of ABHs (Sect. 3). Experiments on plates to show the basic ABH effect serve as a starting point for this (Sect. 2). In addition, essential findings for the integration into more complex material systems (Sect. 4) as well as the applicability to realistic components (Sect. 5) are described.

The effect of ABHs was first described by MIRONOV [1] in 1988. He described mathematically that through a targeted impedance reduction at the end of a beam in form of a thickness reduction, the propagating bending waves become slower in the direction of the beam tip, while the amplitudes increase. With a theoretical thickness

S. Hoffmann (✉) · S. Rothe · S. C. Langer
Institute for Acoustics, Langer Kamp 19, 38106 Braunschweig, Germany
e-mail: steffen.hoffmann@tu-braunschweig.de

S. Rothe
e-mail: sebastian.rothe@tu-braunschweig.de

S. C. Langer
e-mail: s.langer@tu-braunschweig.de

P. Eberhard (ed.), *Calm, Smooth and Smart*, Lecture Notes in Applied
and Computational Mechanics 102, https://doi.org/10.1007/978-3-031-36143-2_9

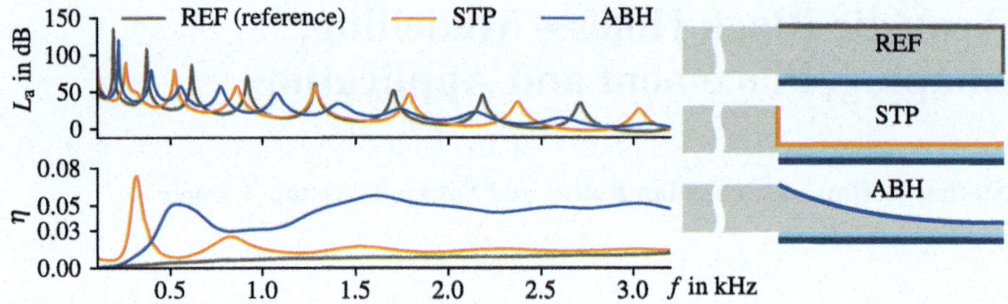

Fig. 1 Mean squared admittance L_a and loss factor η over frequency (simulation results) for three beams with the same applied damping patch but differently shaped ends

reduction down to zero, the wave would stop and not return. This consideration inspired Krylov [2] in 2000 to name this type of material or component weakening an *Acoustic Black Hole* (ABH). Due to the finite thickness reduction that exists in reality, partial reflection of the waves occurs. In order to damp this reflecting wave portion, Krylov combined the ABH with an additional damping layer. Since the increasing amplitudes in the ABH create an ideal region for damping application, a structure carrying structure-borne sound can be damped very efficiently. This becomes clear, for example, when comparing frequency responses (mean squared admittance L_a (whole beam) over frequency f) of beams with different design of the beam end (see Fig. 1).

By using the ABH shape in combination with a damping patch (ABH), the peaks can be damped much more compared to the reference beam with the same amount of damping foil (REF). This effect is also visible in the graph of the resulting system damping (loss factor η). Additionally, the measure results in a desired additional mass reduction. In the results of the design variant with the step at the end of the beam (STP) – high wave reflection due to impedance jump – it is obvious how important the smooth impedance adjustment is for a high damping effect.

Over the last two decades, the effect of ABH has been studied in a number of different structural (see e.g. [3, 4]) and fluid contexts (see e.g. [5]). A good overview of the most important research up to 2020 can be found in Pelat et al. [6]. Nevertheless, it is evident that ABHs have not yet fully established in the industry. The aim of the studies conducted within the DFG founded Priority Program 1897 is to facilitate access to ABHs for mobility applications by providing numerical modelling approaches as well as identifying relations between the expected acoustic effects and design parameters (outer shape, shape function, position, size, etc.). The basis of the investigations are experiments and numerical results based on the finite element method (FEM).

2 Experimental Investigation of ABH Effect on Plates

To demonstrate the theoretical ABH effect described in the introduction on a realistic structure, three different plates are manufactured (see photos in Fig. 2 top right). A 0.495 x 0.395 m and 0.005 m thick aluminium plate is used as a reference structure. In addition, one plate with a circular step and one with a circular ABH (radius: 0.1 m, polynomial degree of ABH $n = 2$, $h_{\mathrm{ABH,min}} = 0.0005$ m) are produced. For this purpose, the one-dimensional shapes shown in Fig. 1 are rotated and integrated into the plate as two-dimensional measures. Circular damping patches with constrained layer (CLD) are applied for every variant in the centre of the ABH respectively at the same position on the reference (REF) and step (STP) plate (radius: 0.05 m).

To measure the plates, they are freely suspended and excited at one point by an electrodynamic shaker (symbolised by red arrow). The force is measured with a force sensor and the surface velocity with a laser scanning vibrometer (LSV) at 755 points on the flat side of the plates. The density of measurement points is doubled in the ABH and step area in order to better scan the locally arising deflection shapes. Excitation and mounting points are chosen according to the minima and maxima of a multiplied superposition of all modes – up to 3200 Hz – in frequency domain

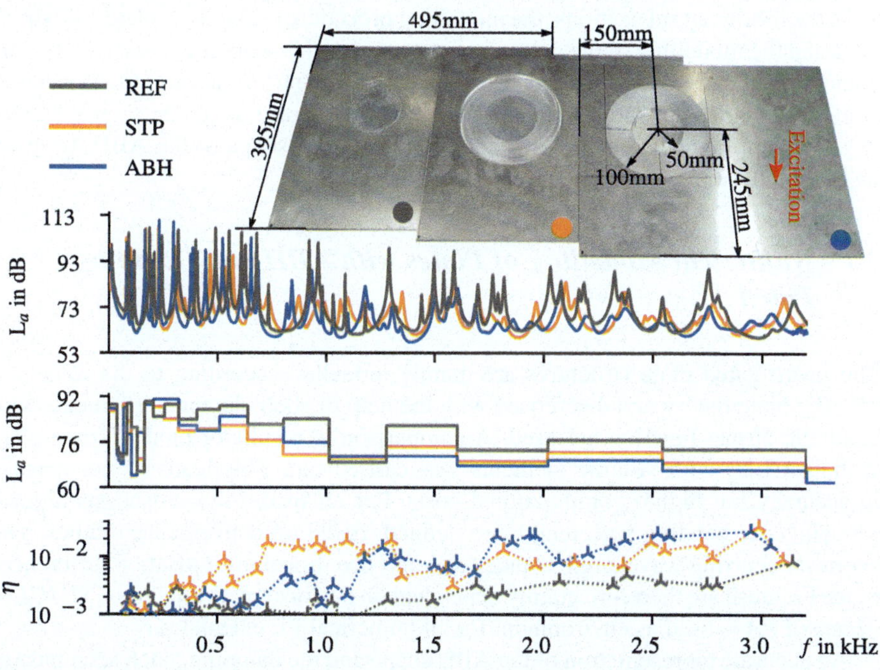

Fig. 2 Comparison of experimental results of reference plate with CLD (grey), plate with step shaped measure (orange) and plate with one ABH (blue): **top**: area-weighted mean squared admittance levels **middle**: mean squared admittance levels L_{a} in third-octave bands **bottom**: loss factors η determined at resonance frequencies

(excitation sensitivity). The experimental results of other plate structures used in the further course of the paper are characterised experimentally using a similar procedure.

The resulting mean squared admittance levels in frequency domain of the three measured plates are shown in Fig. 2. The advantage of the ABH compared to the other variants becomes clear from about 1300 Hz, where the peaks are damped much stronger, even if it is less noticeable than in the beam comparison (compare Fig. 1). Reasons for this include the fact that measurement deviations occur in comparison to simulation results and that optimal comparability of the variants is not possible due to manufacturing inaccuracies. However, the experimental results of the 2D structures serve as a starting point and as a validation basis for the studies and modelling approaches in the following sections.

3 Modelling and Numerical Studies

Models are suitable for investigating the effect of changing system parameters of the ABH on the structure-borne sound behaviour of plate-like structures. When these are transferred into parametric numerical models based on the FEM parameter, changes can be investigated efficiently. However, for reliable predictions, a valid simulation model is essential.

In the following subsections, the modelling procedure of ABHs in plate-like structures and the modelling of the damping patch used in this work are described. Experimental results are used for comparison and assessment of validity. On this basis, numerical studies are carried out subsequently with regard to geometric parameters (e.g. polynomial degree of ABH shape function, outer shape of the ABH, position on the structure).

3.1 Numerical Modelling of Plates with ABH and Damping Patch

The investigated plate structures are mainly modelled according to the *Reissner-Mindlin* plate theory and discretised with the help of shell elements. Elements with quadratic ansatz function are used. A comparison to results of models (base plate without CLD) with volume elements was carried out. This leads to no increase in accuracy but higher computational costs. For all model discretisations at least 15 points per bending wavelength are defined, based on convergence studies. The shortest occurring wavelength is taken as reference. Local discretisation differences of the FE mesh are therefore mainly dependent on the local plate thickness. *SIMULIA Abaqus FEA* is used as environment for all numerical FE calculations.

An accurate representation of the ABH shape and the damping patch accounts for a particular challenge when creating the numerical models. Due to the use of shell elements the shape function of the ABH has to be realised by an element thickness adjustment. In order to realise a sufficient representation of the ABH shape function, a maximum height difference between neighbouring elements must be defined. In all

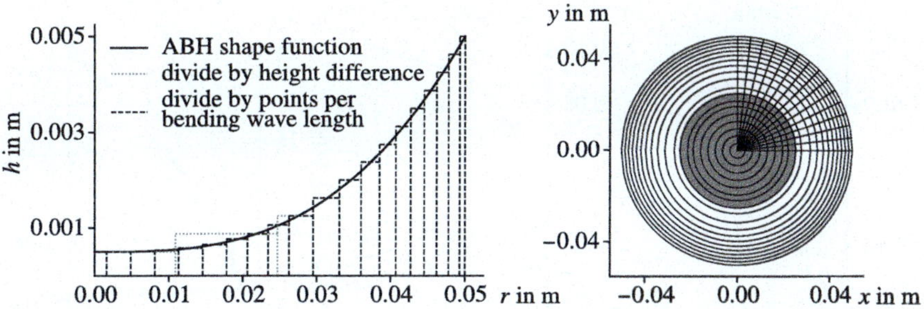

Fig. 3 Schematic representation of the element subdivision in the area of the ABH (**left**: in radial direction; **right**: in peripheral direction)

subsequent calculations a height difference of $h_{\text{diff}} = 0.1 \cdot h_{\text{ABH,min}}$ is used. Herein $h_{\text{ABH,min}}$ describes the minimum thickness of the ABH. This parameter serves as step size for an initial subdivision of the ABH. The second subdivision is made by the required number of points per bending wavelength of each previously defined section and is therefore frequency dependent. The element subdivision in radial direction of the ABH for a maximum thickness of $h_{\text{ABH,max}} = 0.005$ m and a minimum thickness of $h_{\text{ABH,min}} = 0.0005$ m is shown in the left diagram of Fig. 3. The division of the rings in peripheral direction is done by defining the maximum aspect ratio for each element, which is a maximum of five in this case. This is shown schematically on the right side of Fig. 3 as a top view on the ABH.

Based on this partitioning, element sets are created to which the corresponding thicknesses and material properties can be assigned. For the base material (aluminium) linear elastic material is assumed with a Young's modulus $E = 70$ GPa and a density $\varrho = 2700\,\text{kg/m}^3$. The material damping is modelled frequency-dependent and taken into account as *Rayleigh* damping ($\alpha = 0.256$ and $\beta = 4.62\text{e} - 7$; experimental determined acc. to [7]). In the area of the damping patch, considering only an increased damping value is not sufficient, since it neglects the stiffness increase and the frequency-dependent interaction of the sandwich system (base plate and CLD patch). Therefore a more complex modelling is considered, which is explained in more detail hereafter.

In order to be able to represent the influence and effects of the damping patch sufficiently, it is modelled with the help of shell and volume elements. The *M3*™ *Damping foil 2552* [8] used in the experiments in Figs. 2, 10 and 15 serves as the basis. As illustrated in Fig. 1 this are constraining damping measures with multiple layers (damping layer combined with constraining aluminum layer). Since the thin top layer is made of aluminum, it is modeled in the same way as the base structure and with quadratic shell elements. The damping layer is an acrylic that behaves viscoelastically rather than linearly elastically and is discretised by volume elements (20 node elements, quadratic ansatz function).

Since the material parameters of the acrylic adhesive layer are not known, a method for determining these parameters was developed as part of the project (see [7]). For

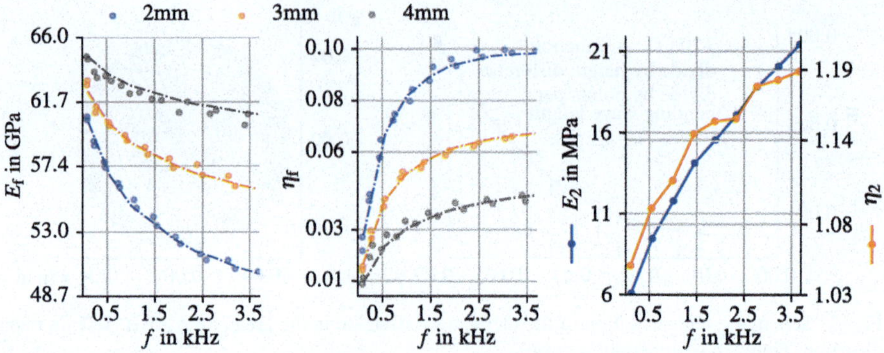

Fig. 4 Experimentally determined homogenised flexural storage moduli (**left**) and loss factors (**middle**) of CLD covered beams of different thickness and curves fitted via RKU [10] (dot-dash line); **right:** inversely determined E-modulus and loss factor of the damping layer [7]

this purpose, aluminum beams of different lengths and thicknesses are treated with the same type of damping foil. The flexural storage modulus E_f (Fig. 4, left) and the flexural loss factor η_f (Fig. 4, middle) of all beams are determined according to DIN EN ISO 6721-3 [9] up to 3.5 kHz.

In order to obtain the parameters of the intermediate damping layer from the homogenising quantities (Fig. 4 left/middle), an inverse method is used. A detailed description of the procedure can be found in [7]. It is based on the approach of *Ross*, *Kervin* and *Ungar* (RKU) [10]. For the complex total bending stiffness $\underline{B}_{\text{tot}}$ of the CLD treated beams the following applies [10, p. 61]:

$$\begin{aligned}\underline{B}_{\text{tot}} = K_2\left[\frac{h_2^2}{12}+H_{21}\right]+K_3\left[\frac{h_3^2}{12}+\frac{gK_1H_{31}^2}{K_1+g(K_1+K_3)}\right] \\ - K_2h_{31}\left[\frac{K_1\left(\frac{H_{21}}{2}+\frac{h_2}{12}\right)+2gK_3H_{21}}{K_1+g\left(K_1+K_3\right)}\right]+K_1\frac{h_1^2}{12}\end{aligned} \tag{1}$$

The tensile stiffness K_s of the individual layer[1] $s = [1, 2, 3]$ are calculated using the respective material (for each layer: ν_s–poisson ratio, E_s–Young's modulus, ϱ_s–density, η_s–loss factor) and geometry parameters (h_s–layer thickness):

$$K_s = E_s h_s\left(1 + j\eta_s\right)$$

Furthermore, the geometry parameters H_{21} and H_{31} are defined as

$$H_{21} = \frac{h_1}{2}+\frac{h_2}{2}, \quad H_{31} = \frac{h_1}{2}+\frac{h_3}{2}+h_2$$

[1] $s = 1$ – base layer, $s = 2$ – intermediate damping layer, $s = 3$ – constraining top layer

as well as the length-related mass m' and the relationship between shear modulus G_2 and elastic modulus E_2 of the damping layer are required, i.e.

$$m' = \rho_1 h_1 + \rho_2 h_2 + \rho_3 h_3, \quad E_2 = 2G_2 (1 + \nu_2) .$$

These quantities can subsequently be used to determine the shear parameter g and the wave number k (B_{tot} indicates the real part of $\underline{B}_{\text{tot}}$, j the complex number):

$$g = \frac{G_2'}{h_2 k^2} \left(\frac{1}{K_1} + \frac{1}{K_3} \right) (1 + j\eta_2) , \quad k = \sqrt[4]{\frac{\Omega^2 m'}{B_{tot}}}$$

The frequency-dependent values for G_2 and η_2 are varied until the E-modulus E_{f} and loss factor η_{f} (ratio of real and imaginary part of the complex bending stiffness $\underline{B}_{\text{tot}}$) fit into the measured curves (dot-dash lines in Fig. 4). The frequency-dependent material parameters G_2 and η_2 obtained from the parameter variation are shown in the right diagram of Fig. 4.

The determined material parameters are used to describe the viscoelastic material behaviour for the volume elements of the adhesive layer. A viscoelastic modelling of the damping layer is not always necessary and must be checked for the individual case. While it has been shown that a viscoelastic calculation of the intermediate layer is appropriate for beam structures with ABH placed at the ends, this does not apply to the present plate structure. As the frequency response functions (FRF) in Fig. 5 show, there seems to be hardly any difference in the stiffness between the model with a linearly elastic intermediate layer (linear) and the model with a viscoelastic modeled damping layer. For the numerical model, the mean values over the frequency range are used (Fig. 4 (right): $E_{2,\text{mean}} = 14$ GPa, $\eta_{2,\text{mean}} = 1.1$).

However, the system damping in the linear model deviates more from the measurement than for the viscoelastic simulation. Nevertheless, since a considerable amount

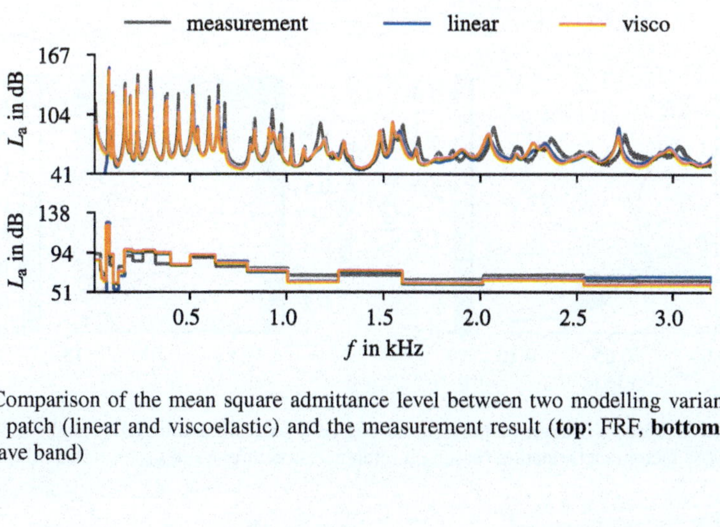

Fig. 5 Comparison of the mean square admittance level between two modelling variants of the damping patch (linear and viscoelastic) and the measurement result (**top**: FRF, **bottom**: FRF as third-octave band)

of computing time can be saved with the linear calculation, a full modelling of the CLD patch with a linear elastic intermediate layer is carried out for the following variations (unless otherwise discussed). These studies focus on crucial parameters in the design and placement of ABH.

3.2 *Analytical Variation of Polynomial Degree and Minimum Height of ABH Shape Function*

To find out which design parameters of the ABH are best for the effect of the damping patch, a variation of the geometric parameters of the ABH shape function h_{ABH} is performed. The shape function of the ABH can be described with the help of the minimum ($h_{\text{ABH,min}}$) and maximum thickness ($h_{\text{ABH,max}}$) of the ABH area, the length of the ABH (ℓ_{ABH}) and the polynomial degree n according to *Mironov* [1] as follows:

$$h_{\text{ABH}} = \frac{h_{\text{ABH,max}} - h_{\text{ABH,min}}}{\ell^n_{\text{ABH}}} x^n + h_{\text{ABH,min}} \quad ; \quad n \geq 2 \tag{2}$$

The origin of the x-coordinate is the end of the ABH shape (beam). Based on the presented RKU approach (Eq. 1) and the determined material parameters (Fig. 4) of the intermediate layer, a variation for the minimum thickness $h_{\text{ABH,min}}$ and the polynomial degree n of the ABH shape function is performed. In Fig. 6 the homogenised loss factor η over x (corresponds to the radius of ABH in plates) in dependency on the frequency f is shown. The results are based only on analytical calculations.

The results in Fig. 6 (left) show why it is reasonable to cover only half of the ABH (from the inside) with CLD, as the resulting loss factor decreases strongly towards the edge. In addition, there is a strong dependence on frequency and thickness.

To obtain a single value for a variation of $h_{\text{ABH,min}}$ and n, the values of the left diagram in Fig. 6 are averaged over the frequency in a first step and afterwards over

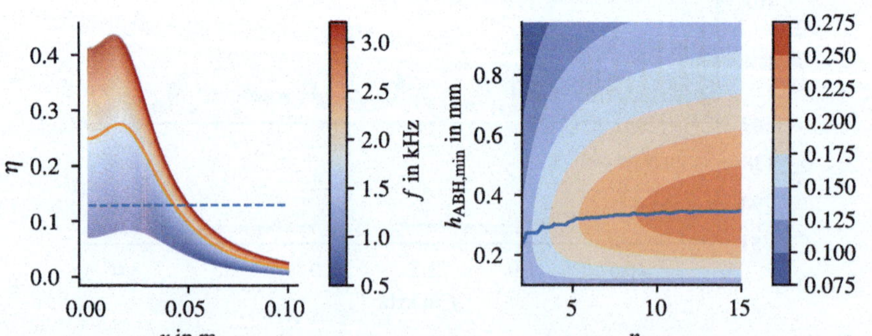

Fig. 6 RKU based variation of polynomial degree n and minimum ABH height $h_{\text{ABH,min}}$ for optimal loss factor determination (based on reference structure: $h_{\text{ABH,max}} = 0.005$ m)

x (radius of the ABH). This spatially- and frequency-averaged loss factor η is plotted in the right diagram for $h_{\text{ABH,min}}$ over n. Here it can be seen that particularly high polynomial degrees are suitable for obtaining high loss factors and that the optimal minimum thickness of the ABH converges to a limit value (≈ 0.35 mm). It can also be seen that for the ideal case of an ABH with $h_{\text{ABH,min}} \approx 0$, considerably lower loss factors can be expected, even if high polynomial degrees are used. This is due to the increasingly constant thickness of the ABH with increasing polynomial degree and the layer structure, in which there is an optimum base layer thickness that leads to an optimal loss factor. Since only the loss factor and no impedance matching is considered here, no general statement for real structures can be made. For this reason, further numerical simulations will be carried out in the following sections.

3.3 *Variation of Polynomial Degree of ABH in Plates*

For the numerical polynomial degree variation, the same ABH plate configuration as described at the beginning of Sect. 2 is taken as a basis. The modelling is done according to the procedure in Sect. 3.1. A total of four different variations are carried out. They differ on the one hand in the modelling of the damping foil and on the other hand in a constant or variable ABH radius. For the first variation, the radius of the ABH is kept constant (const. ABH radius) and the damping patch is only modelled by an increased structural damping (SD) in the application area (mean value based on results in Fig. 4). The second variation, on the other hand, considers the patch as detailed model (FM), as described in Sect. 3.1. For the other two variants, the radius of the ABH is adjusted for each polynomial degree so that the variants have the same mass among each other (constant mass). In this way it can be ensured that only the change in the polynomial degree is assessed and that any mass effects do not falsify the comparison. The radius of the CLD patch remains constant for all variants ($r_{\text{CLD}} = 0.05$ m).

As an assessment quantity for the polynomial degree, the total levels of the mean squared admittance $L_{\text{a,s}}$ are calculated and compared with the total level of the variant with polynomial degree $n = 2$ (see left diagram in Fig. 7). To calculate the total level, the range from 200–3200 Hz is taken into account in each case.

What becomes clear in the comparison is that higher polynomial degrees than $n = 2$ make sense with regard to the reduction of L_{a}. However, there is an optimum. If the polynomial degree is increased further, the reduction of L_{a} decreases. This is also confirmed by the results of the investigations by Rothe et al. [11]. Due to an increasing polynomial degree, the shape function approaches the step shape (STP) and thus no longer enables the smooth impedance matching for the bending waves in the ABH.

An optimal polynomial degree must be determined separately for each application. However, based on the investigations in this paper (see Sect. 3.1), polynomial degrees between $n = 5$ and $n = 12$ can be recommended. They correspond to the minima of the blue and orange curves in Fig. 7 . The FRF as well as its representation in

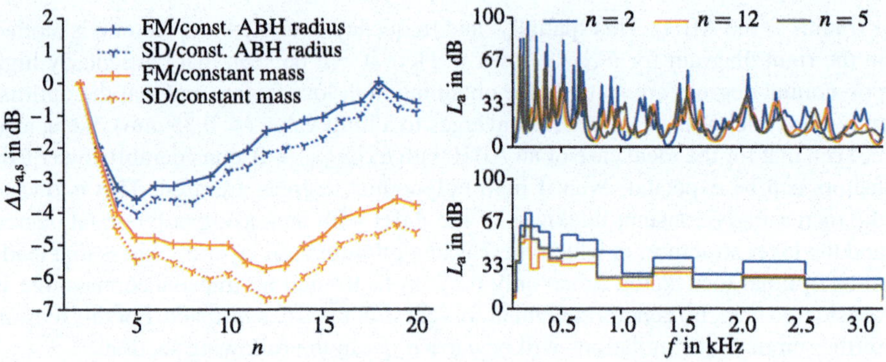

Fig. 7 Results of variation of polynomial degrees of ABH in plates (**left**: difference in the mean squared admittance total level (reference: $n = 2$), **right**: comparison of FRFs of $n = 2$ and the two best polynomial degrees)

third-octave bands is shown in comparison to the variant with $n = 2$ in the right diagrams of Fig. 7. Here the considerable advantage of higher polynomial degrees (over the entire frequency range) becomes clear once again.

3.4 ABH Position Variation on Rectangular Plate

After the suitable polynomial degrees have been determined, the position of the ABH on the previous investigated plate configurations should be examined more closely. The goal is not to determine an optimal position for this structure, which is why no optimization algorithm is used. Rather, the basic structural behaviour with regard to the ABH position and the usefulness of a position optimization in the design process should be evaluated. Due to the previous variation, the degree of the polynomial is set to five, since the curves in Fig. 7 deviates strong for higher degrees. The radius is reduced accordingly with respect to the results in Fig. 7 as a higher admittance decrease is to be expected here. In addition, a larger area of the plate can be used for positional variation without the ABH protruding over the edge of the plate (Fig. 7, left).

The ABH is placed at a total of 400 equidistantly distributed positions on the plate and the total level of the mean squared admittances is calculated for the position assessment in each case. The resulting position rating distribution is shown as a contour plot in Fig. 8 (left).

It becomes clear that positioning the ABH in the area of the force application should be avoided under any circumstances, as this can even result in a worsening of the structure-borne sound behaviour compared to the initial structure without ABH. This also agrees with the findings in [12]. For a better comparison, the FRFs of the best (blue) and worst (grey) position are plotted on the right side of Fig. 8. For

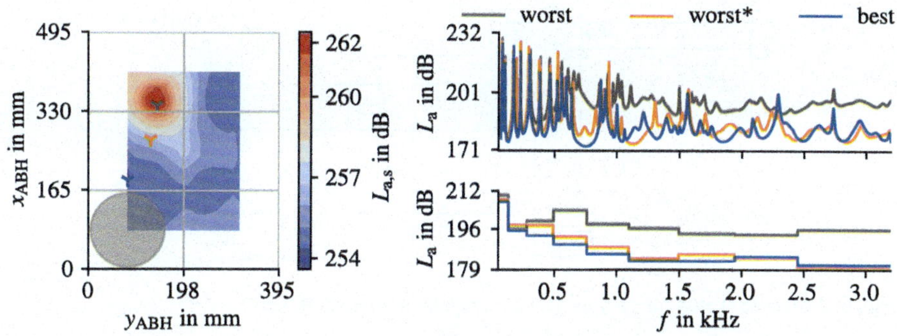

Fig. 8 Results of the ABH position variation (**left**: Position rating distribution of ABH on plate (gray circle represents an exemplary ABH), **right**: comparison of FRFs of worst and best position)

better assessment of the necessity of a position optimisation, the worst (*) position outside the excitation region is shown (orange). These positions are also marked with crosses in the respective colours in the contour plot. In the one-third octave bands, the differences of the position are clearly visible. However, if the positioning of the ABH in the force application area is generally avoided, much smaller differences in the resulting total level are to be expected with different ABH positioning on the plate (≈4 dB).

3.5 *Oval Shaped ABH in Plates*

Beyond the shaping of the ABH cross-section, the optimal external shape of the ABH is also of great interest. Regarding a limited design space, it is obvious to consider not only circular ABH, but also other shapes. The most obvious shape is a special case of the oval, the ellipse. This shape could be well adapted to slim design areas and can "catch" longer bending waves than several small ABHs. Combining the elliptic description with the ABH shape function leads to the following formulation of the material thickness distribution of an elliptic ABH:

$$h_{\mathrm{ABH}}(x_{\mathrm{e}}, y_{\mathrm{e}}) = \left(\frac{h_{\mathrm{ABH,max}} - h_{\mathrm{ABH,min}}}{\left(\sqrt{x_{\mathrm{K}}^2 + y_{\mathrm{K}}^2}\right)^n} \right) \cdot \left(\sqrt{x_{\mathrm{e}}^2 + y_{\mathrm{e}}^2}\right)^n + h_{\mathrm{ABH,min}} \tag{3}$$

The following applies to y_{K} and x_{K}:

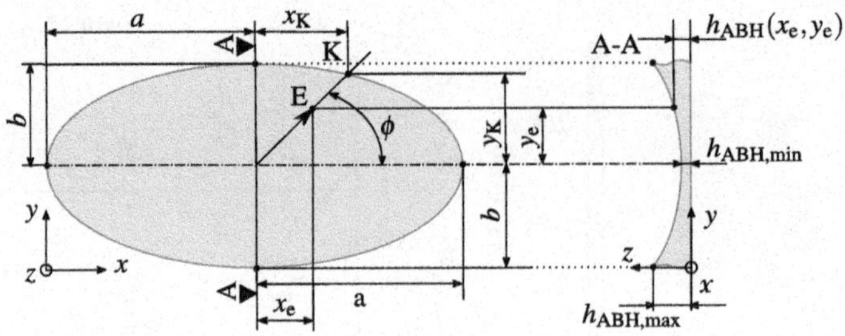

Fig. 9 Principle sketch for geometrical parameters of the oval ABH

$$x_K = \sqrt{\frac{1}{\frac{1}{a^2} + \frac{(x_e \cdot \tan(\phi))^2}{b^2}}} \qquad y_K = \tan(\phi) \cdot x_K$$

The geometric parameters are summarised in the following Fig. 9 as a sketch.

With the help of experiments it is investigated whether the effect of the ABH can be improved by tightening the shape of the ABH (oval ABH) to the aspect ratio of the considered structure respectively design space. For this, an elliptical ABH in the same design area will be compared with three round ABHs.

Assuming that the design space for the application of passive damping measures is limited, a design space with an aspect ratio of 1:3 is defined. This can be seen in Fig. 10 (top, right) slightly framed in black in the photos of the plate under consideration. In one case, three round ABHs are placed. For comparison, an oval ABH (same area as the three round ABHs) is placed on the similar plate with the same design area. The excitation is done at the opposite corner of the ABH positioning. The plates are suspended quasi-freely and their surface velocity is measured at 750 equidistantly distributed points using a laser scanning vibrometer. The area of the ABH is scanned twice as fine in order to sufficiently resolve the local deflection shapes. The CLD patches are placed in the centre of the ABHs and care is taken to apply the same amount of damping foil for both variants.

Figure 10 shows L_a over the frequency as well as the representation in one-third octave bands and the determined loss factor. A clear difference appears especially in the middle frequency range. While the position of the resonances of both structures is similar, the plate with oval ABH shows a strongly increased damping from the fourth resonance on. This effect is noticeable up to 1.5 kHz, above which the loss factors converge and the differences decrease significantly.

The effect is due to the reduced stiffness over a large area, which leads to low-frequency, broadband large deflections in the area of the damping layer. If the bending waves become smaller with increasing frequency and also fit into the round ABHs, the difference between the two structures becomes much smaller. In order to maximise the damping as much as possible with a limited design area, a single connected measure (e.g. oval ABH) should be designed instead of several round ABHs.

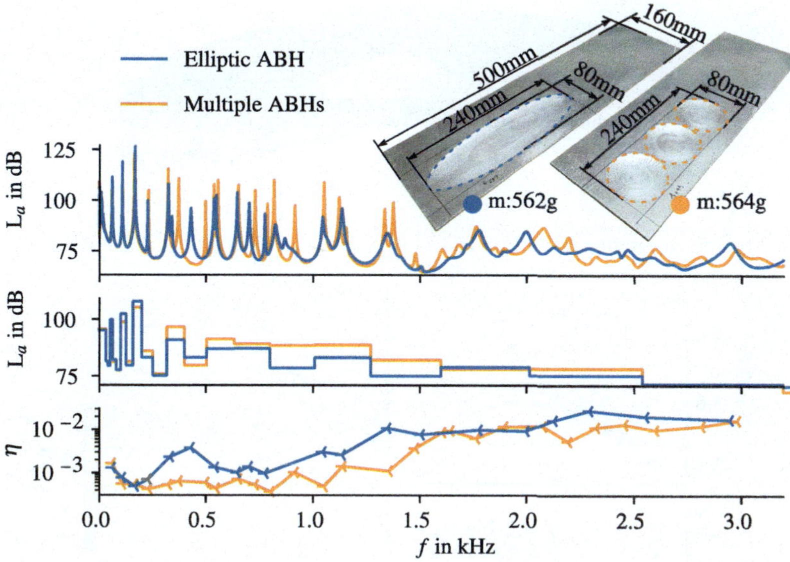

Fig. 10 Comparison of experimental results of plate with multiple ABH (orange) and plate with elliptic ABH (blue) in a defined design space: **top**: area-weighted mean squared admittance levels **middle**: mean squared admittance levels L_a in third-octave bands **bottom**: loss factors η determined at resonance frequencies

In the special case of the elliptical ABH used here, the ABH shape function (see Eq. 2) is fulfilled at every point of the ABH, which is only possible for shapes without undercuts (all ovals). More complex shapes with undercuts would also be possible if not fulfilling the shape function at certain points is accepted.

4 Material Studies

In this section, the possibility of integrating Acoustic Black Holes into non-homogeneous materials is studied. For this purpose, additively manufactured and fibre laminate structures are investigated. Besides numerical studies, mainly experimental investigations on beams and plate samples are done.

4.1 Additively Manufactured Structures

In additive manufacturing, the structures are produced by applying material layer by layer. This enables the realisation of almost any complex geometry and thus also a great variety for the integration of ABH. The additive manufacturing process focused

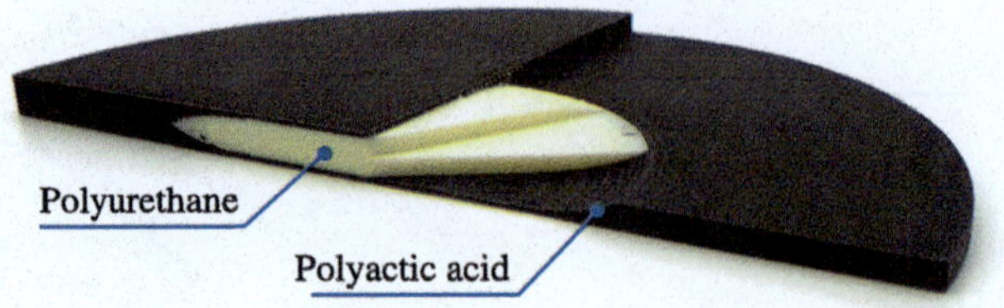

Fig. 11 Possible integration of ABH in additively manufactured structures (demonstrator)

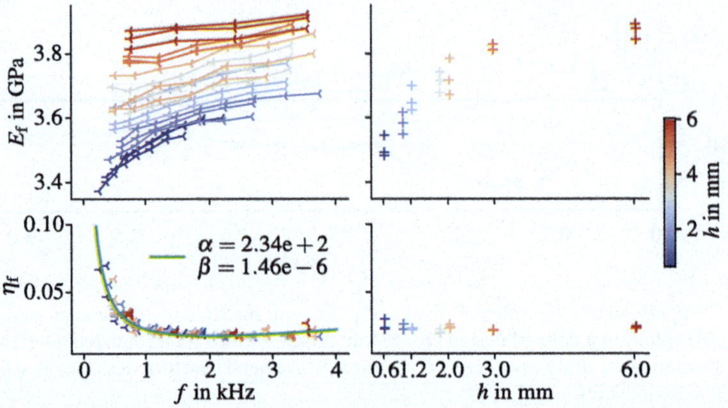

Fig. 12 Material parameters of additively manufactured beams of various thicknesses (0.6, 1, 1.2, 1.8, 2, 3, 6 mm) determined according to Din EN 6721. **Left**: Frequency-dependent flexural storage modulus (top) and loss factor (bottom). **Right**: Flexural storage modulus averaged over frequency range (top) and loss factor (bottom)

on here is fused deposition modelling (FDM). Here, the material, which is available as a filament, is melted with the help of nozzles and continuously applied (material extrusion). A heated print bed serves as the construction base. Thermoplastic polymers are used as materials. In the cases examined here, these are polylactic acid (PLA) and thermoplastic polyurethane (TPU). The following Fig. 11 shows an example of how an ABH can be fully integrated into plate-like structures. The ABH would not be visible from the outside, but would still have its damping-increasing effect.

The realisation of additively manufactured ABH and its effectiveness could already be shown in some project related publications, to which reference is made here [13–15]. One key finding is that the numerical modelling of additively manufactured structures is much more complex. Besides the viscoelastic and anisotropic behaviour, a thickness-dependent effect with regard to the homogenised material properties could be determined (see [16]). This is also shown by the results in Fig. 12.

For this purpose, three beams of PLA (without ABH) with thicknesses of 0.6, 1.0, 1.2, 1.8, 2.0, 3.0, 6.0 mm are additively manufactured and measured at 19 °C in accordance to DIN EN ISO 6721-3 [9]. The influence of shear deformation (especially for thicker beams) is taken into account in the calculation by numerically determined frequency dependent material parameters. The Young's modulus and the loss factor are plotted over the frequency and frequency-averaged values are shown for different thicknesses of the samples.

Comparing the resulting flexural moduli E_f of the beams of different thicknesses h, it becomes clear that the homogenised flexural modulus also increases with increasing thickness. The damping, quantified here via the bending loss factor η_f, is not dependent on the thickness of the beams. It can also be represented quite well via the Rayleigh damping, which is particularly advantageous for more efficient numerical solutions with the help of the finite element method.

4.2 *Laminate Structures*

In contrast to homogeneous structures, layered structures offer the advantage that active (e.g. piezo actuators) or passive (e.g. damping polymers) measures can already be integrated into the layers directly during manufacturing. This is investigated in this section for ABHs. In addition to the often used glass or carbon fibre reinforced materials, laminated woods are particularly suitable due to the usually greater material thicknesses, which is especially advantageous for ABHs. In addition, these are sustainable materials for which there is a drive towards greater use in automotive but also other mobility vehicles and building acoustics.

The challenge here is the fibre orientation to be taken into account and the already high inherent damping of laminated materials (especially wood), which promises only little potential for improvement through ABHs.

After some numerical studies based on the material properties of laminated beams determined according to DIN EN ISO 6721-3 [9], different glued laminated timber (glulam) beams are manufactured with ABH (see Fig. 13). The stepped and ABH-formed beams are provided with the same amount of silicone (white filling of the beams), related to the same length of the measure. The third beam is made with an off-centre ABH within the layered construction and has less damping material. The length of the beam is adjusted to achieve a similar modal density. The excitation is done at one side of the beam, while the surface velocity is measured with a laser scanning vibrometer (double mesh density in the area of the ABH). In addition to the mean square admittance L_a, the homogenised loss factor η of the measured beams in the resonant frequencies are determined with the peak-fit method (see Fig. 13).

The same amount of damping material results in significantly higher damping values due to the ABH shaping (blue curve). By placing the measure outside the neutral axis (red curve), the loss factor can be effectively increased despite much smaller amounts of damping material (damping material is sheared stronger here).

In addition to one-dimensional beam structures, multi-layered plates are also of interest, especially in the context of building acoustics. Among other things, the question arises whether the orientation of an oval ABH adapted to the fibre orientation brings advantages. This is investigated experimentally in the following.

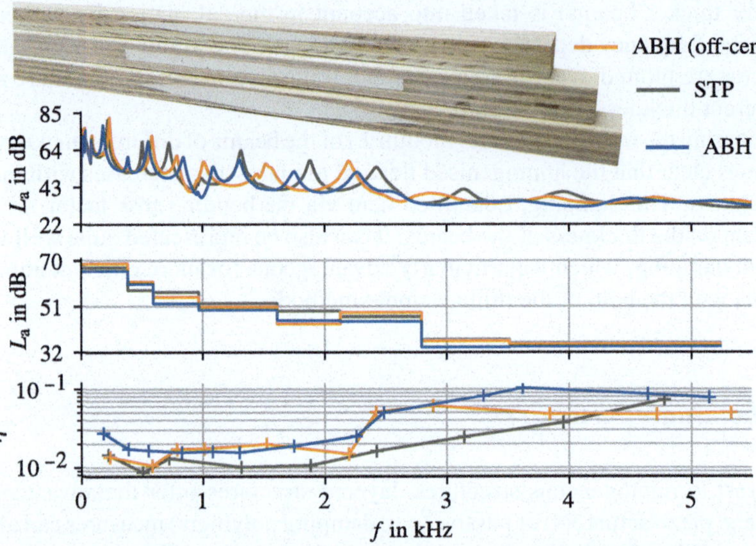

Fig. 13 Glued laminated timber (glulam) beams with damping measures integrated within the layer structure, **top**: area-weighted mean squared admittance levels **middle**: mean squared admittance levels L_a in third-octave bands **bottom**: loss factors η determined at resonance frequencies

The effectiveness of the acoustic measure (oval ABH filled with acrylate, Fig. 14) compared to the plate without ABH can be successfully demonstrated (Fig. 14, light gray). The aspect ratio of the oval ABHs is thereby adapted to the stiffness ratios of the plate in x and y directions. Averaged over the whole frequency range, four times higher loss factors ($\eta = 0.0109...0.0451$) and a maximum level reduction L_a of 15 dB (around 1 kHz) can be achieved. To investigate the influence of the orientation of the ABH, three different plates with an oval ABH are manufactured (see Fig.14, top) and measured.

The results in Fig. 14 highlight that the different orientation of the oval ABH relative to the main fibre orientation ($0\,°, 45\,°, 90\,°$) show no significant effect. Additional numerical studies also did not show a significant advantage of a material stiffness based orientation of the oval ABH, even when the aspect ratios or anisotropies were more significant than in the experimentally studied plates.

5 Application on an Automotive Structure

With the knowledge gained concerning the design and placement of ABHs, several ABHs are integrated into a typical component of an automobile. An oil pan (VAG 038103601NA) made of die-cast aluminium is used as the test object. It is used as a representative of typical thin-walled die-cast aluminium components with a large

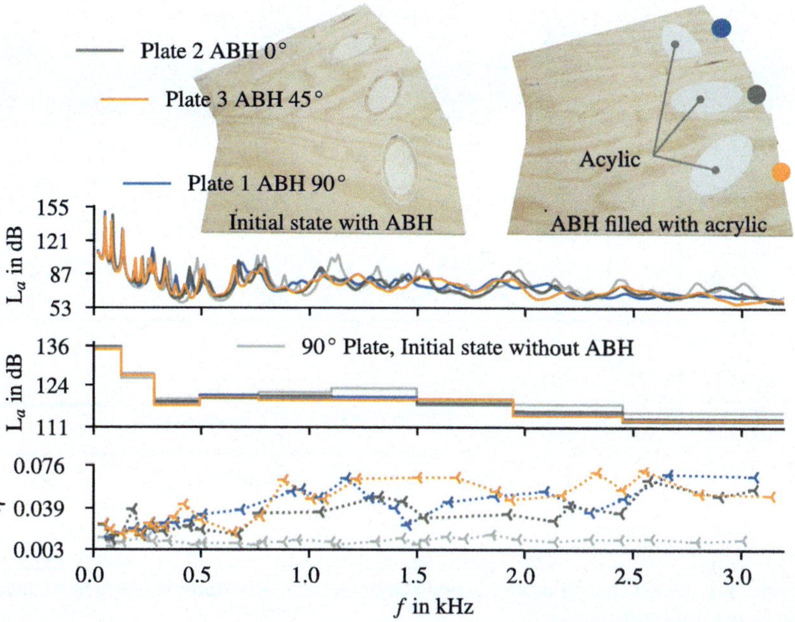

Fig. 14 Glulam plates with damping measures applied on the sample surface, **top**: area-weighted mean squared admittance levels **middle**: mean squared admittance levels L_a in third-octave bands **bottom**: loss factors η determined at resonance frequencies

surface area. The same oil pans from one manufacturer already behave dynamically different in the initial state above 1500 Hz. The component is therefore measured sequentially in three states: in the initial state (Fig. 15 left), with three applied CLD patches (Fig. 15 middle) and finally with three ABHs including CLD patches of same size and position (Fig. 15 right). Unfortunately, a comparison with a step-shaped structure is not possible because the component can only be machined once.

The excitation is done with the help of a shaker at a screw point on the sealing edge (Fig. 15 red arrow). The surface velocity is measured with an laser scanning vibrometer at 700 points. The normal vectors of each measurement point are calculated element by element via the measurement grid in order to compensate the different angles of the individual partial surfaces.

The oval ABHs are milled in the flat surfaces (Fig. 15 black boxed regions in the middle picture). The material thickness of 1.8 mm in the design area is reduced to 0.3 mm to obtain high loss factors with the CLD foil (based on the variation in Fig. 6). The polynomial degree is set to $n = 5$. The size of the ABHs is limited due to the maximal usable area of the flat design spaces.

The amplitude reduction due to the introduction of the ABHs is clearly visible in the curves in Fig. 15. This is particularly noticeable from approx. 500 Hz and is significantly greater compared to CLD patches without ABH. The low effectiveness at low resonances is a frequently occurring effect for ABHs, as here the wavelengths

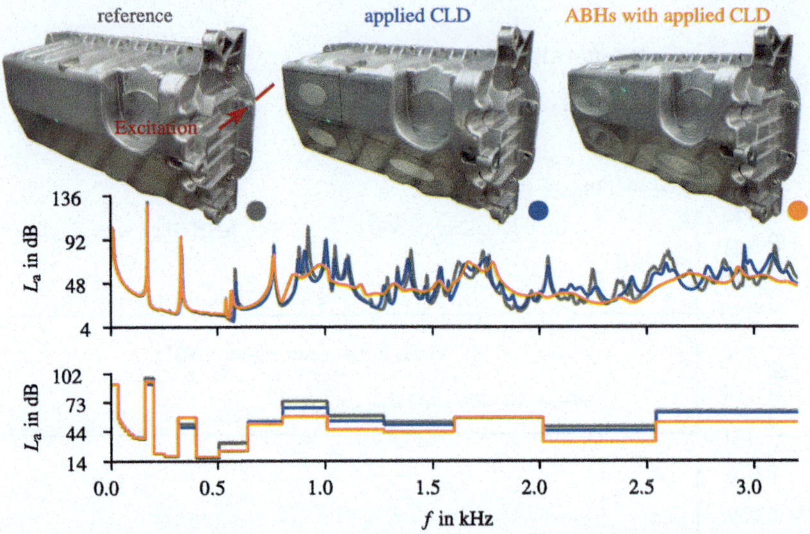

Fig. 15 Comparison of the mean squared admittance levels of an oil pan with different passive measures, **top**: area-weighted mean squared admittance levels **bottom**: mean squared admittance levels L_a in third-octave bands

are too large and the reduced mass for insulation is missing. This can be compensated, for example, by placing point masses in the maxima of the corresponding first two modes without increasing the mass compared to the reference structure [17]. It should nevertheless be noted that the clearly recognisable damping effect cannot be attributed solely to the impedance matching by the ABH shape function, since the comparison with the step function is not possible.

6 Conclusion and Outlook

The studies in the project have identified many sensitive parameters in the design and positioning of ABHs. The results are mainly supported by numerical studies using the finite element method. However, these are always validated by experimental studies. Beams and plates are investigated as generic structures.

A suitable procedure for modelling and discretising (FE mesh) ABHs as well as a CLD patch is proposed. Especially for the plate structures it shows that although a detailed modelling of the layers (damping layer with volume elements and top layer with shell elements) is useful, the consideration of the viscoelastic behaviour of the intermediate layer is not essential for the general vibroacoustic assessment of the investigated structures. To determine the material properties of the intermediate layer, an experimental procedure on beam structures based on the RKU model is proposed.

Furthermore, variations of the polynomial degree of the ABH shape function show that a specific adjustment of the thickness of the base structure can have a considerable influence on the resulting damping by the CLD patch. In general, it can be stated that higher polynomial degrees (here: $n = 5...12$) are more advantageous and lead to a greater reduction of structural vibrations. However, there is an optimum which must be determined for each individual case.

The position of the ABH also has a considerable influence on its mode of action. In any case, positioning in the area of the force application point should be avoided. The variation in position in the remaining area of the plate, on the other hand, showed only small differences in the total level of the mean square admittance ($\Delta \approx 4$ dB).

An adaptation of the outer shape of the ABH can be useful, especially with regard to limited design space in realistic structures. Here, the design of an oval ABH has proven to be advantageous compared to three round ABHs in the same design area.

In addition to the investigations on the geometric design parameters, studies on the integration into other material systems are investigated. Due to the design freedom in additive manufacturing, ABH can be integrated directly into the structure during the manufacturing process (fused deposition modelling). Due to the layer-by-layer build-up, variants that fully enclose the ABH are possible. This could be used, for example, for chemical or physical protection of the damping layer. This is also possible in a similar way when integrating ABH into laminate structures made of wood (glulam), although here the design freedoms is more restricted. Experiments show that the ABH effect can also be used in such structures. However, the valid modelling of the structures is challenging. Here, anisotropic and viscoelastic material properties must be taken into account so that the vibroacoustic behaviour is correctly simulated.

Transferring the findings to a realistic structure – oil pan – on which three oval ABHs are placed, has once again highlighted the high potential of ABH as an acoustic measure. The knowledge gained from the project can help to make ABHs more accessible as a measure in the future and thus enable their standard use in the industry. Nevertheless, the results shown here should be extended by further studies and validated on further realistic structures. Just like the challenge regarding a reduced static load-carrying capacity due to the material thickness weakening.

References

1. Mironov, M.A.: Propagation of a flexural wave in a plate whose thickness decreases smoothly to zero in a finite interval. Soviet Phys. Acoust. USSR **34**(3), 318–319 (1988)
2. Krylov, V.V., Shuvalov, A.L.: Propagation of localised flexural vibrations along plate edges described by a power law. Proc. Inst. Acoust. **22**, 263–270 (2000)
3. Krylov, V.V., Tilman, F.J.B.S.: Acoustic 'black holes' for flexural waves as effective vibration dampers. J. Sound Vib. **274**(3–5), 605–619 (2004)
4. Conlon, S.C., Fahnline, J.B., Semperlotti, F.: Numerical analysis of the vibroacoustic properties of plates with embedded grids of acoustic black holes. J. Acoust. Soc. Am. **137**(1), 447–457 (2015)

5. Ouahabi, A.A.E., Krylov, V.V., O'Boy, D.: Investigation of the acoustic black hole termination for sound waves propagating in cylindrical waveguides. In: Proceedings of ICSV22, Florence (2015)
6. Pelat, A., Gautier, F., Conlon, S.C., Semperlotti, F.: The acoustic black hole: a review of theory and applications. J. Sound Vib. **476**(115316) (2020)
7. Hoffmann, S., Rothe, S., Langer S.C.: Vorgehen zur Charakterisierung und Modellierung von Mehrschichtbelägen. In: Conference proceedings of DAGA, Stuttgart (2022)
8. 3M Industrial Adhesives and Tapes Division. Technical data-damping foil 2552 (November 2017). https://multimedia.3m.com/mws/media/1566784O/iatd-damping-foil-2552-tech-data-sheet.pdf. Accessed 26 Apr 2022
9. DIN EN ISO 6721-3: Kunststoffe-Bestimmung dynamisch-mechanischer Eigenschaften - Teil 3, Deutsches Institut für Normung, Beuth Verlag (2021)
10. Ross, D., Ungar; E.E., Kerwin, E.M.: Damping of plate flexural vibrations by means of viscoelastic laminae, Bolt Beranek and Newmon lnc, Cambridge (1959)
11. Rothe, S., Dorn, M., Langer, S.C.: Numerical study on shape functions for optimal exploitation of the acoustic black hole effect. PAMM **20**(1) (2021)
12. Rothe, S., Ghaffari Mejlej, V., Vietor, T., Langer, S.C.: Optimal adaptation of acoustic black holes by evolutionary optimization algorithms. PAMM **16**(1), 625–626 (2016)
13. Rothe, S., Watschke, H., Langer, S.C.: Study on the producibility of additively manufactured acoustic black holes. In: Proceedings of ICSV24, London (2017)
14. Rothe, S., Blech, C., Watschke, H., Vietor, T., Langer, S.C.: Numerical modelling of additively manufactured acoustic black holes. In: Proceedings of Inter-Noise, Chicago (2018)
15. Rothe, S., Blech, C., Watschke, H., Vietor, T., Langer, S.C.: Layer-effect by additive manufacturing of acoustic black holes. In: Proceedings of Inter-Noise, Madrid (2019)
16. Rothe, S., Blech, C., Watschke, H., Vietor, T., Langer, S.C.: Material parameter identification for acoustic simulation of additively manufactured structures. Materials **14**(1), No. 168 (2021)
17. Hoffmann, S., Rothe, S., Langer, S.C.: Combination of Acoustic Black Holes with point masses. In: Proceedings of Inter-Noise, Glasgow (2022)

Simulation-Free Model Reduction Approaches for Geometric-Nonlinear and Linear-Visco-Elastic Mechanical Systems

Christopher Lerch, Christian Meyer, Daniel J. Rixen, and Boris Lohmann

1 Introduction

1.1 Mechanical Systems

The basic dynamics of mechanical systems are commonly determined by the equilibrium of inertia forces $\boldsymbol{M}\ddot{\boldsymbol{q}}(t)$, damping and internal restoring forces $\widehat{\boldsymbol{f}}(\dot{\boldsymbol{q}}(t), \boldsymbol{q}(t))$ and external forces, i.e. loads, $\widehat{\boldsymbol{F}}(t)$:

$$\boldsymbol{M}\ddot{\boldsymbol{q}}(t) + \widehat{\boldsymbol{f}}(\dot{\boldsymbol{q}}(t), \boldsymbol{q}(t)) = \widehat{\boldsymbol{F}}(t), \qquad \boldsymbol{q}(0) = \boldsymbol{q}_0,\ \dot{\boldsymbol{q}}(0) = \dot{\boldsymbol{q}}_0 \tag{1}$$

with (generalized) displacements $\boldsymbol{q}(t) \in \mathbb{R}^N$, mass matrix $\boldsymbol{M} \in \mathbb{R}^{N \times N}$, $\widehat{\boldsymbol{f}} : \mathbb{R}^N \times \mathbb{R}^N \to \mathbb{R}^N$ and $\widehat{\boldsymbol{F}}(t) \in \mathbb{R}^N$.

From a system-theoretic point of view with the input-output behavior being of importance, the external forces are considered explicitly as a space- and a time-dependent part $\widehat{\boldsymbol{F}}(t) = \boldsymbol{B}\boldsymbol{F}(t)$ where the input matrix $\boldsymbol{B} \in \mathbb{R}^{N \times p}$ contains weights

C. Lerch · B. Lohmann (✉)
Technical University of Munich, Chair of Automatic Control,
Boltzmannstr. 15, 85748 Garching/Munich, Germany
e-mail: lohmann@tum.de

C. Lerch
e-mail: christopher.lerch@tum.de

C. Meyer · D. J. Rixen
Technical University of Munich, Chair of Applied Mechanics,
Boltzmannstr. 15, 85748 Garching/Munich, Germany
e-mail: christian.meyer@tum.de

D. J. Rixen
e-mail: rixen@tum.de

P. Eberhard (ed.), *Calm, Smooth and Smart*, Lecture Notes in Applied
and Computational Mechanics 102, https://doi.org/10.1007/978-3-031-36143-2_10

and allocations to the degrees of freedom of the time-dependent forces, i.e. the input signals $\boldsymbol{F}(t) \in \mathbb{R}^p$ $(p \leq N)$. The corresponding system outputs are given by Eq. (4).

Additionally, the lack of knowledge about the dominating damping mechanisms frequently leads to an assumption of simpler linear viscous damping $\boldsymbol{D}\dot{\boldsymbol{q}}(t)$. Excluding gyroscopic effects allows for writing $\widehat{\boldsymbol{f}}(\dot{\boldsymbol{q}}(t), \boldsymbol{q}(t)) = \boldsymbol{D}\dot{\boldsymbol{q}}(t) + \boldsymbol{f}(\boldsymbol{q}(t))$:

$$\boldsymbol{M}\ddot{\boldsymbol{q}}(t) + \boldsymbol{D}\dot{\boldsymbol{q}}(t) + \boldsymbol{f}(\boldsymbol{q}(t)) = \boldsymbol{B}\boldsymbol{F}(t), \qquad \boldsymbol{q}(0) = \boldsymbol{q}_0, \ \dot{\boldsymbol{q}}(0) = \dot{\boldsymbol{q}}_0 \tag{2}$$

with damping matrix $\boldsymbol{D} \in \mathbb{R}^{N \times N}$ and nonlinear internal restoring forces $\boldsymbol{f}$: $\mathbb{R}^N \to \mathbb{R}^N$.

Sufficiently small displacements around an equilibrium position or initial configuration allows for considering only the linear part of the internal restoring forces $\boldsymbol{f}(\boldsymbol{q}(t)) \approx \boldsymbol{K}\boldsymbol{q}(t)$ resulting in the well-known linear second-order representation

$$\boldsymbol{M}\ddot{\boldsymbol{q}}(t) + \boldsymbol{D}\dot{\boldsymbol{q}}(t) + \boldsymbol{K}\boldsymbol{q}(t) = \boldsymbol{B}\boldsymbol{F}(t), \qquad \boldsymbol{q}(0) = \boldsymbol{q}_0, \ \dot{\boldsymbol{q}}(0) = \dot{\boldsymbol{q}}_0 \tag{3}$$

with stiffness matrix $\boldsymbol{K} \in \mathbb{R}^{N \times N}$ and

$$\boldsymbol{y}(t) = \boldsymbol{C}\boldsymbol{q}(t) \tag{4}$$

with output matrix $\boldsymbol{C} \in \mathbb{R}^{q \times N}$ returning the outputs of interest for the mechanical system if only displacements are considered.

In the Laplace-domain, the transfer behavior from inputs to outputs for zero initial conditions is given by

$$\boldsymbol{G}(s) = \boldsymbol{C}\left(s^2\boldsymbol{M} + s\boldsymbol{D} + \boldsymbol{K}\right)^{-1}\boldsymbol{B} \tag{5}$$

such that $\boldsymbol{Y}(s) = \boldsymbol{G}(s)\boldsymbol{F}(s)$ where $\boldsymbol{Y}(s)$ and $\boldsymbol{F}(s)$ are the Laplace transformed outputs $\boldsymbol{y}(t)$ and inputs $\boldsymbol{F}(t)$, respectively.

Typically, mass ($\boldsymbol{M}$) and stiffness ($\boldsymbol{K}$) matrices are symmetric and positive (semi-) definite for appropriate boundary conditions suppressing rigid body modes. Commonly, linear damping is realized via modal damping. A simple and popular choice is the special case of proportional or Rayleigh damping where $\boldsymbol{D} = \alpha\boldsymbol{M} + \beta\boldsymbol{K}$ with $\alpha, \beta \geq 0$. The damping matrix $\boldsymbol{D}$ is symmetric and positive definite in that case, same as $\boldsymbol{M}$ and $\boldsymbol{K}$.

Further information about mechanical systems and their properties can be found e.g. in [1, 2].

1.2 Model Order Reduction

The computational effort for numerically solving systems (2) or (3) and (4) can be significantly reduced by applying reduced-order models (ROM) that accurately approximate the relevant behavior of the original full-order model (FOM). One classical option to obtain such ROMs is by applying projective model order reduction (MOR).

The full-order displacements $\boldsymbol{q}(t) \in \mathbb{R}^N$ are first approximated as a linear combination of reduced coordinates $\boldsymbol{q}_\mathrm{r}(t) \in \mathbb{R}^n$: $\boldsymbol{q}(t) = \boldsymbol{V}\boldsymbol{q}_\mathrm{r}(t) + \boldsymbol{e}(t)$ where $\boldsymbol{V} \in \mathbb{R}^{N\times n}$, $n \ll N$. Inserting this approximation in (2) or (3) and (4) leads to an overdetermined system with the residuals $\boldsymbol{\varepsilon}(t) \in \mathbb{R}^N$

$$\begin{aligned} \boldsymbol{M}\boldsymbol{V}\ddot{\boldsymbol{q}}_\mathrm{r}(t) + \boldsymbol{D}\boldsymbol{V}\dot{\boldsymbol{q}}_\mathrm{r}(t) + \boldsymbol{f}(\boldsymbol{V}\boldsymbol{q}_\mathrm{r}(t)) &= \boldsymbol{B}\boldsymbol{F}(t) + \boldsymbol{\varepsilon}(t) \qquad \text{or} \\ \boldsymbol{M}\boldsymbol{V}\ddot{\boldsymbol{q}}_\mathrm{r}(t) + \boldsymbol{D}\boldsymbol{V}\dot{\boldsymbol{q}}_\mathrm{r}(t) + \boldsymbol{K}\boldsymbol{V}\boldsymbol{q}_\mathrm{r}(t) &= \boldsymbol{B}\boldsymbol{F}(t) + \boldsymbol{\varepsilon}(t) \qquad \text{and} \\ \boldsymbol{y}_\mathrm{r}(t) &= \boldsymbol{C}\boldsymbol{V}\boldsymbol{q}_\mathrm{r}(t) \end{aligned} \tag{6}$$

Additionally the Petrov-Galerkin conditions $\boldsymbol{W}^\mathrm{T}\boldsymbol{\varepsilon}(t) = \boldsymbol{0}$ are enforced such that the residuals $\boldsymbol{\varepsilon}(t)$ vanish. Premultiplying (6) with $\boldsymbol{W}^\mathrm{T} \in \mathbb{R}^{n\times N}$ leads to the fully determined system

$$\begin{aligned} \boldsymbol{M}_\mathrm{r}\ddot{\boldsymbol{q}}_\mathrm{r}(t) + \boldsymbol{D}_\mathrm{r}\dot{\boldsymbol{q}}_\mathrm{r}(t) + \boldsymbol{f}_\mathrm{r}(\boldsymbol{q}_\mathrm{r}(t)) &= \boldsymbol{B}_\mathrm{r}\boldsymbol{F}(t) \qquad \text{or} \\ \boldsymbol{M}_\mathrm{r}\ddot{\boldsymbol{q}}_\mathrm{r}(t) + \boldsymbol{D}_\mathrm{r}\dot{\boldsymbol{q}}_\mathrm{r}(t) + \boldsymbol{K}_\mathrm{r}\boldsymbol{q}_\mathrm{r}(t) &= \boldsymbol{B}_\mathrm{r}\boldsymbol{F}(t) \qquad \text{and} \\ \boldsymbol{y}_\mathrm{r}(t) &= \boldsymbol{C}_\mathrm{r}\boldsymbol{q}_\mathrm{r}(t) \end{aligned} \tag{7}$$

where the reduced matrices and operators are given by $\{\boldsymbol{M}, \boldsymbol{D}, \boldsymbol{K}\}_\mathrm{r} = \boldsymbol{W}^\mathrm{T}\{\boldsymbol{M}, \boldsymbol{D}, \boldsymbol{K}\}\,\boldsymbol{V}$, $\boldsymbol{B}_\mathrm{r} = \boldsymbol{W}^\mathrm{T}\boldsymbol{B}$, $\boldsymbol{C}_\mathrm{r} = \boldsymbol{C}\boldsymbol{V}$ and $\boldsymbol{f}_\mathrm{r}(\boldsymbol{q}_\mathrm{r}(t)) = \boldsymbol{W}^\mathrm{T}\boldsymbol{f}(\boldsymbol{V}\boldsymbol{q}_\mathrm{r}(t))$. The initial conditions are $\{\boldsymbol{q}, \dot{\boldsymbol{q}}\}_\mathrm{r}(0) = (\boldsymbol{W}^\mathrm{T}\boldsymbol{V})^{-1}\boldsymbol{W}^\mathrm{T}\left\{\boldsymbol{q}_0, \dot{\boldsymbol{q}}_0\right\}$.

The main task of any projective model order reduction technique reduces to finding suitable reduction bases $\boldsymbol{V}, \boldsymbol{W} \in \mathbb{R}^{N\times n}$ that span appropriate subspaces $\mathcal{V} = \mathrm{cspan}(\boldsymbol{V})$ and $\mathcal{W} = \mathrm{cspan}(\boldsymbol{W})$.

Model order reduction in mechanical engineering typically aims at achieving a good approximation quality, the preservation of certain system properties and numerical efficiency. Depending on the application and the characteristic behavior of the FOM, two categories can be distinguished: Initial condition-state based reduction or input-output based reduction. To keep the second-order structure, so-called structure-preserving model reduction is applied. In order not to violate the principle of virtual work, the reduction should be performed by a (orthogonal) Galerkin projection where $\boldsymbol{W} = \boldsymbol{V}$ instead of a two-sided (oblique) Petrov-Galerkin projection.

Further information about model order reduction for mechanical systems and specific algorithms can be found e.g. in [2].

In the following, Sect. 2 presents specific simulation-free model reduction approaches for mechanical systems with geometric nonlinearities which were addressed during the first phase of the DFG priority program 1897. Section 3 continues on simulation-free model reduction approaches for linear mechanical systems with partial visco-elastic material treatments focused on in the second phase. Furthermore, these methods are extended to work on parameterized systems to make the methods usable for applications such as design studies, optimization or sensitivity analyses.

2 Geometrical Nonlinear Mechanical Systems

Model reduction for geometrical nonlinear mechanical systems requires meeting two challenges. First, a reduction basis must be found that is able to capture nonlinear effects originating from large displacements. Classic reduction methods from linear theory are not suitable for this kind of system. Second, a Galerkin projection with a suitable reduction basis is not sufficient to reduce computation time. The reason is that the nonlinear restoring force term must be evaluated and assembled in the full element domain. Therefore, methods are demanded that are able to accelerate this evaluation. These methods are called Hyperreduction.

2.1 *Simulation-Free Reduction Bases*

In contrast to reduction bases obtained from training data, such as the Proper Orthogonal Decomposition of displacement training sets, simulation-free reduction bases do not require time integration of full order models. They are thus much cheaper to compute.

One idea to gain a reduction basis for geometric nonlinear systems is to use simulation-free bases from techniques for linear systems and augment them with special vectors that are able to capture nonlinear effects. A prominent example are the combination of eigenmodes and modal derivatives. They extend the modal truncation reduction basis, that is known from linear model reduction, with their sensitivities in the direction of the modes themselves.

First, the solutions $\boldsymbol{\phi}_i$ to the eigenproblem

$$(\mathbf{K} - \omega_i^2 \mathbf{M})\boldsymbol{\phi}_i = \mathbf{0} \tag{8}$$

describe the modes of the linearized system. Since $\mathbf{K}$ is a function of the displacements $\mathbf{q}$, the eigenproblem can be derived with respect to them. This leads to modal derivatives. However, experience has shown that neglecting the mass matrix for the computation of these derivatives results in basis vectors that lead to more accurate reduced systems. These derivatives are called static modal derivatives [3] $(\nabla_{\boldsymbol{\phi}_j}\boldsymbol{\phi}_i)$ that are determined by solving

$$\mathbf{K}(\nabla_{\boldsymbol{\phi}_j}\boldsymbol{\phi}_i) = -(\nabla_{\boldsymbol{\phi}_j}\mathbf{K})\boldsymbol{\phi}_i. \tag{9}$$

The final reduction basis

$$\mathbf{V} = \left[\boldsymbol{\phi}_1, \ldots, \boldsymbol{\phi}_M, \nabla_{\boldsymbol{\phi}_1}\boldsymbol{\phi}_1, \nabla_{\boldsymbol{\phi}_2}\boldsymbol{\phi}_1, \ldots, \nabla_{\boldsymbol{\phi}_M}\boldsymbol{\phi}_M\right] \tag{10}$$

is then built by stacking some modes and some static modal derivatives into one reduction basis.

The same idea can be applied to other linear reduction techniques, such as the moment matching technique. The linear reduction basis known from moment matching is computed iteratively via

$$\mathbf{K}\mathbf{v}_1 = -\mathbf{B} \tag{11}$$
$$\mathbf{K}\mathbf{v}_2 = -\mathbf{D}\mathbf{v}_1 \tag{12}$$
$$\mathbf{K}\mathbf{v}_j = -\mathbf{D}\mathbf{v}_{j-1} - \mathbf{M}\mathbf{v}_{j-2}. \tag{13}$$

Afterwards, their derivatives can be computed as follows:

$$\mathbf{K}\left(\nabla_{\mathbf{v}_j}\mathbf{v}_1\right) = -(\nabla_{\mathbf{v}_j}\mathbf{K})\,\mathbf{v}_1 \tag{14}$$
$$\mathbf{K}\left(\nabla_{\mathbf{v}_j}\mathbf{v}_2\right) = -\mathbf{D}\left(\nabla_{\mathbf{v}_j}\mathbf{v}_1\right) - (\nabla_{\mathbf{v}_j}\mathbf{K})\,\mathbf{v}_2 \tag{15}$$
$$\mathbf{K}\left(\nabla_{\mathbf{v}_j}\mathbf{v}_i\right) = -\mathbf{D}\left(\nabla_{\mathbf{v}_j}\mathbf{v}_{i-1}\right) - \mathbf{M}(\nabla_{\mathbf{v}_j}\mathbf{v}_{i-2}) - (\nabla_{\mathbf{v}_j}\mathbf{K})\,\mathbf{v}_j \tag{16}$$

where a constant mass and damping matrix according to Eq. 2 is assumed. A case study evaluating the performance of these bases is shown in [4].

2.2 *Simulation-Free Hyper-Reduction*

2.2.1 Polynomial Expansion

One technique that can be considered as hyperreduction is the polynomial expansion. One can show that the nonlinear force term is a polynomial of third order if the system is set up with constitutive laws that are linear in the Green Lagrange strain measure. The nonlinear force term can then be written as

$$\mathbf{f}_i(\mathbf{q}) = \mathbf{K}^1_{ij}\mathbf{q}_j + \mathbf{K}^2_{ijk}\mathbf{q}_j\mathbf{q}_k + \mathbf{K}^3_{ijkl}\mathbf{q}_j\mathbf{q}_k\mathbf{q}_l. \tag{17}$$

The tensors $\mathbf{K}^1$, $\mathbf{K}^2$ and $\mathbf{K}^3$ are costly to store. But they can be reduced by applying a Galerkin projection with a proper reduction basis $\mathbf{V}$ shown in the previous sections such that

$$\bar{\mathbf{K}}^1_{ij}\mathbf{q}_{\mathrm{r}j} + \bar{\mathbf{K}}^2_{ijk}\mathbf{q}_{\mathrm{r}j}\mathbf{q}_{\mathrm{r}k} + \bar{\mathbf{K}}^3_{ijkl}\mathbf{q}_{\mathrm{r}j}\mathbf{q}_{\mathrm{r}k}\mathbf{q}_{\mathrm{r}l} \tag{18}$$

where

$$\bar{\mathbf{K}}^1_{ij} = (\mathbf{V}^T)_{ik}\mathbf{K}^1_{kl}\mathbf{V}_{lj},\ \bar{\mathbf{K}}^2_{ijk} = (\mathbf{V}^T)_{il}\mathbf{K}^2_{lmn}\mathbf{V}_{mj}\mathbf{V}_{nk},\ \bar{\mathbf{K}}^3_{ijkl} = (\mathbf{V}^T)_{im}\mathbf{K}^3_{mnop}\mathbf{V}_{nj}\mathbf{V}_{ok}\mathbf{V}_{pl} \tag{19}$$

and where we use Einstein's summation convention, i.e., indices that appear twice are summed up.

The polynomial expressions allow a very fast evaluation of the nonlinear restoring force term [5]. However, this representation is only feasible for systems that can be described by a low dimensional reduction basis. If a reduction basis of medium or large size is required, other hyperreduction techniques are more suitable, such as the Energy Conserving Sampling and Weighting method (ECSW).

2.2.2 Energy Conserving Sampling and Weighting

The idea of the Energy Conserving Sampling and Weighting method (ECSW) [6] is to not evaluate all elements during the assembly of the restoring force term. Instead, it evaluates only a subset $\tilde{E} \subset E$ of all elements and interpolates their contribution to the full restoring force term

$$\mathbf{V}^T\mathbf{f}(\mathbf{V}\mathbf{q}_\mathrm{r}) = \sum_{e\in E} \mathbf{V}^T\mathbf{L}_e^T\mathbf{f}_e(\mathbf{L}_e\mathbf{V}_e\mathbf{q}_\mathrm{r}) \approx \sum_{e\in\tilde{E}\subset E} \xi_e\mathbf{V}^T\mathbf{L}_e^T\mathbf{f}_e(\mathbf{L}_e\mathbf{V}_e\mathbf{q}_\mathrm{r}) \tag{20}$$

where $\mathbf{L}_e$ describes Boolean localization matrices to map the local elemental degrees of freedom to the global degrees of freedom.

The interpolation is achieved through positive weights ξ_e. These weights and the element set $\tilde{E}$ are chosen by requesting that the virtual work of the restoring force in the direction of all reduction basis vectors is retained in the hyperreduced model for some training sets $\mathbf{q}_{r,\tau}$. This requirement can approximately be formulated as the minimization problem

$$\arg\min_{\xi\in\Phi} \|\mathbf{xi}\|_0 \quad \text{where} \quad \Phi = \{\mathbf{xi} \in \mathbb{R}^{N_e} : \|\mathbf{G}\mathbf{xi} - \mathbf{b}\| \leq \varepsilon_{ECSW}\|\mathbf{b}\| \text{ and } \xi_e \geq 0\} \tag{21}$$

where

$$\mathbf{G} = \begin{bmatrix} \mathbf{g}_{11} & \cdots & \mathbf{g}_{1N_e} \\ \vdots & \ddots & \vdots \\ \mathbf{g}_{N_\tau 1} & \cdots & \mathbf{g}_{N_\tau N_e} \end{bmatrix} \in \mathbb{R}^{nN_\tau \times N_e} \quad \text{and} \quad \mathbf{b} = \begin{bmatrix} \mathbf{b}_1 \\ \vdots \\ \mathbf{b}_{N_\tau} \end{bmatrix} \in \mathbb{R}^{nN_\tau} \tag{22}$$

are built up by the entries

$$\mathbf{g}_{le}(\mathbf{q}_{\mathrm{r},\tau_l}) = \mathbf{V}^T\mathbf{L}_e^T\mathbf{f}_e(\mathbf{L}_e\mathbf{V}\mathbf{q}_{\mathrm{r},\tau_l}) \in \mathbb{R}^n \quad \text{and} \quad \mathbf{b}_l = \mathbf{f}_r(\mathbf{q}_{\mathrm{r},\tau_l}) = \sum_{e=1}^{N_e} \mathbf{g}_{le}(\mathbf{q}_{\mathrm{r},\tau_l}) \,. \tag{23}$$

Here, N_e and N_τ describe the number of elements of the full order model and the number of training sets, respectively.

The training sets $\mathbf{q}_{\tau_l} = \mathbf{V}\mathbf{q}_{\mathrm{r},\tau_l}$ can be gained by obtaining displacement vectors of a full order solution or by the so called Nonlinear Stochastic Krylov Training Sets method.

2.2.3 Nonlinear Stochastic Krylov Training Sets

The idea of these training sets [7] is to build a subspace

$$\mathbf{F}_{kry} = \operatorname{span}\{\mathbf{B}, \mathbf{MK}^{-1}\mathbf{B}, (\mathbf{MK}^{-1})^2\mathbf{B}, \ldots\} = \mathcal{K}(\mathbf{MK}^{-1}, \mathbf{B}) \tag{24}$$

that is able to approximate the nonlinear restoring force $\mathbf{f}$. Afterwards, some random vectors $\mathbf{f}^{\tau}_{NSKTS} \in \mathcal{K}(\mathbf{MK}^{-1}, \mathbf{B})$ are generated and the nonlinear static problems

$$\mathbf{f}(\mathbf{q}^{(k)}_{\tau}) = k\mathbf{f}^{\tau}_{NSKTS}, \quad k \in (0, 1] \tag{25}$$

are solved. The solutions $\mathbf{q}^{(k)}_{\tau}$ are then used as training sets for the ECSW. This procedure avoids costly time integration of the high dimensional model. However, some nonlinear static equations of full dimension must be solved.

2.3 *Extension to Parametric Bases*

2.3.1 Parameterization of Finite Element Models

Design studies or sensitivity analyses require a parameterization of the Finite element model. The parameterization of material data and boundary conditions is quite easy to achieve. But shape parameterization of the mesh is much harder. Classic approaches for parameter studies create a new mesh for each iteration. This is disadvantageous for our applications for two reasons: First, a mesh generation can take large amounts of computation time. Second, the mesh topology can change, which makes already computed reduction bases useless unless one applies special mapping techniques.

Therefore, we use a mesh parameterization approach that avoids both drawbacks: Mesh morphing. The mesh morphing approach just modifies the coordinates of the nodes in the mesh while maintaining the mesh connectivity.

We use Radial Basis functions, more precisely, thin plate splines to move interior nodes and maintain mesh quality. One example is the beam shown in Fig. 1 where a notch's position is parameterized.

After parameterization, the equations of motion (2) become parametric and can be written as

$$\boldsymbol{M}(\boldsymbol{p})\ddot{\boldsymbol{q}}(t) + \boldsymbol{D}(\boldsymbol{p})\dot{\boldsymbol{q}}(t) + \boldsymbol{f}(\boldsymbol{q}, \boldsymbol{p})(t) = \boldsymbol{B}(\boldsymbol{p})\boldsymbol{F}(\boldsymbol{p})(t), \qquad \boldsymbol{q}(0) = \boldsymbol{q}_0, \ \dot{\boldsymbol{q}}(0) = \dot{\boldsymbol{q}}_0. \tag{26}$$

Now, simulation-free reduction basis also depend on the parameters $\boldsymbol{p}$. This implies that a computed reduction basis for a certain parameter value may not be valid for other parameter values as well. Some methods to overcome this burden are summarized in the following sections.

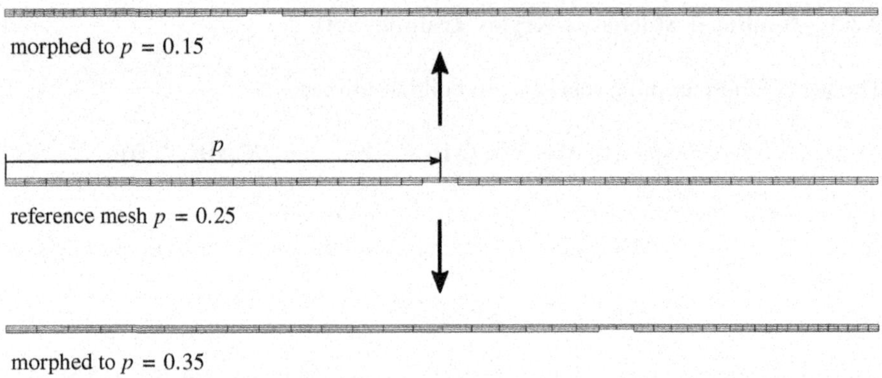

Fig. 1 Mesh morphing for a cantilever beam. The beam has a notch on the bottom side. Middle: The reference mesh. Top: The notch is translated to the left. Bottom: The notch has been morphed to the right. The translation is done via mesh morphing such that only nodal coordinates change while the mesh topology is maintained

2.3.2 Basis Updating

The most simple parametric model reduction technique is to compute a new simulation-free reduction basis for each new parameter value. But since the reduction basis information will not be completely different, information from previous computations can be used to gain new reduction bases with less effort. This procedure is called basis updating. In the typical case where the reduction basis consists of modes and static modal derivatives, this can be done in two steps. First, an inverse free preconditioned Krylov subspace method (IFPKS) is used to update the modal part of the reduction basis. Second, an iterative solver such as the preconditioned conjugate gradient method can be used to gain the static derivative part of the reduction basis. This procedure is described in detail in [8].

2.3.3 Global Reduction Basis Through Sampling

Another idea is to sample the parameter space

$$P_{\text{sample}} = \{\mathbf{p}_1, \mathbf{p}_2, \mathbf{p}_3, \ldots, \mathbf{p}_N\} \tag{27}$$

and compute a simulation-free reduction basis for each sample point

$$\mathbf{V}_{\text{sample}} = \{\mathbf{V}(\mathbf{p}_1), \mathbf{V}(\mathbf{p}_2), \ldots, \mathbf{V}(\mathbf{p}_N)\}. \tag{28}$$

Afterwards, all sampled reduction bases are stacked into one reduction basis. The reduction basis is then deflated such that nearly parallel vectors are removed from the reduction basis. The advantage is that a global reduction basis can be found very easily. But if the reduction basis space changes drastically in the parameter space of interest, the reduction basis can be of high dimension and many sample points could be necessary, which makes other approaches more suitable [9].

2.3.4 Augmentation with Parametric Sensitivities

If only a small parameter space around a certain parameter is of interest, an augmentation by parametric sensitivities is an option [10]. This idea is similar to the ideas of simulation-free reduction bases for geometric nonlinear systems. The basis is built up by stacking the simulation-free reduction basis at a certain point and their parametric derivatives into one basis:

$$\mathbf{V} = \left[\mathbf{v}_1(\mathbf{p}_0), \mathbf{v}_2(\mathbf{p}_0), \ldots, \mathbf{v}_N(\mathbf{p}_0), \nabla_{\mathbf{e}_1}\mathbf{v}_1(\mathbf{p}_0), \nabla_{\mathbf{e}_2}\mathbf{v}_1(\mathbf{p}_0), \ldots \nabla_{\mathbf{e}_P}\mathbf{v}_N(\mathbf{p}_0)\right] \tag{29}$$

where $\mathbf{e}_i$ is the i-th unit vector in the Euclidean space.

2.3.5 Interpolation on Manifolds

The space of reduction bases can be seen as a manifold which enables interpolations between different bases. One option to define this manifold is the Grassmann-manifold. If two reduction bases $\mathbf{V}_1$ and $\mathbf{V}_2$ at sample points p_1 and p_2 have been computed and a reduced model at sample point $\hat{p}$, that is between these two points, is demanded, one can interpolate between the computed reduction bases. One has to perform two steps for this interpolation: First, a singular value decomposition of

$$\left(\mathbf{I}_N - \mathbf{V}_1\mathbf{V}_1^T\right)\mathbf{V}_2\left(\mathbf{V}_1^T\mathbf{V}_2\right)^{-1} = \mathbf{U}\boldsymbol{\Sigma}\mathbf{W}^T \tag{30}$$

is computed. Afterwards, the interpolated reduction basis $\hat{\mathbf{V}}$ at $\hat{p}$ is determined by

$$\hat{\mathbf{V}}(\hat{p}) = \mathbf{V}_1\mathbf{W}\cos\left[\left(\frac{\hat{p}-p_1}{p_2-p_1}\right)\tan^{-1}(\boldsymbol{\Sigma})\right] + \mathbf{U}\sin\left[\left(\frac{\hat{p}-p_1}{p_2-p_1}\right)\tan^{-1}(\boldsymbol{\Sigma})\right]. \tag{31}$$

This method works very well even for systems whose modes show high parametric dependencies [11].

2.4 Parametric Hyper-Reduction

Similar to the global reduction basis approach, the hyperreduction problem can also be globalized [12]. First, we compute $\mathbf{q}_\tau(\mathbf{p}_s)$ (using NSKTS) for each sample point $\mathbf{p}_s \in P_{\text{sample}}$. Then, the quantities

$$\mathbf{G}_{\text{global}} = \begin{bmatrix} \tilde{\mathbf{g}}_{11} & \cdots & \tilde{\mathbf{g}}_{1N_e} \\ \vdots & \ddots & \vdots \\ \tilde{\mathbf{g}}_{N_S 1} & \cdots & \tilde{\mathbf{g}}_{N_S N_e} \end{bmatrix} \in \mathbb{R}^{k \cdot \beta \cdot N_S \times N_e} \quad \text{and} \quad \mathbf{b}_{\text{global}} = \begin{bmatrix} \tilde{\mathbf{b}}_1 \\ \vdots \\ \tilde{\mathbf{b}}_{N_S} \end{bmatrix} \in \mathbb{R}^{k \cdot \beta \cdot N_S} \tag{32}$$

are built up by the entries

$$\tilde{\mathbf{g}}_{se} = \begin{bmatrix} \mathbf{V}^T(\mathbf{p}_s) \cdot \mathbf{L}_e^T \mathbf{f}_e(\mathbf{p}_s, \mathbf{L}_e \mathbf{V}(\mathbf{p}_s)\, \mathbf{q}_{\tau_{1,1}}(\mathbf{p}_s)) \\ \vdots \\ \mathbf{V}^T(\mathbf{p}_s) \cdot \mathbf{L}_e^T \mathbf{f}_e(\mathbf{p}_s, \mathbf{L}_e \mathbf{V}(\mathbf{p}_s)\, \mathbf{q}_{\tau_{1,k}}(\mathbf{p}_s)) \\ \mathbf{V}^T(\mathbf{p}_s) \cdot \mathbf{L}_e^T \mathbf{f}_e(\mathbf{p}_s, \mathbf{L}_e \mathbf{V}(\mathbf{p}_s)\, \mathbf{q}_{\tau_{2,1}}(\mathbf{p}_s)) \\ \vdots \\ \vdots \\ \mathbf{V}^T(\mathbf{p}_s) \cdot \mathbf{L}_e^T \mathbf{f}_e(\mathbf{p}_s, \mathbf{L}_e \mathbf{V}(\mathbf{p}_s)\, \mathbf{q}_{\tau_{\beta,k}}(\mathbf{p}_s)) \end{bmatrix} \in \mathbb{R}^{k \cdot \beta \cdot N_S} \tag{33}$$

$$\tilde{\mathbf{b}}_s = \sum_{e=1}^{N_e} \tilde{\mathbf{g}}_{se} \tag{34}$$

where $N_S = |P_{\text{sample}}|$ is the number of sample points, β is the number of NSKTS vectors per sample, k is the number of load increments (Eq. 25) that are computed for each NSKTS vector and $\mathbf{V}(\mathbf{p}_s)$ are local reduction bases that are computed at sampling point $\mathbf{p}_s$. The matrix $\mathbf{G}_{\text{global}}$ and the vector $\mathbf{b}_{\text{global}}$ are used to compute a global set of weights and elements for the ECSW as described by Eq. (21).

2.5 *Case Study*

We want to show a simple case study showing the performance of the most promising approaches. A notched cantilever beam with hexagonal elements is set up as shown in Fig. 1. The beam is parameterized with the position of the notch. The left end is fixed and an harmonic excitation force is applied at the tip. Since the beam is very slender, a highly nonlinear behavior is expected for sufficiently large excitation forces.

Two approaches are compared to find a suitable reduction basis for $p = 0.25$. First, the interpolation method is tested by interpolating between two reduction bases at $p = 0.24$ and $p = 0.26$. Second, the parametric sensitivity approach is tested by computing a reduction basis at $p = 0.24$ and its parametric sensitivities. The tip displacement errors for the different bases are depicted in Fig. 2. One can conclude that both approaches lead to a good reduced model. The parametric sensitivity approach performs best and gives almost the accuracy of a model with direct computation of a reduction basis at $p = 0.25$.

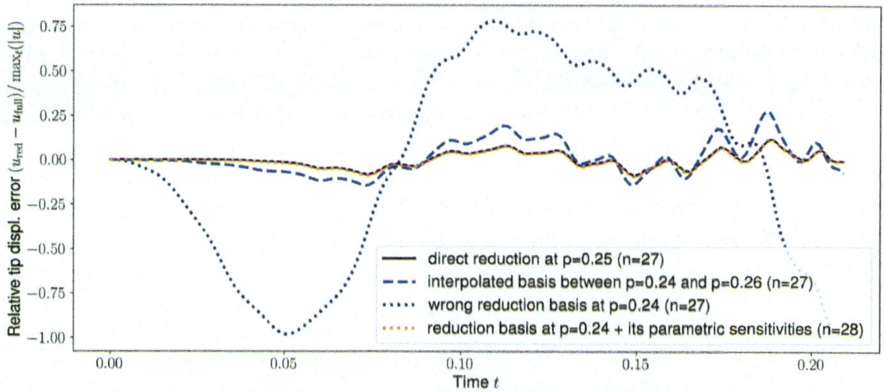

Fig. 2 Tip displacement error for a model at $p = 0.25$ over time for different reduction bases. The parametric sensitivity approach at $p = 0.24$ (4 modes + static modal derivatives + parametric sensitivities = 28 basis vectors) gives almost the same accuracy as the directly computed reduction basis for $p = 0.25$ (6 modes + static modal derivatives = 27 basis vectors). The interpolation of the basis between $p = 0.24$ and $p = 0.26$ (6 modes + static modal derivatives = 27 basis vectors) also gives good accuracy. The reduction basis at $p = 0.24$ (6 modes + static modal derivatives) is not suitable for reduction

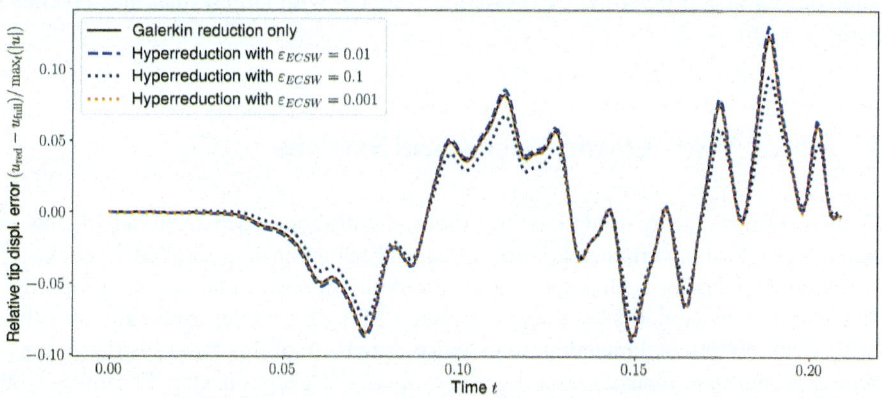

Fig. 3 Tip displacement errors of hyperreduced models compared to a model that is reduced without hyperreduction. The smallest tolerance leads to an accuracy compared to the non-hyperreduced model. A very broad tolerance gives best performance due to a softening effect by using too few integration points

Furthermore, a hyperreduction with the ECSW and NSKTS is conducted with the reduction that performed best, namely the parametric sensitivity approach. Figure 3 shows the tip displacement error for the hyperreduced model with different tolerances for the ECSW. One can see that a very tight tolerance leads to a model that is as accurate as a reduced order model without hyperreduction. However, a very high tolerance leads to a model that is more accurate compared to the non-reduced solution.

Table 1 Number of evaluated elements, simulation times for reduced order models and speedup factor compared to full order model. Row one shows the full order model and rows 2–4 shows hyperreduced models with different tolerances for ECSW weight generation according to Fig. 3

Type	ε_{ECSW}	no. of elements $\lvert\tilde{E}\rvert$	sim. time online	speedup
Full order model	–	248	3155	1.0
ECSW hyperreduced model	0.1	85	1581	2.0
ECSW hyperreduced model	0.01	114	1762	1.8
ECSW hyperreduced model	0.001	142	1871	1.7

This probably originates from a softening effect because too few integration points (i.e. selected finite elements) are used.

Table 1 lists simulation times in seconds, measured with Python's function `process_time()`, and speedup factors for the hyperreduced models compared to the full order model. The simulations are conducted on a machine with Intel Xeon CPU E3-1270 v5 (3.6 GHz) with 32 GB RAM. The table also contains the number of elements that are evaluated for the nonlinear restoring force vector. For this small academical problem, the hyperreduced model with largest tolerance reaches a speedup factor of 2.0.

3 Linear Visco-Elastic Mechanical Systems

Many materials that are used to damp structural vibrations, such as rubber-like layers placed on plate-like structures, show a material behavior that is called viscoelastic. Viscoelastic behavior is characterized by a mixture of elastic and viscous properties. These are often modeled by the Generalized Maxwell model especially if a time domain simulation is demanded. One major drawback of this model is that it introduces internal state variables that must be evaluated in each timestep. For this reason, the computational effort is drastically increased for such models and, thus, model reduction is highly desired.

3.1 Modeling Aspects

3.1.1 Generalized Maxwell Model

Figure 4 illustrates the Generalized Maxwell model that is used to model viscoelastic materials in time domain. It consists of several Maxwell elements built from linear elastic springs and viscous dashpots.

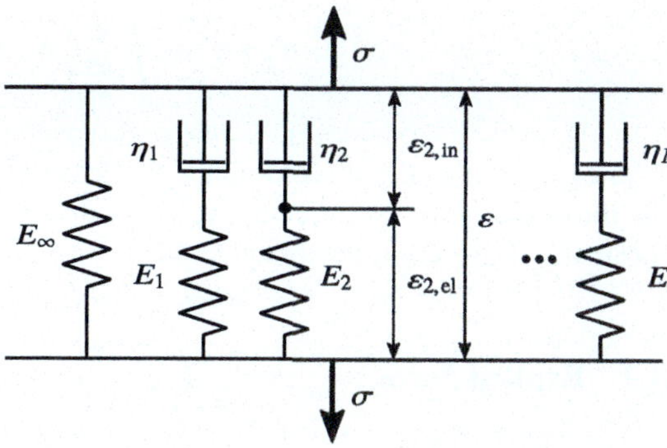

Fig. 4 Generalized Maxwell Model consisting of M Maxwell elements. Each Maxwell element is composed of a spring and a dashpot. A single spring with stiffness E_∞ is added to model long time elastic behavior

The constitutive equations

$$\varepsilon_{m,\text{el}} = \frac{\sigma_m}{E_m}, \qquad \dot{\varepsilon}_{m,\text{in}} = \frac{\sigma_m}{\eta_m} \tag{35}$$

for the dashpot and the spring and the kinematic relation

$$\varepsilon = \varepsilon_{m,\text{el}} + \varepsilon_{m,\text{in}} \tag{36}$$

leads to the equation

$$\sigma_m = E_m \varepsilon_{m,\text{el}} = \eta_m \dot{\varepsilon}_{m,\text{in}} \quad \Leftrightarrow \quad E_m(\varepsilon - \alpha) = \eta_m \dot{\alpha} \tag{37}$$

where $\varepsilon_{m,\text{in}}$ has been replaced by α. We call α an internal variable. This leads to two equations for the constitutive law:

$$\frac{\eta_m}{E_m}\dot{\alpha} + \alpha = \varepsilon\,, \qquad \sigma_m = E_m(\varepsilon - \alpha) \tag{38}$$

When a step load in strain with amplitude ε_0 is applied to a Maxwell element, its response is

$$\sigma_m(t) = \varepsilon_0 E_m \,\mathrm{e}^{-\frac{t}{\theta_m}} \tag{39}$$

where we use the definition $\theta_m := \frac{\eta_m}{E_m}$. The full response of all Maxwell elements is the sum

$$\sigma(t) = \sigma_\infty + \sum_{m=1}^{M} \sigma_m(t) = \varepsilon_0 \left(E_\infty + \sum_{m=1}^{M} E_m \,\mathrm{e}^{-\frac{t}{\theta_m}} \right) \tag{40}$$

Therefore, the constitutive equation can also be expressed by a Duhamel integral

$$\sigma(t) = E_\infty \varepsilon(t) + \int_0^t \sum_{m=1}^{M} E_m \, \mathrm{e}^{-\frac{t-s}{\theta_m}} \, \dot{\varepsilon}(s) \, \mathrm{d}s \tag{41}$$

An extension to the three dimensional case is easy to achieve by introducing a split into volumetric and deviatoric parts and using an internal variable for each coordinate of the strain tensors [13, 14].

3.1.2 Explicit Form

The explicit state form contains the internal variables in its system state vector $\mathbf{x} = [\mathbf{u}, \boldsymbol{\alpha}]$. This allows to apply model reduction techniques for linear systems because this representation leads to a system

$$\begin{bmatrix} \mathbf{M}_{uu} & \mathbf{0} \\ \mathbf{0} & \mathbf{0} \end{bmatrix} \begin{bmatrix} \ddot{\mathbf{u}} \\ \ddot{\boldsymbol{\alpha}} \end{bmatrix} + \begin{bmatrix} \mathbf{D}_{uu} & \mathbf{0} \\ \mathbf{0} & \mathbf{D}_{\alpha\alpha} \end{bmatrix} \begin{bmatrix} \dot{\mathbf{u}} \\ \dot{\boldsymbol{\alpha}} \end{bmatrix} + \begin{bmatrix} \mathbf{K}_{uu} & \mathbf{K}_{u\alpha} \\ \mathbf{K}_{\alpha u} & \mathbf{K}_{\alpha\alpha} \end{bmatrix} \begin{bmatrix} \mathbf{u} \\ \boldsymbol{\alpha} \end{bmatrix} = \begin{bmatrix} \mathbf{F} \\ \mathbf{0} \end{bmatrix} \tag{42}$$

that is linear.

The matrices $\mathbf{M}_{uu}$, $\mathbf{D}_{uu}$, $\mathbf{K}_{uu}$ are similar to those from classic Finite Element models containing linear materials. The matrices $\mathbf{D}_{\alpha\alpha}$ and $\mathbf{K}_{\alpha\alpha}$ are fully diagonal. The coupling between the internal states $\boldsymbol{\alpha}$ and the displacements $\mathbf{u}$ happens in the stiffness matrix through the blocks $\mathbf{K}_{u\alpha}$ and $\mathbf{K}_{\alpha u}$.

3.2 Model Reduction via Decoupling into Subsystems

In a first naiv approach, classical model order reduction approaches, like moment matching or modal truncation, are applied the fully coupled system (42). With this approach different physics, displacements $\boldsymbol{u}$ and partial stresses (internal variables) $\boldsymbol{\alpha}$ are mixed in the reduced coordinates $\boldsymbol{x}_\mathrm{r}$, thereby losing their physical meaning and limiting the reduction process.

To avoid those limitations, displacement variables and internal partial stress variables are treated separately. For this, the visco-elastic structural system (42) $\mathcal{S}$ is treated as a coupled respectively closed-loop system: The purely elastic structural subsystem $\mathcal{S}_1$ with N_u degrees of freedom $\boldsymbol{u}$ is coupled to the viscous subsystem $\mathcal{S}_2$ with N_α degrees of freedom $\boldsymbol{\alpha}$ via the interface equations $\mathcal{I}$

$$
\mathcal{S}: \begin{cases}
\mathcal{S}_1: \begin{cases} \boldsymbol{M}_{uu}\ddot{\boldsymbol{u}}(t) + \boldsymbol{D}_{uu}\dot{\boldsymbol{u}}(t) + \boldsymbol{K}_{uu}\boldsymbol{u}(t) = -\boldsymbol{K}_{u\alpha}\boldsymbol{u}_{12}(t) + \boldsymbol{B}\boldsymbol{F}(t) \\ \boldsymbol{y}_1(t) = \boldsymbol{C}_1\boldsymbol{u}(t) \\ \boldsymbol{y}_{12}(t) = \mathbf{I}\boldsymbol{u}(t) \end{cases} \\
\mathcal{S}_2: \begin{cases} \boldsymbol{M}_{\alpha\alpha}\ddot{\boldsymbol{\alpha}}(t) + \boldsymbol{D}_{\alpha\alpha}\dot{\boldsymbol{\alpha}}(t) + \boldsymbol{K}_{\alpha\alpha}\boldsymbol{\alpha}(t) = -\boldsymbol{K}_{u\alpha}^T\boldsymbol{u}_{21}(t) \\ \boldsymbol{y}_2(t) = \boldsymbol{C}_2\boldsymbol{\alpha}(t) \\ \boldsymbol{y}_{21}(t) = \mathbf{I}\boldsymbol{\alpha}(t) \end{cases} \\
\mathcal{I}: \begin{cases} \boldsymbol{u}_{12}(t) - \boldsymbol{y}_{21}(t) = \boldsymbol{0} \\ \boldsymbol{u}_{21}(t) - \boldsymbol{y}_{12}(t) = \boldsymbol{0} \end{cases}
\end{cases}
$$

With regard to model reduction, this way both systems can be treated separately. Thus, for each subsystem, well established first- or second-order methods like first-/second-order moment matching, modal truncation with complex eigenvectors, second-order balanced truncation and others can be applied. The reduced and re-coupled system will have the dimension $n = n_1 + n_2$. Additionally, the interface equations are also reduced, i.e. (internal) in- and outputs are reduced.

3.2.1 Second-Order Moment Matching

Transfer function (5) of the full and reduced system are represented by Taylor series around the shift $s_0 \in \mathbf{C}$:

$$
\boldsymbol{G}(s) = \boldsymbol{C}(s^2\boldsymbol{M} + s\boldsymbol{D} + \boldsymbol{K})^{-1}\boldsymbol{B} = \sum_{i=0}^{\infty} \boldsymbol{M}_{s_0,i}(s - s_0)^i \tag{43}
$$

$$
\boldsymbol{G}_{\mathrm{r}}(s) = \boldsymbol{C}_{\mathrm{r}}(s^2\boldsymbol{M}_{\mathrm{r}} + s\boldsymbol{D}_{\mathrm{r}} + \boldsymbol{K}_{\mathrm{r}})^{-1}\boldsymbol{B}_{\mathrm{r}} = \sum_{i=0}^{\infty} \boldsymbol{M}_{\mathrm{r},s_0,i}(s - s_0)^i \tag{44}
$$

where $\boldsymbol{M}_{s_0,i}$ and $\boldsymbol{M}_{\mathrm{r},s_0,i}$ $\forall i = 0, \dots, \infty$ are called the moments of the full and reduced system, respectively.

The basic ansatz consists in making a specified amount of moments match around s_0:

$$
\boldsymbol{M}_{\mathrm{r},s_0,i} \overset{!}{=} \boldsymbol{M}_{s_0,i} \qquad \forall i = 0, \dots, q_0 \tag{45}
$$

This can implicitly and numerically efficiently be achieved by using second-order Krylov subspaces as reduction bases [15]. The necessary Krylov subspace is defined as

$$
\mathcal{K}_n(\boldsymbol{M}_1, \boldsymbol{M}_2, \boldsymbol{V}) = \mathrm{colspan}\,\{\boldsymbol{P}_0, \boldsymbol{P}_1, \dots, \boldsymbol{P}_{n-1}\} \tag{46}
$$

where $\boldsymbol{P}_0 = \boldsymbol{V}$, $\boldsymbol{P}_1 = \boldsymbol{M}_1\boldsymbol{V}$ and $\boldsymbol{P}_i = \boldsymbol{M}_1\boldsymbol{P}_{i-1} + \boldsymbol{M}_2\boldsymbol{P}_{i-2}$.

In general with one-sided moment matching q_0 moments can be matched if $\mathbf{V}$ is chosen such that it includes the input Krylov subspace:

$$\mathcal{K}_{q_0}(\boldsymbol{K}_{s_0}^{-1}\boldsymbol{D}_{s_0}, \boldsymbol{K}_{s_0}^{-1}\boldsymbol{M}, \boldsymbol{K}_{s_0}^{-1}\boldsymbol{B}) \subseteq \text{colspan}\, \boldsymbol{V} \tag{47}$$

where $\boldsymbol{K}_{s_0} = s_0^2\boldsymbol{M} + s_0\boldsymbol{D} + \boldsymbol{K}$ and $\boldsymbol{D}_{s_0} = 2s_0\boldsymbol{M} + \boldsymbol{D}$.

To achieve moment matching around different shifts $(s_0, q_0), (s_1, q_1), \ldots$ the appropriate Krylov subspaces simply need to be augmented.

3.2.2 Reduction of Coupling Blocks

The coupling block $\boldsymbol{K}_{u\alpha}$ and $\boldsymbol{K}_{\alpha u}$ are reduced via singular value decomposition and only considering the dominant singular values:

$$\boldsymbol{K}_{ab} = \boldsymbol{U}_{ab}\boldsymbol{\Sigma}_{ab}\boldsymbol{V}_{ab}^{\mathrm{T}} \approx \sum_{i=0}^{n_{ab}} \boldsymbol{U}_{ab}[:, i]\boldsymbol{\Sigma}_{ab}[i, i]\boldsymbol{V}_{ab}^{\mathrm{T}}[i, :] \tag{48}$$

Thus, internal in- and output matrices for both subsystems can be defined

$$\boldsymbol{B}_{ab} = \boldsymbol{U}_{ab}[:, : n_{ab}]\sqrt{\boldsymbol{\Sigma}_{ab}}[: n_{ab}, : n_{ab}] \tag{49}$$

$$\boldsymbol{C}_{ba} = \sqrt{\boldsymbol{\Sigma}_{ab}}[: n_{ab}, : n_{ab}]\boldsymbol{V}_{ab}^{\mathrm{T}}[: n_{ab}, :] \tag{50}$$

and included in the moment matching process for each subsystem guaranteeing moment matching for the fully re-coupled system [16].

3.3 Schur Complement

The diagonality of $\mathbf{D}_{\alpha\alpha}$ and $\mathbf{K}_{\alpha\alpha}$ can be exploited to condense the internal states in the equations of motion (42). The equations of motion in frequency domain

$$\left(\begin{bmatrix}\mathbf{K}_{uu} & \mathbf{K}_{u\alpha}\\ \mathbf{K}_{\alpha u} & \mathbf{K}_{\alpha\alpha}\end{bmatrix} + \mathrm{i}\Omega\begin{bmatrix}\mathbf{D}_{uu} & \mathbf{0}\\ \mathbf{0} & \mathbf{D}_{\alpha\alpha}\end{bmatrix} - \Omega^2\begin{bmatrix}\mathbf{M}_{uu} & \mathbf{0}\\ \mathbf{0} & \mathbf{0}\end{bmatrix}\right)\begin{bmatrix}\mathbf{U}\\ \mathbf{A}\end{bmatrix} = \begin{bmatrix}\mathbf{F}\\ \mathbf{0}\end{bmatrix} \tag{51}$$

consist of two blocks of rows. The second block can be transformed

$$(\mathbf{K}_{\alpha\alpha} + \mathrm{i}\Omega\mathbf{D}_{\alpha\alpha})\mathbf{A} = -\mathbf{K}_{\alpha u}\mathbf{U} \quad \Rightarrow \quad \mathbf{A} = -(\mathbf{K}_{\alpha\alpha} + \mathrm{i}\Omega\mathbf{D}_{\alpha\alpha})^{-1}\mathbf{K}_{\alpha u}\mathbf{U} \tag{52}$$

such that the states $\mathbf{A}$ can be inserted into the first block. This results in

$$(\underbrace{\mathbf{K}_{uu} + \mathrm{i}\Omega\mathbf{D}_{uu} - \mathbf{K}_{u\alpha}(\mathbf{K}_{\alpha\alpha} + \mathrm{i}\Omega\mathbf{D}_{\alpha\alpha})^{-1}\mathbf{K}_{\alpha u}}_{\underline{\mathbf{K}}_{\text{schur}}(\mathrm{i}\Omega)} - \Omega^2\mathbf{M}_{uu})\mathbf{U} = \mathbf{F}. \tag{53}$$

This procedure is a Schur complement of the dynamics stiffness matrix. It can be computed very cheaply because $(\mathbf{K}_{\alpha\alpha} + \mathrm{i}\Omega\mathbf{D}_{\alpha\alpha})$ is diagonal and its inverse is very cheap to evaluate. The same procedure can also be done in time domain if a certain time integration scheme is applied but we stick to the frequency domain for simplicity.

The Schur complement is an exact procedure, i.e. it produces no procedural error. It reduces the number of degrees of freedom to the number of displacement degrees of freedom. This can be a large reduction, especially if large regions of the model are considered viscoelastic or if viscoelastic materials have many Maxwell elements.

3.3.1 Modal Reduction

The Schur complement is a good starting point to reduce the degrees of freedom of the equations of motion, but we want to go further. One idea is to apply a modal reduction after the application of the Schur complement. One can use the eigenvectors of the eigenvalue problem

$$(\mathbf{K}_{uu} - \omega_i^2 \mathbf{M}_{uu})\phi_i^0 = \mathbf{0} \tag{54}$$

and stack these modes into a reduction basis such that

$$\mathbf{V} = [\phi_1, \phi_2, \ldots, \phi_n] \tag{55}$$

The reduced system can then be expressed as

$$(\underbrace{\mathbf{V}^T \mathbf{K}_{\text{schur}}(\mathrm{i}\Omega)\mathbf{V}}_{\underline{\mathbf{K}}_{\text{red}}(\mathrm{i}\Omega)} - \Omega^2 \mathbf{V}^T \mathbf{M}_{uu}\mathbf{V}) \cdot \mathbf{Q} = \mathbf{V}^T \mathbf{F}, \qquad \mathbf{U} = \mathbf{V}\mathbf{Q}. \tag{56}$$

However, we will see that these modes are not a good choice for the reduction. The reason is that these modes are the modes of an elastic system where all dampers of Maxwell elements are blocked ($\boldsymbol{\alpha} = \mathbf{0}$). This results in a system that is stiffer and leads too high eigenfrequencies in the reduced system. A better approach is to use the eigenmodes of the elastic system where no Maxwell element is active, i.e. only the long-time elastic behavior is considered ($\dot{\boldsymbol{\alpha}} = \mathbf{0}$). These modes are computed by solving the eigenproblem

$$(\underline{\mathbf{K}}_{\text{schur}}(i0) - \omega_i^2 \mathbf{M}_{uu})\boldsymbol{\phi}_i^\infty = (\mathbf{K}_{uu} - \mathbf{K}_{u\alpha}(\mathbf{K}_{\alpha\alpha})^{-1}\mathbf{K}_{\alpha u} - \omega_i^2 \mathbf{M}_{uu})\boldsymbol{\phi}_i^\infty = \mathbf{0}. \tag{57}$$

An extension to this idea is the augmentation by the static response to loads that are generated by the imaginary part of the stiffness matrix when the structure is deformed according to a mode [17, 18]. The reduction basis then reads

$$\mathbf{V} = \left[\phi_1^\infty, \phi_2^\infty, \ldots, \phi_n^\infty, \mathbf{K}_\infty^{-1}\Im(\mathbf{K}_{\text{schur}}(\mathrm{i}\omega_1^\infty))\boldsymbol{\phi}_1^\infty, \mathbf{K}_\infty^{-1}\Im(\mathbf{K}_{\text{schur}}(\mathrm{i}\omega_2^\infty))\boldsymbol{\phi}_2^\infty \ldots\right] \tag{58}$$

with $\mathbf{K}_\infty = \mathbf{K}_{\text{schur}}(\mathrm{i}0)$.

3.4 Numerical Example: Plate with Acoustic Black Hole and Visco-Elastic Constrained Layer Damping

The proposed methods are illustrated by a model of an aluminium plate that contains a so called acoustic black hole (ABH). ABHs are regions where the plate thickness is decreased with a special shape function. The theory claims that bending waves traveling through the plate decrease their travel velocity when the plate thickness is decreased. Therefore, the waves stay longer in regions with decreased thickness and can damped more effectively in these regions.

The proposed model, that is taken from [19], is depicted in Fig. 5. It consists of an aluminium plate with a circular ABH and a constrained layer damper treatment placed in the ABH region. The constrained layer damper treatment consists of a viscoelastic rubber-like layer and a constrained layer made of CFRP. The Finite Element model has 14,769 displacement degrees of freedom $\mathbf{u}$ and 14,280 internal states $\boldsymbol{\alpha}$.

Figure 6 shows a part of the frequency response function evaluated at a point on the plate for different reduction methods. When modes ϕ^0 are used for reduction, one can see that the solution is similar to a full solution where viscoelasticity is neglected, i.e. the system $\{\mathbf{M}_{uu}, \mathbf{D}_{uu}, \mathbf{K}_{uu}\}$. The reason is that the modes are computed for a very stiff constrained layer damper (all dashpots of Maxwell elements are blocked) and they do not activate the viscoelastic layer. A modal reduction with the modes ϕ^∞ gives better results. But modes computed with ϕ^∞ and the augmentation according to Eq. 58 gives best results. However, these perform worse in the higher frequency domain if the dimension of the reduction basis is equal for all methods (this is not illustrated in the Figure). Due to the augmentation only half of the modes ϕ^∞ can be used to keep the same dimension. The moment matching approach is not able to approximate the frequency response fir a comparable reduced dimension of 50. However, moment matching can approximate the FRF very well if a dimension of about 300 is used which is also shown in [19].

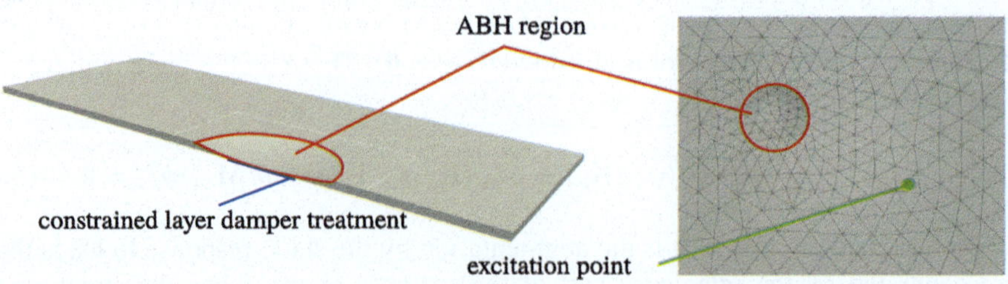

Fig. 5 Free aluminium plate with acoustic black hole (ABH) and constrained layer damper treatment (CLD). The CLD is placed on the backside in the ABH's region. The plate is excited at the marked point

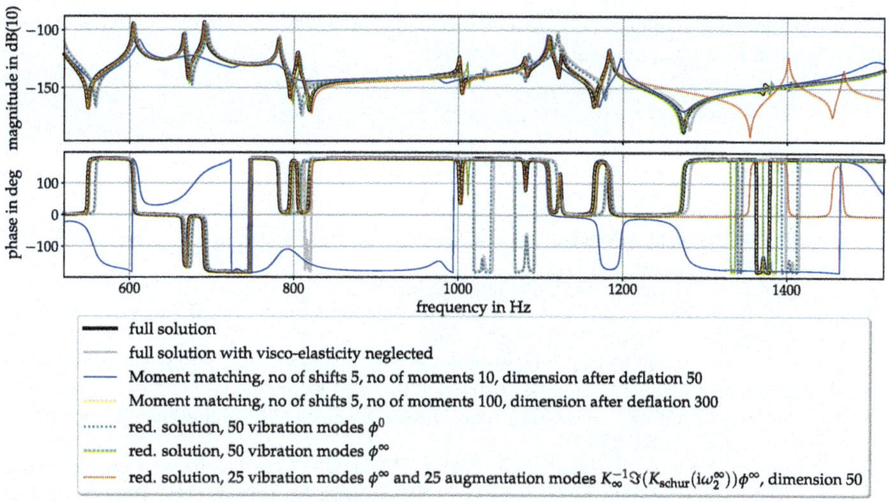

Fig. 6 Frequency response function computed with different reduced order models compared to the full order model. The Schur complement gives exact results. Modal reduction with modes ϕ^0 approximate a full solution where viscoelasticity is neglected. The reduced order model with ϕ^∞ give good results while the augmentation according to Eq. (58) gives best accuracy for reduction bases with dimension 50. The moment matching approach needs a higher number of basis vectors to give a good accuracy

4 Conlusion and Outlook

Accurate simulation models to motivate design decisions in early product development phases is a challenge. Especially models containing viscoelastic materials or undergoing large deformations can lead to high computation times. It is desired to reduced these times to accelerate simulations in the concept phase. A promising method to achieve this goal is model reduction.

We have shown different methods to reduce Finite element models of mechanical structures undergoing large deformations. The challenge here is to find reduction bases that are able to capture nonlinear effects and parametric dependencies. Furthermore a hyperreduction must be applied to accelerate evaluation of the nonlinear restoring force term. A cantilever beam case study illustrates the potential of the methods. Furthermore, we have introduced how viscoelasticity is modeled in Finite Element models. The equations of motion contain many internal states that can be considered as additional system states that can increase the system dimension drastically. Some reduction bases are proposed to reduce these models. A case study on a plate with an acoustic black hole illustrates the performance of the reduction bases. Further research is necessary to also apply the parametric methods from the geometric nonlinear reduced order models to the viscoelastic systems. Hyperreduction methods can also be a potential candidate to reduce the evaluation costs of internal states in viscoelastic models.

Acknowledgements Funding: This research was funded by the Deutsche Forschungsgemeinschaft (DFG, German Research Foundation)—314987946.

References

1. Géradin, M., Rixen, D.J.: Mechanical Vibrations: Theory and Application to Structural Dynamics. John Wiley & Sons (2014)
2. Benner, P. et al.: Model Order Reduction. Applications, vol. 3 De Gruyter (2021)
3. Slaats, P., de Jongh, J., Sauren, A.: Model reduction tools for nonlinear structural dynamics. Int. J. Solids Struct. **38**(10), 2131–2147 (2001)
4. Lerch, C., Meyer, C.H.: Modellordnungsreduktion für parametrische nichtlineare mechanische Systeme mittels erweiterter simulationsfreier Basen und Hyperreduktion–6. In Lohmann, B.; Roppenecker, G. (Hrsg.): Methoden und Anwendungen der Regelungstechnik, pp. 67–86, Shaker Verlag (1. Aufl.) (2017)
5. Lohmann, B., Bechtold, T., Eberhard P., Fehr, J., Rixen, D.J., Cruz Varona M., Lerch, C., Yuan, C.D., Rudnyi, E.B., Fröhlich, B., Holzwarth, P., Grunert, D., Meyer, C.H., Rutzmoser, J.B.: Model Order Reduction in Mechanical Engineering. Model Order Reduction–Applications, De Gruyter, vol. 3, pp. 33–74 (2020)
6. Farhat, C., Chapman, T., Avery, P.: Structure-preserving, stability, and accuracy properties of the energy-conserving sampling and weighting method for the hyper reduction of nonlinear finite element dynamic models. Int. J. Numeric. Methods Eng. **102**(5), 1077–1110 (2015)
7. Rutzmoser, J.B., Rixen, D.J.: A lean and efficient snapshot generation technique for the Hyper-Reduction of nonlinear structural dynamics. Comput. Methods Appl. Mech. Eng. **325**, 330–349 (2017)
8. Meyer, C.H., Lerch, C., Karamooz Mahdiabadi, M., Rixen, D.: Efficient basis updating for parametric nonlinear model order reduction. PAMM **18**(1) (2018)
9. Meyer, C.H., Rixen, D.: Global proper orthogonal decomposition for parametric model reduction of geometrically nonlinear structures. PAMM **19**(1) (2019)
10. Rothe, M.: Parametrische Basen für nichtlineare Modellreduktion von mechanischen Finite Elemente Modellen. Technische Universität München, Bachelorarbeit (2018)
11. Meyer C.H., Rixen D.J.: Simulation-free reduction basis interpolation to reduce parametrized dynamic models of geometrically non-linear structures. In: Kerschen G., Brake M.R., Renson L. (Eds.) Nonlinear Structures & Systems. Conference Proceedings of the Society for Experimental Mechanics Series, vol. 1. Springer, Cham (2021)
12. Meyer, C.H., Rixen, D.J.: Simulation-lean training-sets for hyper-reduction of parametric geometric non-linear structures. In: Proceedings of the International Conference on Structural Dynamic, EURODYN. vol. 1, pp. 80–92. Athens, Greece (2020)
13. Rust, W.: Nichtlineare Finite-Elemente-Berechnungen: Kontakt, Kinematik, Material, p. 156. Springer (2016)
14. Wriggers, P.: Nichtlineare Finite-Element-Methoden, pp. 66–67. Springer (2001)
15. Salimbahrami, B., Lohmann, B.: Order reduction of large scale second-order systems using Krylov subspace methods. Linear Algebra Appl. **415**, 385–405 (2006)
16. Lutowska, A.: Model order reduction for coupled systems using low-rank approximations. Technische Universiteit Eindhoven (2012)
17. Rouleau, L., Deü, J.-F., Legay, A.: A comparison of model reduction techniques based on modal projection for structures with frequency-dependent damping. Mech. Syst. Signal Process. **90**, 110–125 (2017)
18. Plouin, A., Balmès, E.: Steel/viscoelastic/steel sandwich shells computational methods and experimental validations. In: International Modal Analysis Conference, pp. 384–390 (2000)
19. Meyer, C.H., Lerch, C., Rothe, S., Hoffmann, S., Langer S.C., Lohmann, B., Rixen, D.J.: Study on model reduction for position optimization of acoustic black holes. In: PAMM 21 (2021). (Submitted for publication)

Structure-Preserving Model Reduction for Dissipative Mechanical Systems

Rebekka S. Beddig, Peter Benner, Ines Dorschky, Timo Reis, Paul Schwerdtner, Matthias Voigt, and Steffen W. R. Werner

1 Introduction

We consider model order reduction of dynamical systems arising from modeling of mechanical systems, which have the property of *dissipativity*. That is, energy is only consumed and not produced by the system. In the particular focus of this work are *linear second-order systems*

$$\begin{aligned} M\ddot{q}(t) + D\dot{q}(t) + Kq(t) &= B_{\mathrm{u}}u(t), \\ y(t) &= C_{\mathrm{p}}q(t) + C_{\mathrm{v}}\dot{q}(t) \end{aligned} \tag{1}$$

R. S. Beddig
Institute of Mathematics, Hamburg University of Technology, Am Schwarzenberg-Campus 3 Gebäude E, 21073 Hamburg, Germany
e-mail: rebekka.beddig@tuhh.de

P. Benner
Max Planck Institute for Dynamics of Complex Technical Systems, Sandtorstraße 1, 39106 Magdeburg, Germany
e-mail: peter.benner@ovgu.de; benner@mpi-magdeburg.mpg.de

Faculty of Mathematics, Otto von Guericke University Magdeburg, Universitätsplatz 2, 39106 Magdeburg, Germany

I. Dorschky
Capgemini Engineering, Olof-Palme-Straße 14, 81829 Munich, Germany
e-mail: ines.dorschky@uni-hamburg.de

T. Reis
Institute of Mathematics, Technische Universität Ilmenau, Weimarer Straße 25, 98693 Ilmenau, Germany
e-mail: timo.reis@tu-ilmenau.de

P. Eberhard (ed.), *Calm, Smooth and Smart*, Lecture Notes in Applied and Computational Mechanics 102, https://doi.org/10.1007/978-3-031-36143-2_11

with $M, D, K \in \mathbb{R}^{n\times n}$, $B_{\mathrm{u}} \in \mathbb{R}^{n\times m}$, and $C_{\mathrm{p}}, C_{\mathrm{v}} \in \mathbb{R}^{p\times n}$. These occur naturally by modeling mechanical systems via force balances, in which the second derivative of the position vector $q(t)$ at time $t \in \mathbb{R}$ occurs by Newton's second law. Hereby, the matrices M, D and K are respectively called *mass matrix*, *damping matrix* and *stiffness matrix*. The function $t \mapsto u(t)$ expresses the *input* to the system (external forces), a function that can be chosen by the operator (or, alternatively called, the "user") of the system. Moreover, the model contains an *output* $t \mapsto y(t)$ that contains some linear combinations of the state variables and its first derivative, which are of particular interest. The typical situation is, especially for systems of high complexity, that the position vector $q(t)$ evolves in a high-dimensional space, that is, the number n is large. In contrast to that, the input and output spaces are low-dimensional, i.e., $m \ll n$ and $p \ll n$. Since the number n of position variables is a significant measure for the difficulty of the numerical simulation of (1), there is a need for efficient and reliable methods for model reduction, i.e., the approximation of such systems by ones whose solutions can be computed with significantly less effort. In this context, "reliable" means that the output of the reduced-order system is (mathematically proven to be) close to the output of the original system for the same input signal, whereas "efficient" means that the determination of the reduced-order system comes with as little effort as possible. Another important demand on model order reduction methods is that they preserve inherent properties such as stability and the second-order structure of the system (to mention only a few). By the latter, we mean that the reduced-order model is of the form

$$\begin{aligned} \widehat{M}\ddot{\hat{q}}(t) + \widehat{D}\dot{\hat{q}}(t) + \widehat{K}\hat{q}(t) &= \widehat{B}_{\mathrm{u}}u(t), \\ \hat{y}(t) &= \widehat{C}_{\mathrm{p}}\hat{q}(t) + \widehat{C}_{\mathrm{v}}\dot{\hat{q}}(t) \end{aligned} \tag{2}$$

with $\widehat{M}, \widehat{D}, \widehat{K} \in \mathbb{R}^{r\times r}$, $\widehat{B}_{\mathrm{u}} \in \mathbb{R}^{r\times m}$ and $\widehat{C}_{\mathrm{p}}, \widehat{C}_{\mathrm{v}} \in \mathbb{R}^{p\times r}$, and with $r \ll n$. Moreover, models of mechanical systems have the property that the mass and stiffness matrices are symmetric positive definite, whereas the negative of the damping matrix is *dissipative*, that is, $D + D^{\mathsf{T}}$ is positive semi-definite. These properties are requested to be preserved as well by the reduced-order system (2).

P. Schwerdtner
Institut für Mathematik, Technische Universität Berlin, Straße des 17. Juni 136, 10623 Berlin, Germany
e-mail: schwerdt@math.tu-berlin.de

M. Voigt
UniDistance Suisse, Schinerstrasse 18, 3900 Brig-VS, Switzerland
e-mail: matthias.voigt@fernuni.ch

S. W. R. Werner (✉)
Courant Institute of Mathematical Sciences, New York University, 251 Mercer Street, New York, NY 10012, USA
e-mail: steffen.werner@nyu.edu

Meanwhile, model order reduction is an established discipline within applied mathematics and is subject of textbooks and collections, see [1–5]. In particular, for first-order systems

$$\begin{aligned}\dot{x}(t) &= \mathcal{A}x(t) + \mathcal{B}u(t),\\ y(t) &= \mathcal{C}x(t),\end{aligned}$$

there exists a rich theory for their approximation by reduced-order systems of low state-space dimension; see [1] for an overview. These methods are indeed applicable to first-order representations of second-order systems like

$$\begin{bmatrix} I_n & 0 \\ 0 & M \end{bmatrix} \frac{\mathrm{d}}{\mathrm{d}t} \begin{bmatrix} q(t) \\ \dot{q}(t) \end{bmatrix} = \begin{bmatrix} 0 & I_n \\ -K & -D \end{bmatrix} \begin{bmatrix} q(t) \\ \dot{q}(t) \end{bmatrix} + \begin{bmatrix} 0 \\ B_{\mathrm{u}} \end{bmatrix} u(t),$$
$$y(t) = \begin{bmatrix} C_{\mathrm{p}} & C_{\mathrm{v}} \end{bmatrix} \begin{bmatrix} q(t) \\ \dot{q}(t) \end{bmatrix}.$$

However, the problem with this is that the reduced-order system is again of first order, and, in general, it does not have a physical interpretation as a mechanical system. The structure-preserving model order reduction problem of second-order systems is therefor a problem on its own and new techniques have to be developed.

The model order reduction problem for linear time-invariant systems can also be considered in the frequency domain. More precisely, the *transfer function* mapping inputs to outputs in frequency domain can be considered, which for (1) is given by

$$H(s) = (C_{\mathrm{p}} + sC_{\mathrm{v}})(s^2M + sD + K)^{-1}B_{\mathrm{u}} = \begin{bmatrix} C_{\mathrm{p}} & C_{\mathrm{v}} \end{bmatrix} \begin{bmatrix} sI_n & -I_n \\ K & sM + D \end{bmatrix}^{-1} \begin{bmatrix} 0 \\ B_{\mathrm{u}} \end{bmatrix}.$$

Plancherel's theorem [1, Prop. 5.1] provides a link between the time and frequency domain in a way that—very roughly speaking—"the better the transfer function of the reduced-order system approximates that of the original system, the better the outputs of original and reduced-order systems coincide". Two important measures for the distance between transfer functions are the *$\mathcal{H}_\infty$-norm*, which for asymptotically stable systems expresses the supremal distance between the transfer functions on the imaginary axis; and the so-called *gap metric* [6], which applies to arbitrary, possibly unstable, systems and can be expressed by the $\mathcal{H}_\infty$-norm of certain stable factorizations of transfer functions. Whereas in the time domain, the $\mathcal{H}_\infty$-norm expresses the L^2-norm differences of the outputs of the original and reduced-order system, the gap metric can be seen as a quantitative measure for the distance of the dynamics of systems.

Besides considering arbitrary linear outputs $y(t) = C_{\mathrm{p}}q(t) + C_{\mathrm{v}}\dot{q}(t)$, in our considerations special emphasis is put on *co-located* velocity outputs $y(t) = B_{\mathrm{u}}^{\mathsf{T}}\dot{q}(t)$, which corresponds to measurements of velocities directly at the force actuators forming the input. This special input-output configuration has the additional property that it provides an energy balance, namely

$$\forall t \geq 0: \quad \tfrac{1}{2}\left(\dot{q}(t)^{\mathsf{T}} M \dot{q}(t) + q(t)^{\mathsf{T}} K q(t)\right) - \tfrac{1}{2}\left(\dot{q}(0)^{\mathsf{T}} M \dot{q}(0) + q(0)^{\mathsf{T}} K q(0)\right)$$
$$= \int_0^t y(\tau)^{\mathsf{T}} u(\tau)\,\mathrm{d}\,\tau - \int_0^t \dot{q}(\tau)^{\mathsf{T}} D \dot{q}(\tau)\,\mathrm{d}\,\tau.$$

The expressions $\frac{1}{2}\dot{q}(t)^{\mathsf{T}} M \dot{q}(t)$ and $\frac{1}{2}q(t)^{\mathsf{T}} K q(t)$ respectively stand for the kinetic and potential energies of the system at time t, whereas $\int_0^t \dot{q}(\tau)^{\mathsf{T}} D \dot{q}(\tau)\,\mathrm{d}\,\tau$ is the dissipated energy, and $\int_0^t y(t)^{\mathsf{T}} u(t)\,\mathrm{d}\,t$ is the energy put into the system at the actuators within the time interval $[0, t]$. In particular, since D is dissipative and the mass and stiffness matrices are positive definite, in the case where the system is in a standstill at $t = 0$, i.e., $\dot{q}(0) = q(0) = 0$, this energy balance reduces to

$$\forall t \geq 0: \quad 0 \leq \int_0^t y(\tau)^{\mathsf{T}} u(\tau)\,\mathrm{d}\,\tau.$$

Systems with this property are called *passive*, a property which is further desired to be preserved by the reduced-order model. Note that the frequency domain pendant of passivity is *positive realness*, i.e., the transfer function $H(s)$ has no poles in the open right complex half plane and, additionally, $-H(s)$ is dissipative in the open right complex half plane.

For linear time-invariant systems, there are (among others) three "prominent" techniques for model order reduction, namely *modal-based*, *balancing-based*, and *interpolation-based* approaches. The modal-based methods consider eigenvalue problems associated with the potential poles of the transfer function to retain chosen poles from the original in the reduced-order model. Balancing-based methods use energy considerations to figure out parts of the state only contributing marginally to the input-output behavior, which are truncated to obtain a reduced-order system of a priori known quality by providing error bounds, e.g., in the $\mathcal{H}_\infty$-norm or gap metric. The main cost in the determination of reduced-order models by balanced truncation is the numerical solution of matrix equations of Lyapunov or Riccati type. Interpolation-based methods use certain projections of the state space, which guarantee exactness of the transfer function of the reduced-order system at some prescribed frequencies.

For second-order systems, the general ideas of balancing-based model order reduction are subject to various contributions [7–11]; see also [12] for an overview. Some progress has been made in preservation of certain physical properties like passivity in model order reduction of second-order systems with co-located inputs and outputs [10–12], but none of these methods are provided with an error bound. Besides these, there exist interpolatory methods, which succeed either in preserving the second-order structure [13–16] or deliver a posteriori $\mathcal{H}_\infty$ error bounds [17]. However, all the approaches mentioned lack a combination of the two.

In the project "Structure-Preserving Model Reduction for Dissipative Mechanical Systems" of the German Research Foundation (DFG) Priority Program "Calm,

Smooth, and Smart—Novel Approaches for Influencing Vibrations by Means of Deliberately Introduced Dissipation" (SPP1897), several new approaches for model order reduction of second-order systems have been developed. An extract of this work can be found in [18], as well as in the dissertations of Werner [19] and Dorschky [20].

The structure of this report is as follows. In Sect. 2, we present our results on a dominant pole algorithm for modally damped mechanical systems. In Sect. 3, a novel balancing-based approach for second-order systems is presented, which considers the dominant behavior of the system on some prescribed time and frequency intervals. In Sect. 4, we consider an alternative balancing-based method for second-order systems with co-located inputs and outputs. We prove that an error bound in the gap metric holds and that, under some additional assumptions, a special state-space transformation leads to a second-order system realization. Section 5 is devoted to an interpolation-based model order reduction method for second-order systems, based on an optimization-based technique, which generates a sequence of reduced-order models of descending error in the $\mathcal{H}_\infty$-norm. The report is concluded in Sect. 6.

2 A Dominant Pole Algorithm for Modally Damped Mechanical Systems

One of the oldest model order reduction approaches, which also directly translates into a structure-preserving setting for second-order systems (1), is the *modal truncation method* [21]. Thereby, the projection basis for the reduced-order model only consists of the left and right eigenvectors corresponding to the desired eigenvalues. In case of second-order systems like (1), the corresponding quadratic eigenvalue problem

$$\left(\lambda_i^2 M + \lambda_i D + K\right) x_i = 0, \quad y_i^{\mathsf{H}} \left(\lambda_i^2 M + \lambda_i D + K\right) = 0$$

has to be considered for the left and right eigenvectors $y_i, x_i \in \mathbb{C}^n \setminus \{0\}$ corresponding to the eigenvalue $\lambda_i \in \mathbb{C}$. Hereby, y^{H} stands for the conjugate transpose of y.

For this model order reduction method, the choice of the eigenvalues that remain in the reduced-order system is critical. Classical choices are, e.g., taking the rightmost eigenvalues in the complex plane or the eigenvalues with smallest absolute values. A significant drawback of those simple choices is the neglection of the input and output matrices, which have a significant influence on the actual input-to-output behavior of the system. The extension of the classical modal truncation method to a more sophisticated choice of eigenvalues is the *dominant pole algorithm* [22]. Here, the eigenvalues with the strongest influence on the system behavior are computed and then chosen for constructing the reduced-order model. Adaptations of the dominant pole algorithm to the case of general second-order systems have been suggested in [23] for single-input single-output systems and in [24] for multi-input multi-output systems.

A modeling approach used very often for mechanical structures results in modally damped second-order systems. Hereby, for the second-order system (1) it is assumed that $C_v = 0$, M, D, K are symmetric positive definite and, additionally, it holds that $DM^{-1}K = KM^{-1}D$, i.e., the system can be rewritten into modal coordinates and completely decouples into independent mechanical systems of order 1; see, e.g., [25]. Classical damping approaches, like Rayleigh and critical damping, fall into this category. For this type of mechanical systems, the idea of the dominant pole algorithm can be reformulated. As shown in [25], choosing X as a scaled eigenvector basis gives

$$X^{\mathsf{T}}MX = \Omega^{-1} \quad \text{and} \quad X^{\mathsf{T}}KX = \Omega \tag{3}$$

with $\Omega = \operatorname{diag}(\omega_1, \ldots, \omega_n) \in \mathbb{R}^{n\times n}$ and $X = [x_1, \ldots, x_n] \in \mathbb{R}^{n\times n}$. By the modal damping assumption, we further get

$$X^{\mathsf{T}}DX = 2\Xi \tag{4}$$

with $\Xi = \operatorname{diag}(\xi_1, \ldots, \xi_n) \in \mathbb{R}^{n\times n}$. Combining (3) and (4), the transfer function of (1) can be written in a structured pole-residue form

$$\begin{aligned} H(s) &= C_{\mathrm{p}}(s^2M + sD + K)^{-1}B_{\mathrm{u}} \\ &= C_{\mathrm{p}}X(s^2\Omega^{-1} + 2s\Xi + \Omega)^{-1}X^{\mathsf{T}}B_{\mathrm{u}} \\ &= \sum_{k=1}^{n} \frac{\omega_k \left(C_{\mathrm{p}}x_k\right)\left(x_k^{\mathsf{T}}B_{\mathrm{u}}\right)}{(s-\lambda_k^+)(s-\lambda_k^-)}, \end{aligned} \tag{5}$$

where the eigenvalue pairs λ_k^+, λ_k^- are given by

$$\lambda_k^{\pm} = -\omega_k\xi_k \pm \omega_k\sqrt{\xi_k^2 - 1}.$$

The pole-residue formulation (5) is now used to derive a new dominant pole algorithm for modally damped second-order systems. With (5), an appropriate extension of classical dominant poles as in [24] to pole pairs reads as: The pole pair $(\lambda_k^+, \lambda_k^-)$ is called dominant if it satisfies

$$\frac{\|\omega_k(C_{\mathrm{p}}x_k)(x_k^{\mathsf{T}}B_{\mathrm{u}})\|_2}{|\operatorname{Re}(\lambda_k^+)(\operatorname{Im}(\lambda_k^+)\mathrm{i} - \lambda_k^-)|} > \frac{\|\omega_j(C_{\mathrm{p}}x_j)(x_j^{\mathsf{T}}B_{\mathrm{u}})\|_2}{|\operatorname{Re}(\lambda_j^+)(\operatorname{Im}(\lambda_j^+)\mathrm{i} - \lambda_j^-)|} \quad \text{for all } j \neq k. \tag{6}$$

Note the difference to the dominance measure in [26], for which the two poles of a pair are considered as independent quantities. The corresponding dominant pole algorithm then computes the r most dominant pole pairs (6) and the corresponding eigenvectors, such that the reduced-order model's transfer function is given by

$$\widehat{H}(s) = \sum_{k=1}^{r} \frac{\omega_k \left(C_{\mathrm{p}} x_k\right) \left(x_k^{\mathsf{T}} B_{\mathrm{u}}\right)}{(s - \lambda_k^{+})(s - \lambda_k^{-})} \approx H(s),$$

with an appropriate ordering of the terms in (5) with respect to (6). The projection basis is then given by the r eigenvectors corresponding to the chosen pole pairs. The basic ideas of the resulting algorithm are published in [19, 26]. Additionally, we published an implementation of this new algorithm for large-scale sparse systems as MATLAB and GNU Octave toolbox [27].

Remark 1 A big advantage of this new approach, compared to the methods in [23, 24], is the restriction to one-sided projections. This preserves the system and eigenvalue structure in each single step such that the resulting eigenvector basis will be real and no additional unrelated Ritz values, which usually disturb the resulting approximation, are introduced in the reduced-order model.

As an illustrative example, we consider the butterfly gyroscope benchmark from [28] with $n = 17\,361$, $m = 1$ and $p = 12$. The used Rayleigh damping for the D $(= 10^{-6} \cdot K)$ matrix belongs to the class of modal damping. We are using the implementation from [27] to compute a reduced-order model with the first 10 dominant pole pairs by the criterion (6). By construction, the resulting reduced-order model has order 10. Figure 1 shows the results in the frequency domain with the frequency response behavior of the original and the reduced-order model and the pointwise relative error of the approximation. Up to a frequency of about 10^6 rad/s, the behavior of the original system is well reproduced, while later, the two lines begin to diverge slightly. Additionally, Fig. 2 shows the position of the computed dominant poles as projection onto the imaginary axis and the corresponding transfer function behavior.

Comparisons to other model reduction methods and further examples using this new dominant pole algorithm can be found in [19].

3 Second-Order Frequency- and Time-Limited Balanced Truncation

A global approximation of the system behavior in frequency or time domain is often not required in practice. The second-order limited balanced truncation approaches are a suitable tool for model order reduction restricted to certain frequency and time ranges. Thereby, the ideas from the first-order frequency- and time-limited balanced truncation methods [29] are combined with different second-order balanced truncation approaches [7, 8, 12]. A first version of these methods can be found in [18, 26, 30, 31] and the complete theory with applications to large-scale sparse systems is contained in [19, 32].

The idea of the method is based on the first companion form realization of (1), which is given by

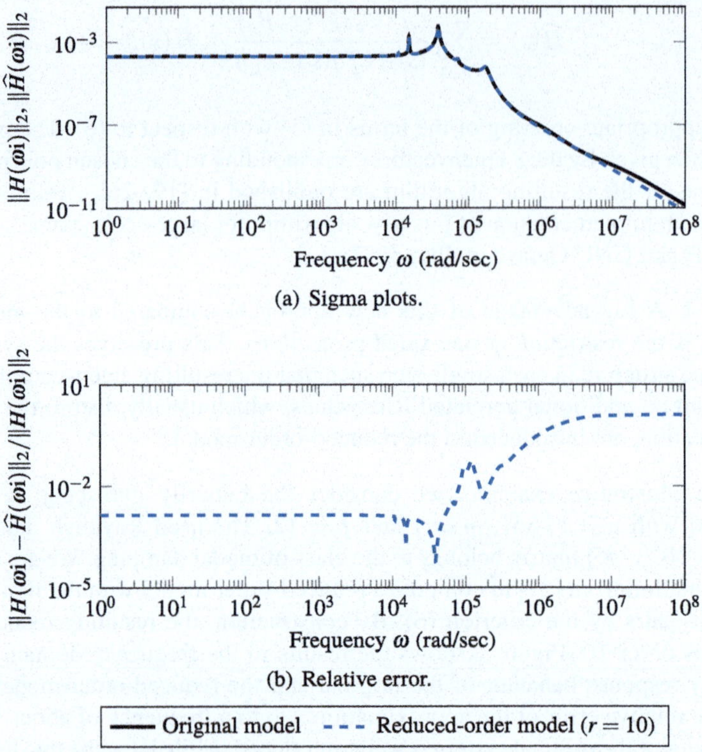

(a) Sigma plots.

(b) Relative error.

Fig. 1 Modally damped dominant pole algorithm results for the butterfly gyroscope example

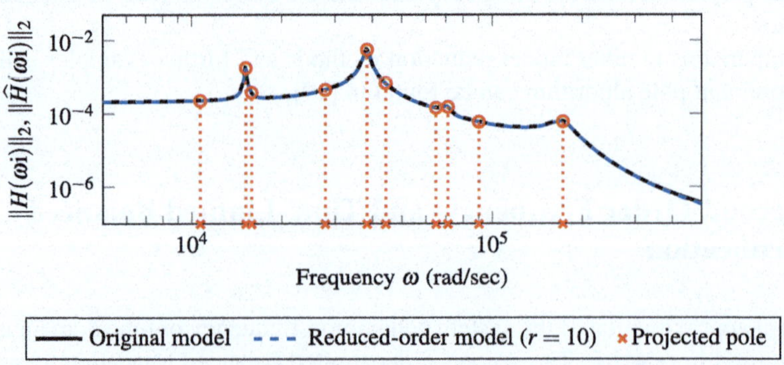

Fig. 2 Projection of the computed dominant poles on the frequency axis compared with the transfer functions for the butterfly gyroscope example

$$\underbrace{\begin{bmatrix} I_n & 0 \\ 0 & M \end{bmatrix}}_{=:\mathcal{E}} \dot{x}(t) = \underbrace{\begin{bmatrix} 0 & I_n \\ -K & -D \end{bmatrix}}_{=:\mathcal{A}} x(t) + \underbrace{\begin{bmatrix} 0 \\ B_{\mathrm{u}} \end{bmatrix}}_{=:\mathcal{B}} u(t),$$
$$y(t) = \underbrace{\begin{bmatrix} C_{\mathrm{p}} & C_{\mathrm{v}} \end{bmatrix}}_{=:C} x(t). \tag{7}$$

For (7), the classical controllability and observability Gramians are defined and can be limited as in [29]. Therefore, the frequency-limited Gramians P_Ω and Q_Ω of (7) are the unique symmetric positive semi-definite solutions of the potentially indefinite Lyapunov equations

$$\mathcal{A}P_\Omega\mathcal{E}^\mathsf{T} + \mathcal{E}P_\Omega\mathcal{A}^\mathsf{T} + \mathcal{B}_\Omega\mathcal{B}^\mathsf{T} + \mathcal{B}\mathcal{B}_\Omega^\mathsf{T} = 0,$$
$$\mathcal{A}^\mathsf{T} Q_\Omega\mathcal{E} + \mathcal{E}^\mathsf{T} Q_\Omega\mathcal{A} + C_\Omega^\mathsf{T} C + C^\mathsf{T} C_\Omega = 0,$$

for a specified frequency range $\Omega = [\omega_1, \omega_2] \cup [-\omega_2, -\omega_1]$; see [29, 33]. The right-hand side matrices contain matrix functions, which are given by

$$\mathcal{B}_\Omega = \mathrm{Re}\left(\frac{\mathrm{i}}{\pi}\ln\left((\mathcal{A}+\omega_2\mathrm{i}\mathcal{E})(\mathcal{A}+\omega_1\mathrm{i}\mathcal{E})^{-1}\right)\right)\mathcal{B},$$
$$C_\Omega = C\mathrm{Re}\left(\frac{\mathrm{i}}{\pi}\ln\left((\mathcal{A}+\omega_1\mathrm{i}\mathcal{E})^{-1}(\mathcal{A}+\omega_2\mathrm{i}\mathcal{E})\right)\right).$$

Analogously, the time-limited Gramians P_Θ and Q_Θ of (7) are given as the unique symmetric positive semi-definite solutions of the (potentially) indefinite Lyapunov equations

$$\mathcal{A}P_\Theta\mathcal{E}^\mathsf{T} + \mathcal{E}P_\Theta\mathcal{A}^\mathsf{T} + \mathcal{B}_{t_0}\mathcal{B}_{t_0}^\mathsf{T} - \mathcal{B}_{t_\mathrm{f}}\mathcal{B}_{t_\mathrm{f}}^\mathsf{T} = 0,$$
$$\mathcal{A}^\mathsf{T} Q_\Theta\mathcal{E} + \mathcal{E}^\mathsf{T} Q_\Theta\mathcal{A} + C_{t_0}^\mathsf{T} C_{t_0} - C_{t_\mathrm{f}}^\mathsf{T} C_{t_\mathrm{f}} = 0,$$

where the time-dependent right-hand sides are defined as

$$\mathcal{B}_{t_0} = \mathrm{e}^{\mathcal{A}\mathcal{E}^{-1}t_0}\mathcal{B}, \quad \mathcal{B}_{t_\mathrm{f}} = \mathrm{e}^{\mathcal{A}\mathcal{E}^{-1}t_\mathrm{f}}\mathcal{B}, \quad C_{t_0} = C\mathrm{e}^{\mathcal{E}^{-1}\mathcal{A}t_0}, \quad C_{t_\mathrm{f}} = C\mathrm{e}^{\mathcal{E}^{-1}\mathcal{A}t_\mathrm{f}}$$

on the time interval $\Theta = [t_0, t_\mathrm{f}]$; see [29, 34]. Using those Gramians for the different second-order balanced truncation approaches [7, 8, 12] leads to the second-order limited balanced truncation methods as described in [19, 32]. The dense version of the resulting methods is contained in the current version of the MORLAB toolbox [35, 36], and an implementation for large-scale sparse systems as MATLAB and GNU Octave toolbox can be found in [37].

As numerical example for the frequency-limited approach, we consider the triple chain oscillator example as in [18]. We reduce the original model ($n = 1\,201$) by the second-order frequency-limited balanced truncation method in the interval $[5 \cdot 10^{-3}, 5 \cdot 10^{-2}]$ rad/s using the eight different second-order balancing formulas

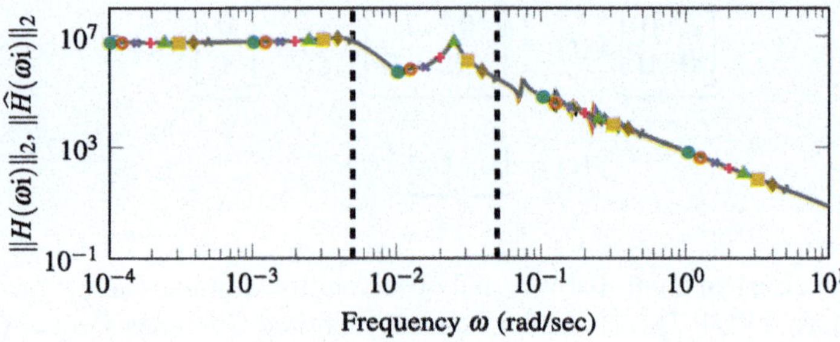

(a) Sigma plots.

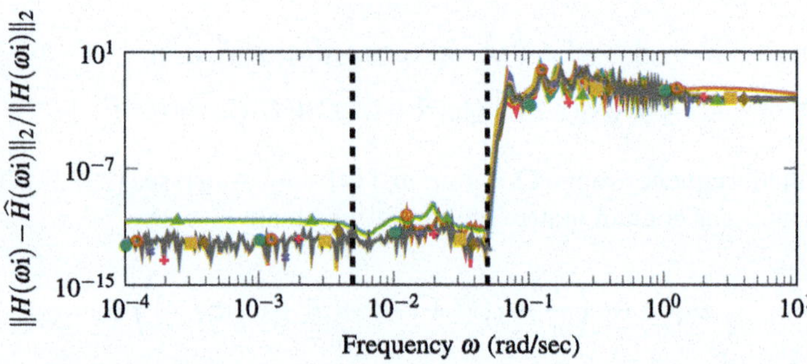

(b) Relative errors.

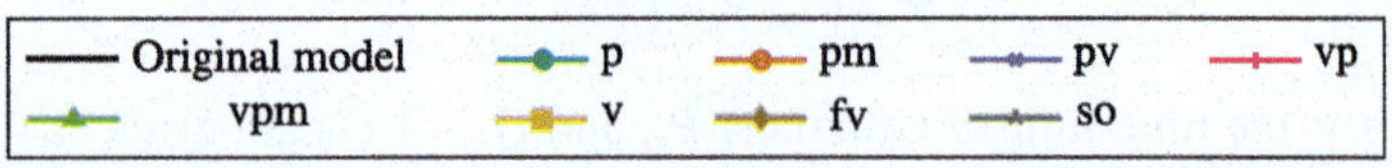

Fig. 3 Frequency-limited balanced truncation results for the triple chain oscillator example

from [32] to the order $r = 34$. The computations are done using the dense implementation of the second-order frequency-limited balanced truncation method from the latest version of the MORLAB toolbox [35, 36]. The results can be seen in Fig. 3 with the transfer functions and the relative approximation errors. The computed reduced-order models are denoted according to the used balancing formulas from [32]. We clearly see the good approximation behavior in the frequency range of interest. Only the system computed by the fv formula is stable, while all others become unstable. The preservation of stability by the fv formula follows directly from the preservation of the symmetric positive definiteness of the system matrices via the underlying one-sided projection. However, note that in general for any balancing formula in the classical second-order balanced truncation method, there are counter-examples for the stability preservation [12].

To illustrate the time-limited balanced truncation method, we use the single chain oscillator example as described in [32]. Here, we use the implementation for large-scale sparse mechanical systems from [37] to reduce the original system ($n = 12\,000$) to order $r = 3$ in the time interval $[0, 20]$ s. The results for the second output entry

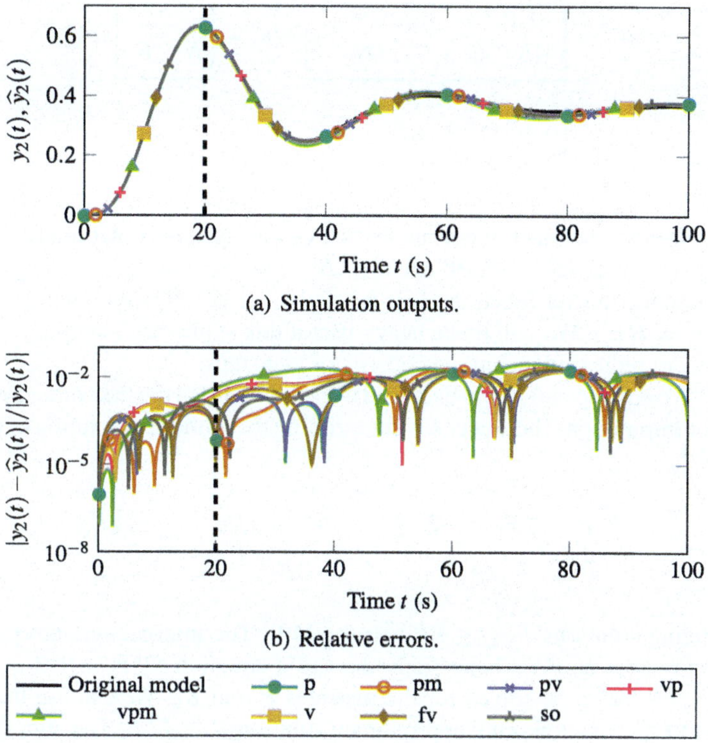

Fig. 4 Time-limited balanced truncation results for the single chain oscillator example

can be seen in Fig. 4, where in the time region of interest, the original system is nicely approximated by the reduced-order models. For all balancing formulas, the resulting systems are stable.

More detailed comparisons of the different balancing formulas and further examples can be found in [19, 32].

4 Positive Real Balanced Truncation for Second-Order Systems

In this section, we consider second-order systems of the form (1), where $M, K > 0$, $D \geq 0$, and either we exclusively measure positions, i.e., $C_{\mathrm{v}} = 0$, or velocities, i.e., $C_{\mathrm{p}} = 0$. We start with the second case and additionally assume co-located inputs and outputs, which means $B_{\mathrm{u}} = C_{\mathrm{v}}^{\mathsf{T}}$. This case is treated in [38]. Using the Cholesky factorizations $K = GG^{\mathsf{T}}$ and $M = NN^{\mathsf{T}}$, the system can be rewritten in first-order form as

$$\dot{x}(t) = \underbrace{\begin{bmatrix} 0 & G^{\mathsf{T}}N^{-\mathsf{T}} \\ -N^{-1}G & -N^{-1}DN^{-\mathsf{T}} \end{bmatrix}}_{=:\mathcal{A}} x(t) + \underbrace{\begin{bmatrix} 0 \\ N^{-1}B_{\mathrm{u}} \end{bmatrix}}_{=:\mathcal{B}} u(t),$$
$$y(t) = \underbrace{\begin{bmatrix} 0 & B_{\mathrm{u}}^{\mathsf{T}}N^{-\mathsf{T}} \end{bmatrix}}_{=:\mathcal{C}} x(t). \tag{8}$$

Besides passivity, the most important feature of this system is that it has an internal symmetry structure $\mathcal{A}\mathcal{S}_n = \mathcal{S}_n\mathcal{A}^{\mathsf{T}}$ and $\mathcal{C} = \mathcal{B}^{\mathsf{T}} = \mathcal{B}^{\mathsf{T}}\mathcal{S}_n$, where $\mathcal{S}_n := \operatorname{diag}(-I_n, I_n)$. In particular, its transfer function $H(s) = \mathcal{C}(sI_{2n} - \mathcal{A})^{-1}\mathcal{B}$ is symmetric, i.e., it fulfills $H(s)^{\mathsf{T}} = H(s)$. We will make heavy use of this symmetry structure.

The model reduction technique consists of two steps:

Step 1: We apply positive real balanced truncation [39] to the first-order system (8). Here, the computational bottleneck is determining the numerical solution of the Lur'e equations

$$\begin{aligned} \mathcal{A}^{\mathsf{T}}P + P\mathcal{A} &= -K_{\mathrm{c}}^{\mathsf{T}}K_{\mathrm{c}}, & \mathcal{A}Q + Q\mathcal{A}^{\mathsf{T}} &= -K_{\mathrm{o}}^{\mathsf{T}}K_{\mathrm{o}}, \\ P\mathcal{B} - \mathcal{C}^{\mathsf{T}} &= 0, & Q\mathcal{C}^{\mathsf{T}} - \mathcal{B} &= 0, \end{aligned}$$

for stabilizing solutions $P, Q \in \mathbb{R}^{2n\times 2n}$; see [40]. The internal symmetry structure of (8) yields $Q = \mathcal{S}_n P\mathcal{S}_n$, hence only the Lur'e equation $\mathcal{A}^{\mathsf{T}}P + P\mathcal{A} = -K_{\mathrm{c}}^{\mathsf{T}}K_{\mathrm{c}}$, $P\mathcal{B} - \mathcal{C}^{\mathsf{T}} = 0$ has to be solved for the matrices P and K_{c}. We can use the method from [41] to obtain a low-rank approximate solution $P \approx L^{\mathsf{T}}L$. The sign symmetry of the first-order system (8) yields that its positive real characteristic values (which are defined to be the positive square roots of the eigenvalues of PQ) can in a certain sense be allocated to the symmetry structure of the system (8). More precisely, the positive real characteristic values are the absolute values of the eigenvalues of the symmetric matrix $L\mathcal{S}_nL^{\mathsf{T}}$. By truncating equally many states corresponding to positive and negative eigenvalues of $L\mathcal{S}_nL^{\mathsf{T}}$, it is shown that the resulting first-order model is—without any further computational effort – of the form

$$\begin{aligned} \dot{\hat{x}}(t) &= \begin{bmatrix} 0 & 0 & 0 & 0 & 0 & \mathcal{A}_{16} \\ 0 & 0 & 0 & 0 & \mathcal{A}_{25} & \mathcal{A}_{26} \\ 0 & 0 & \mathcal{A}_{33} & \mathcal{A}_{34} & 0 & \mathcal{A}_{36} \\ 0 & 0 & -\mathcal{A}_{34}^{\mathsf{T}} & \mathcal{A}_{44} & 0 & \mathcal{A}_{46} \\ 0 & -\mathcal{A}_{25}^{\mathsf{T}} & 0 & 0 & 0 & 0 \\ -\mathcal{A}_{16}^{\mathsf{T}} & -\mathcal{A}_{26}^{\mathsf{T}} & -\mathcal{A}_{36}^{\mathsf{T}} & \mathcal{A}_{46}^{\mathsf{T}} & 0 & \mathcal{A}_{66} \end{bmatrix} \hat{x}(t) + \begin{bmatrix} 0 \\ 0 \\ 0 \\ 0 \\ 0 \\ \mathcal{B}_6 \end{bmatrix} u(t), \\ \hat{y}(t) &= \begin{bmatrix} 0 & 0 & 0 & 0 & 0 & \mathcal{B}_6^{\mathsf{T}} \end{bmatrix} \hat{x}(t), \end{aligned} \tag{9}$$

where the block sizes from left to right and from top to bottom are m, ℓ, p, p, ℓ, m, with $r = p + m + \ell$. Note that, if $\mathcal{A}_{33}$ is zero, then—by merging some of the blocks—(9) has the form

$$\begin{aligned}\dot{\hat{x}}(t) &= \begin{bmatrix} 0 & \widehat{G}^\mathsf{T} \\ -\widehat{G} & -\widehat{D} \end{bmatrix} \hat{x}(t) + \begin{bmatrix} 0 \\ \widehat{B}_\mathrm{u} \end{bmatrix} u(t), \\ \hat{y}(t) &= \begin{bmatrix} 0 & \widehat{B}_\mathrm{u}^\mathsf{T} \end{bmatrix} \hat{x}(t),\end{aligned} \tag{10}$$

which results in a reduced-order model in second-order form (2) with $\widehat{M} = I_r$, $\widehat{K} = \widehat{G}\widehat{G}^\mathsf{T}$, $\widehat{C}_\mathrm{p} = 0$, and $\widehat{C}_\mathrm{v} = \widehat{B}_\mathrm{u}^\mathsf{T}$. This is regrettably not the case in general, which is why we apply the following step.

Step 2: We apply a state-space transformation to (9) such that the matrix $\mathcal{A}_{33}$ vanishes. More precisely, we first intend to find some invertible $T \in \mathbb{R}^{2p\times 2p}$ that preserves the symmetry structure, i.e., it fulfills $T^\mathsf{T} \mathcal{S}_p T = \mathcal{S}_p$ and

$$T^{-1} \begin{bmatrix} \mathcal{A}_{33} & \mathcal{A}_{34} \\ -\mathcal{A}_{34}^\mathsf{T} & \mathcal{A}_{44} \end{bmatrix} T = \begin{bmatrix} 0 & \widehat{\mathcal{A}}_{34} \\ -\widehat{\mathcal{A}}_{34}^\mathsf{T} & \widehat{\mathcal{A}}_{44} \end{bmatrix}.$$

Then, a state-space transformation with $V = \operatorname{diag}(I_{m+\ell}, T, I_{\ell+m})$ leads to a system, which is indeed of the form (10) and can then be rewritten as a second-order system. Such a transformation is based on techniques from indefinite linear algebra [42], and can be computed without any remarkable computational effort. The existence of such a transformation is linked to the $2k$ real eigenvalues of $\begin{bmatrix} \mathcal{A}_{33} & \mathcal{A}_{34} \\ -\mathcal{A}_{34}^\mathsf{T} & \mathcal{A}_{44} \end{bmatrix}$. Under the assumption of semi-simplicity, these eigenvalues can be assigned a certain signature structure according to the sign of $-v_i^\mathsf{T} \mathcal{S}_k v_i$, where v_i is the eigenvector of the i-th eigenvalue. In particular, we have k eigenvalues $\mu_1^- \leq \ldots \leq \mu_k^-$ of negative type and k eigenvalues $\mu_1^+ \leq \ldots \leq \mu_k^+$ of positive type. It is then shown that such a transformation is possible, if $\mu_i^- < \mu_i^+$ for all $i = 1, \ldots, k$. A sufficient criterion on the original system for the existence of such a transformation is that it is *overdamped*, that is $\left(v^\mathsf{T} D v\right)^2 > 4\left(v^\mathsf{T} M v\right)\left(v^\mathsf{T} K v\right)$ for all $v \in \mathbb{R}^n$.

The resulting second-order system in particular fulfills $\widehat{M}, \widehat{K} > 0$, and it is further shown that $\widehat{D} = \widehat{D}^\mathsf{T}$ has at most m negative eigenvalues. If the original system is overdamped, then even $\widehat{D} > 0$. Moreover, the gap metric distance between the transfer functions $H(s)$ of the original and $\widehat{H}(s)$ of the reduced-order system is shown to be bounded from above by twice the sum of the truncated positive real characteristic values.

Considering a system dilation, we are able to extend the previously presented reduction to second-order systems with velocity measurements, which are not necessarily co-located to the input. This can be done by considering the extended system

$$\begin{aligned} M\ddot{q}(t) + D\dot{q}(t) + Kq(t) &= \begin{bmatrix} B_\mathrm{u} & C_\mathrm{v}^\mathsf{T} \end{bmatrix} \begin{bmatrix} u_1(t) \\ u_2(t) \end{bmatrix}, \\ \begin{bmatrix} y_1(t) \\ y_2(t) \end{bmatrix} &= \begin{bmatrix} B_\mathrm{u}^\mathsf{T} \\ C_\mathrm{v} \end{bmatrix} \dot{q}(t). \end{aligned}$$

This system is again positive real and from the algorithm above we thus obtain a reduced-order system for the extended system as

$$\begin{aligned} \widehat{M}\ddot{\hat{q}}(t) + \widehat{D}\dot{\hat{q}}(t) + \widehat{K}\hat{q}(t) &= \begin{bmatrix} \widehat{B}_{\mathrm{u}} & \widehat{C}_{\mathrm{v}}^{\mathsf{T}} \end{bmatrix} \begin{bmatrix} u_1(t) \\ u_2(t) \end{bmatrix}, \\ \begin{bmatrix} \hat{y}_1(t) \\ \hat{y}_2(t) \end{bmatrix} &= \begin{bmatrix} \widehat{B}_{\mathrm{u}}^{\mathsf{T}} \\ \widehat{C}_{\mathrm{v}} \end{bmatrix} \dot{\hat{q}}(t), \end{aligned} \tag{11}$$

where $\widehat{B}_{\mathrm{u}} \in \mathbb{R}^{r\times m}$ and $\widehat{C}_{\mathrm{v}} \in \mathbb{R}^{p\times r}$. From that we obtain a reduced-order system for (1) as

$$\begin{aligned} \widehat{M}\ddot{\hat{q}}(t) + \widehat{D}\dot{\hat{q}}(t) + \widehat{K}\hat{q}(t) &= \widehat{B}_{\mathrm{u}} u_1(t), \\ \hat{y}_2(t) &= \widehat{C}_{\mathrm{v}}^{\mathsf{T}} \dot{\hat{q}}(t). \end{aligned} \tag{12}$$

As the transfer function of (12) is a submatrix of the transfer function of the system (11), and the same holds for the original system, the reduction error in the gap metric is bounded by the one of the extended system.

We illustrate the performance of the reduction method above with an example of three coupled mass-spring-damper chains; see [43, Ex. 2]. The triple chain consists of three rows that are coupled via a mass m_0, which is connected to the fixed base with a spring with stiffness k_0. Each row contains g masses, $g+1$ springs and one damper, which is attached to a wall; see Fig. 5. One can write the free system as

$$M\ddot{q}(t) + D\dot{q}(t) + Kq(t) = 0,$$

where M, D, and K are defined as $M = \operatorname{diag}(m_1, \ldots, m_1, m_2, \ldots, m_2, m_3, \ldots, m_3)$, $D = \alpha M + \beta K + \nu_1 e_1 e_1^{\mathsf{T}} + \nu_{g+1} e_{g+1} e_{g+1}^{\mathsf{T}} + \nu_{2g+1} e_{2g+1} e_{2g+1}^{\mathsf{T}}$ and

$$K = \begin{bmatrix} K_{11} & & & -\kappa_1 \\ & K_{22} & & -\kappa_2 \\ & & K_{33} & -\kappa_3 \\ -\kappa_1^{\mathsf{T}} & -\kappa_2^{\mathsf{T}} & -\kappa_3^{\mathsf{T}} & k_1 + k_2 + k_3 + k_0 \end{bmatrix}, \quad K_{ii} = k_i \begin{bmatrix} 2 & -1 & & & \\ -1 & 2 & -1 & & \\ & \ddots & \ddots & \ddots & \\ & & -1 & 2 & -1 \\ & & & -1 & 2 \end{bmatrix}$$

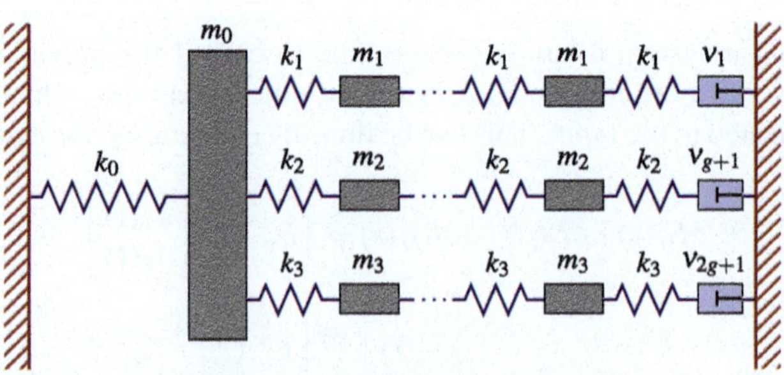

Fig. 5 (3g + 1)-mass (triple chain) oscillator with three dampers [43]

with $\kappa_i = \begin{bmatrix} 0 \ldots 0 \ k_i \end{bmatrix}^\mathsf{T} \in \mathbb{R}^{1\times g}$ and $K_{ii} \in \mathbb{R}^{g\times g}$ for $i = 1, 2, 3$. We choose the input $B_\mathrm{u} = \begin{bmatrix} 1 \ldots 1 \end{bmatrix}^\mathsf{T}$ and equally measure the velocities such that $C_\mathrm{v} = B_\mathrm{u}^\mathsf{T}$. The second-order control system reads

$$M\ddot{q}(t) + D\dot{q}(t) + Kq(t) = B_\mathrm{u}u(t), \quad y(t) = B_\mathrm{u}^\mathsf{T}\dot{q}(t).$$

We consider the triple chain with $g = 500$, thus the number of differential equations is $n = 3g + 1 = 1\,501$. The stiffness and mass parameters are set as

$$\begin{aligned} k_0 &= 50, & k_1 &= 10, & k_2 &= 20, & k_3 &= 1, \\ m_0 &= 1, & m_1 &= 1, & m_2 &= 2, & m_3 &= 3, \end{aligned}$$

and the damping parameters $\alpha = \beta = 0.002$ and $\nu_1 = \nu_{g+1} = \nu_{2g+1} = 5$.

Following the previously presented theory, we first compute a reduced-order model in first-order form of order $2r = 200$ and then recover the structure of a second-order model of order $r = 100$. The latter has again the form

$$\widehat{M}\ddot{\hat{q}}(t) + \widehat{D}\dot{\hat{q}}(t) + \widehat{K}\hat{q}(t) = \widehat{B}_\mathrm{u}u(t), \quad \hat{y}(t) = \widehat{B}_\mathrm{u}^\mathsf{T}\hat{q}(t),$$

with symmetric $\widehat{M}, \widehat{D}, \widehat{K} \in \mathbb{R}^{r\times r}$ and $\widehat{B}_\mathrm{u} \in \mathbb{R}^{r\times m}$, where $\widehat{M} = I_r$, $\widehat{K} > 0$ and $\widehat{D} = \widehat{D}^\mathsf{T}$. The plot of the absolute value of the original and reduced-order transfer functions can be found in Fig. 6a, whereas Fig. 6b displays the relative error of the transfer function on the imaginary axis, respectively. With a maximum relative error of approximately $3.1 \cdot 10^{-2}$ we obtain a good match between the original and the reduced-order system.

5 $\mathcal{H}_\infty$-Optimal Rational Approximation

In this section, we briefly describe an interpolatory $\mathcal{H}_\infty$ model order reduction scheme for systems with symmetric mass, damping, and stiffness matrices and co-located inputs and position outputs, i.e., we have $M, D, K > 0$, $B_\mathrm{u} = C_\mathrm{p}^\mathsf{T}$ and $C_\mathrm{v} = 0$ in order to be able to preserve symmetry and asymptotic stability by an appropriate choice of the projection spaces. More precisely, we construct a sequence of reduced-order transfer functions of the form

$$\widehat{H}_j(s) := B_\mathrm{u}^\mathsf{T}V_j\left(s^2V_j^\mathsf{T}MV_j + sV_j^\mathsf{T}DV_j + V_j^\mathsf{T}KV_j\right)^{-1}V_j^\mathsf{T}B_\mathrm{u}, \quad j = 1, 2, \ldots$$

for appropriately chosen projection matrices V_j. Our method aims to iteratively reduce the $\mathcal{H}_\infty$-norm of the error transfer function $\mathcal{E}_j(s) := H(s) - \widehat{H}_j(s)$. To do so, we compute $\|\mathcal{E}_j\|_{\mathcal{H}_\infty} := \max_{\omega\in\mathbb{R}\cup\{\infty\}} \|\mathcal{E}_j(\omega\mathrm{i})\|_2$. The point where $\|\mathcal{E}_j\|_{\mathcal{H}_\infty}$ is attained is denoted by $\omega_j\mathrm{i}$. We then choose $V_{j+1} :=$

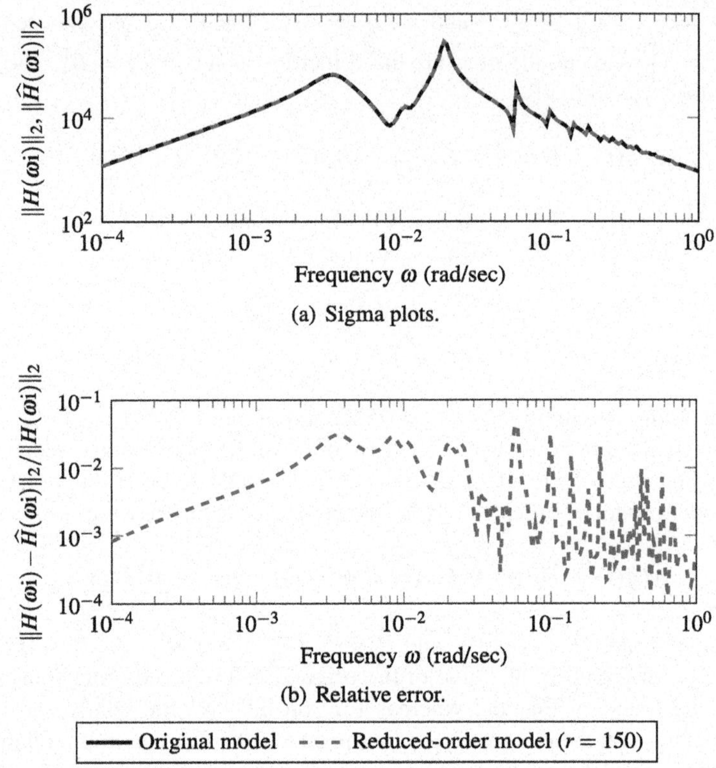

Fig. 6 Positive real balanced truncation results for the triple chain oscillator

$\left[V_j \; (-\omega_j^2 M + \omega_j \mathrm{i} D + K)^{-1} B_{\mathrm{u}}\right]$. For numerical reasons, V_{j+1} (and all other projection matrices appearing in this section) are orthogonalized. This choice guarantees Hermite interpolation properties between the original and reduced-order transfer functions $H(s)$ and $\widehat{H}_j(s)$ at the interpolation points $\omega_1\mathrm{i}, \omega_2\mathrm{i}, \ldots, \omega_j\mathrm{i}$, such that the error near these points becomes small. This procedure is repeated until a specified error tolerance is met.

The main computational cost of this algorithm is the repeated computation of the $\mathcal{H}_\infty$-norm of the error transfer function, which is expensive to evaluate since the error system is of large dimension. However, these computations have been made possible by the methods presented in [44, 45], which we use here and which work as follows.

Assume that a transfer function is given by $H(s) = C(sE - A)^{-1}B$ with the regular matrix pencil $sE - A \in \mathbb{R}[s]^{n \times n}$, $B \in \mathbb{R}^{n \times m}$, and $C \in \mathbb{R}^{p \times n}$ with $n \gg m, p$. Then, the algorithm determines a sequence of reduced-order transfer functions of the form $H_j(s) = CV_j(sW_j^{\mathsf{H}}EV_j - W_j^{\mathsf{H}}AV_j)^{-1}W_j^{\mathsf{H}}B$, where $V_j, W_j \in \mathbb{C}^{n \times k_j}$ and $k_j \ll n$ for $j = 1, 2, \ldots$, and where $\left\|H_j\right\|_{\mathcal{L}_\infty} := \max_{\omega \in \mathbb{R} \cup \{\infty\}} \left\|H_j(\omega \mathrm{i})\right\|_2$ converges to $\|H\|_{\mathcal{H}_\infty}$. Since the matrices defining the reduced-order transfer functions $H_j(s)$, are of small dimensions, $\|H_j\|_{\mathcal{L}_\infty}$ can be efficiently computed using well-established methods such as [46, 47]. Assume first that $m = p$. Further suppose that j interpolation points $\omega_1\mathrm{i}, \ldots, \omega_j\mathrm{i} \in \mathrm{i}\mathbb{R}$ are already given. In this case, we choose

$$V_j = \left[(\omega_1\mathrm{i}E - A)^{-1}B \ \ldots \ (\omega_j\mathrm{i}E - A)^{-1}B\right],$$
$$W_j = \left[(\omega_1\mathrm{i}E - A)^{-\mathsf{H}}C^{\mathsf{H}} \ \ldots \ (\omega_j\mathrm{i}E - A)^{-\mathsf{H}}C^{\mathsf{H}}\right],$$

which amounts to the Hermite interpolation conditions

$$H(\omega_k\mathrm{i}) = H_j(\omega_k\mathrm{i}), \quad H'(\omega_k\mathrm{i}) = H_j'(\omega_k\mathrm{i}), \quad k = 1, \ldots, j,$$

that carry over directly to the functions $\sigma(s) := \|H(s)\|_2$ and $\sigma_j(s) := \|H_j(s)\|_2$. These Hermite interpolation conditions are then used to prove a superlinear rate of convergence to a local maximum of $\sigma(\mathrm{i}\cdot)$ provided that the algorithm converges. The situation is more difficult if $m \neq p$ since then, V_j and W_j would have different dimensions and the pencil $sW_j^{\mathsf{H}}EV_j - W_j^{\mathsf{H}}AV_j$ would be singular. This situation also occurs if V_j or W_j do not have full column rank. Thus, an alternative choice for V_j and W_j, which is outlined in [44], and QR factorizations can be used to obtain the matrices V_j, W_j such that the pencil $sW_j^{\mathsf{H}}EV_j - W_j^{\mathsf{H}}AV_j$ is regular.

In Fig. 7, we illustrate the effectiveness of our algorithm using the triple chain oscillator benchmark example [43] (see Fig. 5) of order $n = 1\,000$ and an $\mathcal{H}_\infty$-error bound of 10^{-6}. Compared with the second-order balanced truncation (SOBT, vp version) approach described in [12, Alg. 3.2], the greedy interpolation approach has a slightly larger maximal error for the same reduced order $r = 29$. However, in contrast to SOBT, we obtain full information on the current error and may terminate whenever the reduced-order model satisfies a prescribed error bound. Moreover, the method also allows for an easy adaptation to frequency-limited reduction. We currently investigate post-processing strategies to improve the performance of the greedy interpolation such as an additional optimization of the interpolation points.

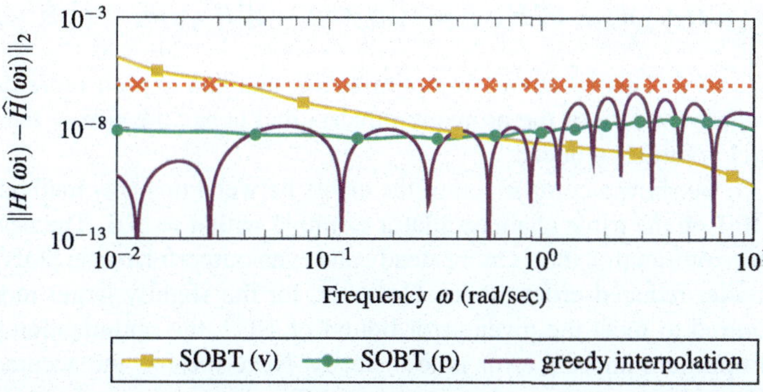

Fig. 7 Comparison of absolute errors of different methods for second-order model reduction. The interpolation points of our greedy approach are plotted as red crosses

Furthermore, our approach may be combined with the subspaces obtained from (SO)BT to initialize the first projection space.

The previously described algorithm leads to an error function $\mathcal{E}_j(s)$, which is zero at the interpolation points in exact arithmetics. This results in the spiky shape of the error maximum singular value function that can be observed in Fig. 7. Such a behavior is generally unwanted, since this indicates that our reduced-order model approximates the given model at a few frequencies much better than at others. In this way, accuracy is "wasted" in a few regions that could be used to improve the overall accuracy of the reduced-order model. This is an inherent problem of a greedy approximation strategy with interpolation on the imaginary axis.

To smoothen the error maximum singular value function and reach a better approximation with respect to the $\mathcal{H}_\infty$-norm, we use direct numerical optimization. In particular, after a new interpolation point has been chosen according to the previously described greedy algorithm, we vary the interpolation points such that the $\mathcal{H}_\infty$-error is locally minimized. This requires the solution of a nonsmooth, nonconvex and nonlinear optimization problem. We use the method described in [48], implemented in the software package GRANSO.[1] This iterative optimization requires the repeated evaluation of the $\mathcal{H}_\infty$-norm of the error system, which is high-dimensional. For that, we again apply the method described in [44], which is well-suited for this task.

It is important to note that the gradient of $\|\mathcal{E}_j\|_{\mathcal{H}_\infty}$ with respect to the interpolation points can be computed analytically. Furthermore, the direct optimization strategy is not limited to just the interpolation points. On top of that, we can optimize *tangential directions* of the interpolation as well. In case of tangential interpolation, the projection matrix can be chosen as

$$V_j = \left[(s_1^2 M + s_1 D + K)^{-1} B_{\mathrm{u}} b_1 \ \dots \ (s_j^2 M + s_j D + K)^{-1} B_{\mathrm{u}} b_j\right],$$

where $b_k \in \mathbb{C}^m$ and $s_k \in \mathbb{C}$ for $k = 1, \dots, j$ are not in the spectrum of $s^2 M + sD + K$. In this way, the interpolation condition is relaxed such that we now only have *tangential* Hermite interpolation between $H(s)$ and $\widehat{H}_j(s)$, that is

$$H(s_k)b_k = \widehat{H}_j(s_k)b_k, \quad b_k^{\mathsf{H}} H(s_k) = b_k^{\mathsf{H}} \widehat{H}_j(s_k), \quad b_k^{\mathsf{H}} H'(s_k)b_k = b_k^{\mathsf{H}} \widehat{H}_j'(s_k)b_k,$$

for $k = 1, \dots, j$. This results in an optimization that can exploit more degrees of freedom, while the size of the projection matrix and hence the size of the reduced-order model is further reduced.

In Fig. 8, we show a comparison of the errors between this new method and (the faster) SOBT on the triple chain oscillator example with $n = 301$. The new method leads to an error function that is more steady and even outperforms the SOBT method for the smaller reduced-order model. However, for the slightly larger model order that is required to meet the given error bound of 10^{-6}, the optimization got stuck in a local optimum and the error is less steady. Nevertheless, the accuracy is still comparable with the accuracy obtained by SOBT.

[1] Available at http://www.timmitchell.com/software/GRANSO.

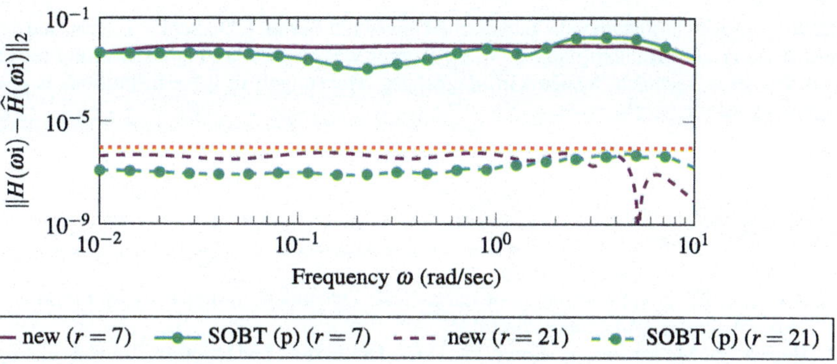

Fig. 8 Comparison of the (absolute) errors of SOBT (vp) with the new, optimization-based method

In order to improve the behavior of the error and make it globally more steady we have also recently developed a new optimization framework. There, we do not optimize the $\mathcal{H}_\infty$-error itself but rather parametrize the reduced-order model and then minimize the sum of squares of the error evaluated at certain sampling points on the imaginary axis [49].

6 Conclusions

We have presented an overview of recently developed structure-preserving model order reduction methods for second-order systems. We have started with an adaptation of the dominant pole algorithm for modally damped mechanical systems and, afterwards, have introduced extensions of the frequency- and time-limited balanced truncation methods for second-order systems in various ways. We have presented an approach for structure recovery of second-order systems based on positive real balanced truncation, which also yields an a priori error bound in the gap metric, and concluded with an $\mathcal{H}_\infty$ greedy interpolation approach yielding an $\mathcal{H}_\infty$-error optimal approximation. Numerical examples for all the presented approaches have illustrated their effectiveness.

Additionally to the approaches for linear systems summarized here, we were able to develop model order reduction techniques for (parametric) mechanical systems with special nonlinearities, namely bilinear control systems and quadratic-bilinear systems. These techniques are described in [19, 50, 51].

Acknowledgements This work has been supported by the German Research Foundation (DFG) Priority Program 1897: “Calm, Smooth and Smart – Novel Approaches for Influencing Vibrations by Means of Deliberately Introduced Dissipation”. P. Benner and S. W. R. Werner were also supported by the German Research Foundation (DFG) Research Training Group 2297 “Mathematical Complexity Reduction (MathCoRe)”, Magdeburg. This work has been carried out while R. S. Beddig and M. Voigt were affiliated with Technische Universität Berlin, I. Dorschky, T. Reis, and M. Voigt were at Universität Hamburg and S. W. R. Werner was at the Max Planck Institute for Dynamics of Complex Technical Systems in Magdeburg. The support of these institutions is gratefully acknowledged.

References

1. Antoulas, A.C.: Approximation of Large-Scale Dynamical Systems, vol. 6. SIAM, Philadelphia, PA (2005). (Adv. Des. Control.)
2. Benner, P., Mehrmann, V., Sorensen, D.C.: Dimension Reduction of Large-Scale Systems. Lecture Notes in Computational Science and Engineering, vol. 45. Springer, Berlin, Heidelberg (2005)
3. Quarteroni, A., Rozza, G.: Reduced Order Methods for Modeling and Computational Reduction. MS&A–Modeling, Simulation and Applications, vol. 9. Springer, Cham (2014)
4. Benner, P., Cohen, A., Ohlberger, M., Willcox, K.: Model Reduction and Approximation: Theory and Algorithms. Computational Science & Engineering. SIAM, Philadelphia, PA (2017)
5. Benner, P., Schilders, W., Grivet-Talocia, S., Quarteroni, A., Rozza, G., Silveira, L.M.: Model Order Reduction. System- and Data-Driven Methods and Algorithms, vol. 1. De Gruyter, Berlin, Boston (2021)
6. Georgiou, T.T., Smith, M.C.: Optimal robustness in the gap metric. IEEE Trans. Autom. Control **35**(6), 673–686 (1990)
7. Meyer, D.G., Srinivasan, S.: Balancing and model reduction for second-order form linear systems. IEEE Trans. Autom. Control **41**(11), 1632–1644 (1996)
8. Chahlaoui, Y., Lemonnier, D., Vandendorpe, A., Van Dooren, P.: Second-order balanced truncation. Linear Algebra Appl. **415**(2–3), 373–384 (2006)
9. Hartmann, C., Vulcanov, V.M., Schütte, C.: Balanced truncation of linear second-order systems: a Hamiltonian approach. Multiscale Model. Simul. **8**(4), 1348–1367 (2010)
10. Benner, P., Saak, J.: Efficient balancing-based MOR for large-scale second-order systems. Math. Comput. Model. Dyn. Syst. **17**(2), 123–143 (2011)
11. Benner, P., Kürschner, P., Saak, J.: An improved numerical method for balanced truncation for symmetric second-order systems. Math. Comput. Model. Dyn. Syst. **19**(6), 593–615 (2013)
12. Reis, T., Stykel, T.: Balanced truncation model reduction of second-order systems. Math. Comput. Model. Dyn. Syst. **14**(5), 391–406 (2008)
13. Bai, Z., Su, Y.: Dimension reduction of large-scale second-order dynamical systems via a second-order Arnoldi method. SIAM J. Sci. Comput. **26**(5), 1692–1709 (2005)
14. Freund, R.W.: Padé-type model reduction of second-order and higher-order linear dynamical systems. In Benner, P., Mehrmann, V., Sorensen, D.C. (eds.) Dimension Reduction of Large-Scale Systems. Lecture Notes in Computational Science and Engineering, vol. 45, pp. 173–189. Springer, Berlin, Heidelberg (2005)
15. Salimbahrami, B., Lohmann, B.: Order reduction of large scale second-order systems using Krylov subspace methods. Linear Algebra Appl. **415**, 385–405 (2006)
16. Beattie, C.A., Gugercin, S.: Interpolatory projection methods for structure-preserving model reduction. Syst. Control Lett. **58**(3), 225–232 (2009)
17. Panzer, H.K.F., Wolf, T., Lohmann, B.: H_2 and H_∞ error bounds for model order reduction of second order systems by Krylov subspace methods. In: Proceedings of the 2013 European Control Conference, pp. 4484–4489 (2013)
18. Beddig, R.S., Benner, P., Dorschky, I., Reis, T., Schwerdtner, P., Voigt, M., Werner, S.W.R.: Model reduction for second-order dynamical systems revisited. Proc. Appl. Math. Mech. **19**(1), e201900224 (2019)
19. Werner, S.W.R.: Structure-Preserving Model Reduction for Mechanical Systems. Dissertation, Otto-von-Guericke-Universität, Magdeburg, Germany (2021)

20. Dorschky, I.: Balancing-Based Structure Preserving Model Order Reduction of Second Order Systems. Dissertation, Universität Hamburg, Germany (2021)
21. Davison, E.J.: A method for simplifying linear dynamic systems. IEEE Trans. Autom. Control **11**(1), 93–101 (1966)
22. Martins, N., Lima, L.T.G., Pinto, H.J.C.P.: Computing dominant poles of power system transfer functions. IEEE Trans. Power Syst. **11**(1), 162–170 (1996)
23. Rommes, J., Martins, N.: Computing transfer function dominant poles of large-scale second-order dynamical systems. IEEE Trans. Power Syst. **21**(4), 1471–1483 (2006)
24. Benner, P., Kürschner, P., Tomljanović, Z., Truhar, N.: Semi-active damping optimization of vibrational systems using the parametric dominant pole algorithm. ZAMM Z. Angew. Math. Mech. **96**(5), 604–619 (2016)
25. Beattie, C., Benner, P.: $\mathcal{H}_2$-Optimality Conditions for Structured Dynamical Systems. Preprint MPIMD/14-18, Max Planck Institute for Dynamics of Complex Technical Systems Magdeburg (2014)
26. Saak, J., Siebelts, D., Werner, S.W.R.: A comparison of second-order model order reduction methods for an artificial fishtail. at-Automatisierungstechnik **67**(8), 648–667 (2019)
27. Benner, P., Werner, S.W.R.: SOMDDPA–Second-Order Modally-Damped Dominant Pole Algorithm (version 2.0) (2021)
28. Oberwolfach Benchmark Collection: Butterfly Gyroscope. Hosted at MORwiki–Model Order Reduction Wiki (2004)
29. Gawronski, W., Juang, J.N.: Model reduction in limited time and frequency intervals. Int. J. Syst. Sci. **21**(2), 349–376 (1990)
30. Haider, K., Ghafoor, A., Imran, M., Malik, F.M.: Frequency interval Gramians based structure preserving model reduction for second-order systems. Asian J. Control **20**(2), 790–801 (2018)
31. Haider, K., Ghafoor, A., Imran, M., Malik, F.M.: Time-limited Gramian-based model order reduction for second-order form systems. Trans. Inst. Meas. Control **41**(8), 2310–2318 (2019)
32. Benner, P., Werner, S.W.R.: Frequency- and time-limited balanced truncation for large-scale second-order systems. Linear Algebra Appl. **623**, 68–103 (2021). (Special issue in honor of P. Van Dooren, Edited by F. Dopico, D. Kressner, N. Mastronardi, V. Mehrmann, and R. Vandebril)
33. Benner, P., Kürschner, P., Saak, J.: Frequency-limited balanced truncation with low-rank approximations. SIAM J. Sci. Comput. **38**(1), A471–A499 (2016)
34. Kürschner, P.: Balanced truncation model order reduction in limited time intervals for large systems. Adv. Comput. Math. **44**(6), 1821–1844 (2018)
35. Benner, P., Werner, S.W.R.: MORLAB–Model Order Reduction LABoratory (version 5.0) (2019). https://www.mpi-magdeburg.mpg.de/projects/morlab
36. Benner, P., Werner, S.W.R.: MORLAB—The Model Order Reduction LABoratory. In: Benner, P., Breiten, T., Faßbender, H., Hinze, M., Stykel, T., Zimmermann, R. (eds.) Model Reduction of Complex Dynamical Systems. International Series of Numerical Mathematics, vol. 171, pp. 393–415. Birkhäuser, Cham (2021)
37. Benner, P., Werner, S.W.R.: SOLBT–Limited Balanced Truncation for Large-Scale Sparse Second-Order Systems (version 3.0) (2021)
38. Dorschky, I., Reis, T., Voigt, M.: Balanced truncation model reduction for symmetric second order systems–a passivity-based approach. SIAM J. Matrix Anal. Appl. **42**(4), 1602–1635 (2021)
39. Guiver, C., Opmeer, M.R.: Error bounds in the gap metric for dissipative balanced approximations. Linear Algebra Appl. **439**(12), 3659–3698 (2013)
40. Reis, T.: Lur'e equations and even matrix pencils. Linear Algebra Appl. **434**(1), 152–173 (2011)
41. Poloni, F., Reis, T.: A deflation approach for large-scale Lur'e equations. SIAM J. Matrix Anal. Appl. **33**(4), 1339–1368 (2012)
42. Gohberg, I., Lancaster, P., Rodman, L.: Matrices and Indefinite Scalar Products. Birkhäuser, Basel, Bosten, Stuttgart (1983)
43. Truhar, N., Veselić, K.: An efficient method for estimating the optimal dampers' viscosity for linear vibrating systems using Lyapunov equation. SIAM J. Matrix Anal. Appl. **31**(1), 18–39 (2009)

44. Aliyev, N., Benner, P., Mengi, E., Schwerdtner, P., Voigt, M.: Large-scale computation of $\mathcal{L}_\infty$-norms by a greedy subspace method. SIAM J. Matrix Anal. Appl. **38**(4), 1496–1516 (2017)
45. Schwerdtner, P., Voigt, M.: Computation of the $\mathcal{L}_\infty$-norm using rational interpolation. IFAC-PapersOnline. In: 9th IFAC Symposium on Robust Control Design ROCOND 2018, vol. 51, No. 25, pp. 84–89. Florianópolis, Brazil (2018)
46. Boyd, S., Balakrishnan, V.: A regularity result for the singular values of a transfer matrix and a quadratically convergent algorithm for computing its L_∞-norm. Syst. Control Lett. **15**(1), 1–7 (1990)
47. Benner, P., Sima, V., Voigt, M.: $\mathcal{L}_\infty$-norm computation for continuous-time descriptor systems using structured matrix pencils. IEEE Trans. Autom. Control **57**(1), 233–238 (2012)
48. Curtis, F.E., Mitchell, T., Overton, M.L.: A BFGS-SQP method for nonsmooth, nonconvex, constrained optimization and its evaluation using relative minimization profiles. Optim. Methods Softw. **32**(1), 148–181 (2017)
49. Schwerdtner, P., Voigt, M.: SOBMOR: structured optimization-based model order reduction. SIAM J. Sci. Comput. **45**(2), A502–A529 (2023)
50. Benner, P., Gugercin, S., Werner, S.W.R.: Structure-preserving interpolation for model reduction of parametric bilinear systems. Automatica J. IFAC **132**, 109799 (2021)
51. Benner, P., Gugercin, S., Werner, S.W.R.: Structure-preserving interpolation of bilinear control systems. Adv. Comput. Math. **47**(3), 43 (2021)

Numerical and Experimental Assessment of Acoustic Radiation Damping

Suhaib Koji Baydoun and Steffen Marburg

1 Introduction

In the light of the current environmental crisis, lightweight design becomes an increasingly important aspect in various engineering disciplines. Appropriate application of lightweight materials and structures can significantly reduce the energy consumption during manufacturing, operation and recycling of a product. Lightweight structures are characterized by a high stiffness relative to their mass, which however makes them prone to noise, vibration and harshness. In the current engineering practice, vibration mitigation is usually achieved by adding damping treatments in a late stage of the design process or even after manufacturing. The resulting increase in mass and volume can be significant and deteriorates the ecological footprint of the product. Recent approaches for a deliberate introduction of energy dissipation includes materials with high inherent damping [1] and the use of special devices such as particle dampers [2], electrorheological valves [3] and acoustic black holes [4].

This contribution deals with another, often neglected damping phenomenon, commonly denoted as *acoustic radiation damping*. That is the energy dissipation in vibrating structures by virtue of far-field sound radiation, accounting for a significant share in the overall damping of lightweight structures. Although radiation damping is not to be understood as a deliberately introduced damping mechanism, it nevertheless requires accurate quantification. In fact, deliberately introduced treatments require that the extent of added damping is at least within the same order of magnitude as radiation damping in order to be effective [5]. Furthermore, including the effect of radiation damping facilitates more reliable vibroacoustic simulations.

Early attempts to estimate radiation damping are based on the principles of power flow and assume a sufficiently high modal density [6, 7]. While these expressions are accurate approximations in the high frequency range above coincidence [8],

S. K. Baydoun (✉) · S. Marburg
TUM School of Engineering and Design, Chair of Vibroacoustics of Vehicles and Machines, Technical University of Munich, Boltzmannstraße 15, 85748 Garching, Germany
e-mail: suhaib.baydoun@tum.de

P. Eberhard (ed.), *Calm, Smooth and Smart*, Lecture Notes in Applied and Computational Mechanics 102, https://doi.org/10.1007/978-3-031-36143-2_12

they are rather deficient at low frequencies due to well-separated modes. The latter implies that radiation damping is strongly dependent on the actual excitation and mounting condition of the structure at hand. The applicability of the above mentioned theoretical expressions to complex geometries and to inhomogeneous materials is also difficult to judge. Alternatively, experimental assessment of radiation damping is possible based on sound power measurements inside reverberation chambers and anechoic rooms [9]. This allows to consider geometries that are more complex and also to accurately reflect material properties. If the measurements are conducted in the actual mounting condition, even the effect of the boundary conditions is realistically included. However, in the low frequency range, reliable acoustic measurements are hardly possible due to the modal behavior of the rooms [10].

This contribution presents a framework for a low frequency assessment of radiation damping by acoustic simulations based on the boundary element method (BEM) [11]. The use of BEM yields clear advantages compared to other approaches for evaluating sound radiation and associated damping. Where domain-based discretization methods such as the finite element method (FEM) [12] necessitate special treatments for truncating the far-field sound radiation [13], modeling in BEM is restricted to the radiating surface of the structure. Moreover, compared to the above mentioned experimental approaches, which require special facilities with limited scope of application, the applicability of BEM is more versatile allowing to cover a large range of frequencies and geometrical configurations. This contribution combines BEM with both a structural simulation model [8] and with an experimental characterization of structural mobility [14, 15]. Excitation-dependent radiation loss factors as well as modal loss factors that are associated with certain structural modes are derived.

The remainder of the contribution is organized as follows. Section 2 presents the methodological tools for evaluating radiation damping, including BEM, FEM and structural mobility measurements. Section 3 presents an application of these methods to a honeycomb sandwich panel. Finally, the main content and results are summarized in Sect. 4.

2 Methods

This section describes the framework for a numerical and an experimental-numerical assessment of radiation damping. In the upcoming derivations, a harmonic time dependency of $e^{-\mathrm{i}\omega t}$ is assumed, where $\mathrm{i} = \sqrt{-1}$, the angular frequency is $\omega = 2\pi f$ and t denotes the time.

2.1 Boundary Element Method for the Evaluation of Sound Radiation

In this work, the propagation of acoustic waves in the fluid surrounding the structure is modeled by the inhomogeneous Helmholtz equation [16]

$$\Delta p(\mathbf{x}) + \frac{\omega^2}{c^2} p(\mathbf{x}) = -q, \tag{1}$$

in which c is the speed of sound and q refers to the source. The sound pressure field $p(\mathbf{x})$ is the sum of the scattered field p^{s} and the incident field p^{i}. The same applies to the fluid particle velocity v_{f}, i.e.

$$p(\mathbf{x}) = p^{\mathrm{s}}(\mathbf{x}) + p^{\mathrm{i}}(\mathbf{x}) \tag{2}$$

$$v_{\mathrm{f}}(\mathbf{x}) = v_{\mathrm{f}}^{\mathrm{s}}(\mathbf{x}) + v_{\mathrm{f}}^{\mathrm{i}}(\mathbf{x}). \tag{3}$$

Reformulation of Eq. (1) by the Kirchhoff integral theorem and collocation discretization using boundary elements yields the linear system of equations [16]

$$\mathbf{H}(\omega)\mathbf{p} = \mathbf{G}(\omega)\left(\mathbf{v}_{\mathrm{f}} - \mathbf{v}_{\mathrm{f}}^{\mathrm{i}}\right) + \mathbf{H}(\omega)\mathbf{p}^{\mathrm{i}} \tag{4}$$

for the description of the acoustic field. Therein, $\mathbf{p}$ is the unknown vector containing the sound pressure solution at the nodes. The frequency dependent coefficient matrices $\mathbf{H}(\omega)$ and $\mathbf{G}(\omega)$ relate the fluid particle velocity $\mathbf{v}_{\mathrm{f}}$ to the sound pressure. Acoustically rigid baffles can be taken into account by evaluating $\mathbf{H}(\omega)$ and $\mathbf{G}(\omega)$ using a half-space formulation with a modified Green's function [17]. The incident sound pressure field is denoted by $\mathbf{p}^{\mathrm{i}}$ and the corresponding incident particle velocity is $\mathbf{v}_{\mathrm{f}}^{\mathrm{i}}$.

After solving Eq. (4), the time-averaged radiated sound power can be obtained from

$$P = \frac{1}{2}\mathrm{Re}\left(\mathbf{p}^{\mathrm{T}}\mathbf{\Theta}\mathbf{v}_{\mathrm{f}}^{*}\right), \tag{5}$$

where $(\cdot)^*$ denotes the conjugate complex, $(\cdot)^{\mathrm{T}}$ is the transpose, and $\mathbf{\Theta}$ is the boundary mass matrix, which is obtained by integration of the boundary element interpolation functions. Note that only the real part $\mathrm{Re}\,(\cdot)$ of the complex sound power is associated with radiation damping. The imaginary part of the sound power corresponds to near-field sound radiation, which has a mass-like effect on the structure and hence, does not dissipate energy.

2.2 *Numerical Assessment of Radiation Damping Using Finite and Boundary Elements*

The fluid particle velocity $\mathbf{v}_\mathrm{f}$ in Eqs. (4) and (5) is determined by the oscillations of the solid structure that is submerged by the fluid. The underlying equations of linear time-harmonic elasticity are discretized by finite elements, which yields [18]

$$\left(\mathbf{K} - \mathrm{i}\omega\mathbf{D} - \omega^2\mathbf{M}\right)\mathbf{u} = \mathbf{f}_\mathrm{s} + \mathbf{f}_\mathrm{f}. \tag{6}$$

The vector $\mathbf{u}$ contains the unknown displacement degrees of freedom (DOFs). The matrices $\mathbf{K}, \mathbf{D}, \mathbf{M}$ are related to the stiffness, damping and mass of the structure. The excitation is a combination of structural forces $\mathbf{f}_\mathrm{s}$ and fluid forces $\mathbf{f}_\mathrm{f}$. The latter acts by virtue of the surrounding acoustic field. This coupling condition and the continuity condition can be expressed as

$$\mathbf{f}_\mathrm{f} = \mathbf{C}_\mathrm{sf}\mathbf{p}, \quad \text{and} \quad \mathbf{v}_\mathrm{f} = -\mathrm{i}\omega\mathbf{C}_\mathrm{fs}\mathbf{u}, \tag{7}$$

in which the matrices $\mathbf{C}_\mathrm{sf}$ and $\mathbf{C}_\mathrm{fs}$ establish the coupling between the structural and the acoustic subdomains [19]. Combining Eqs. (4), (6) and (7) yields the fully coupled structural acoustic system

$$\begin{bmatrix} \mathbf{K} - \mathrm{i}\omega\mathbf{D} - \omega^2\mathbf{M} & -\mathbf{C}_\mathrm{sf} \\ \mathrm{i}\omega\mathbf{G}(\omega)\mathbf{C}_\mathrm{fs} & \mathbf{H}(\omega) \end{bmatrix} \begin{bmatrix} \mathbf{u} \\ \mathbf{p} \end{bmatrix} = \begin{bmatrix} \mathbf{f}_\mathrm{s} \\ -\mathbf{G}(\omega)\mathbf{v}_\mathrm{f}^\mathrm{i} + \mathbf{H}(\omega)\mathbf{p}^\mathrm{i} \end{bmatrix}. \tag{8}$$

Forming the Schur complement of $\mathbf{H}(\omega)$ with respect to the block system matrix in Eq. (8) and thereby eliminating the pressure DOFs from Eq. (8) yields

$$\left[\mathbf{K} - \mathrm{i}\omega\mathbf{D} - \omega^2\mathbf{M} + \mathrm{i}\omega\mathbf{C}_\mathrm{sf}\mathbf{H}^{-1}(\omega)\mathbf{G}(\omega)\mathbf{C}_\mathrm{fs}\right]\mathbf{u} = \underbrace{\mathbf{f}_\mathrm{s} + \mathbf{C}_\mathrm{sf}\left(\mathbf{p}^\mathrm{i} - \mathbf{H}^{-1}(\omega)\mathbf{G}(\omega)\mathbf{v}_\mathrm{f}^\mathrm{i}\right)}_{\mathbf{f}_\mathrm{t}}, \tag{9}$$

which can be interpreted as a structural equation with an additional term corresponding to the mass and damping contributions of the fluid. The total force vector $\mathbf{f}_\mathrm{t}$ in Eq. (9) comprises both, structural loading $\mathbf{f}_\mathrm{s}$ and acoustic loading due to the incident acoustic field characterized by $\mathbf{p}^\mathrm{i}$ and $\mathbf{v}_\mathrm{f}^\mathrm{i}$. Efficient strategies for solving equations of form (9) are discussed in e.g. [20–22].

2.2.1 Harmonic Radiation Loss Factors

Harmonic radiation loss factors are a result of a frequency-wise response analysis and thus depend on the type of the excitation. They are obtained by relating the time-averaged radiated sound power P obtained by Eq. (5) to the time-averaged total energy E_tot of the vibration [23], i.e.

$$\eta_{\mathrm{r}} = \frac{P}{|\omega E_{\mathrm{tot}}|}. \tag{10}$$

In harmonic problems, the time-averaged total vibrational energy in Eq. (10) is equivalent to twice the time-averaged potential energy. This assumption gives [18]

$$E_{\mathrm{tot}} = \frac{1}{2}\mathbf{u}^{\mathrm{T}}\mathbf{K}\mathbf{u}^{*} - \frac{1}{2}\mathbf{f}_{\mathrm{t}}^{\mathrm{H}}\mathbf{u}. \tag{11}$$

The first term in Eq. (11) is the potential energy due to the elastic straining and the second term corresponds to the work done by external forces. The superscript $(\cdot)^{\mathrm{H}}$ indicates the conjugate transpose of a matrix. In view of the balance of forces in Eq. (9), the time-averaged total vibrational energy can alternatively be expressed as twice the sum of time-averaged kinetic and dissipated energies of the structural acoustic system, i.e.

$$E_{\mathrm{tot}} = \frac{1}{2}\mathbf{u}^{\mathrm{T}}\left(\omega^{2}\mathbf{M} + \mathrm{i}\omega\mathbf{D} - \mathrm{i}\omega\mathbf{C}_{\mathrm{sf}}\mathbf{H}^{-1}(\omega)\mathbf{G}(\omega)\mathbf{C}_{\mathrm{fs}}\right)\mathbf{u}^{*}. \tag{12}$$

Besides the inertial and damping terms corresponding to the finite element matrices **M** and **D**, Eq. (12) also includes energy contributions of the acoustic field. The imaginary part of $\mathbf{H}^{-1}(\omega)\mathbf{G}(\omega)$ is associated with the additional mass effect of the fluid, and its real part corresponds to radiation damping.

Since the energy expressions in Eqs. (11) and (12) are fully equivalent to each other, either of them can be employed to evaluate radiation damping by Eq. (10) in the context of a coupled FEM-BEM analysis. In practice, the expression based on the potential energy in Eq. (11) is more efficient to evaluate since it only comprises sparse coefficient matrices.

2.2.2 Modal Radiation Loss Factors

While the harmonic loss factor given by Eq. (10) depends on the type of the excitation, an alternative measure for radiation damping can be derived based on a modal analysis of the structural acoustic system. Modal analyses of closed acoustic cavities can be solved using FEM [24] or frequency independent BEM formulations [25]. These techniques yield linear eigenvalue problems (EVPs) for which well-established algorithms are available [26]. The situation is different in the context of exterior acoustic domains, which are the relevant case for the analysis of radiation damping. When using the coupled FEM-BEM formulation as described in this contribution, the underlying EVP is obtained by setting the right-hand side of Eq. (9) to zero, i.e.

$$\underbrace{\left[\mathbf{K} - \mathrm{i}\tilde{\omega}\mathbf{D} - \tilde{\omega}^{2}\mathbf{M} + \mathrm{i}\tilde{\omega}\mathbf{C}_{\mathrm{sf}}\mathbf{H}^{-1}(\tilde{\omega})\mathbf{G}(\tilde{\omega})\mathbf{C}_{\mathrm{fs}}\right]}_{\mathbf{A}(\tilde{\omega})}\boldsymbol{\Psi} = \mathbf{0}, \tag{13}$$

in which the vector $\boldsymbol{\Psi}$ denotes a fluid-loaded displacement mode. Equation (13) is a nonlinear EVP, since the coefficient matrices $\mathbf{H}(\tilde{\omega})$ and $\mathbf{G}(\tilde{\omega})$ implicitly depend on the eigenvalue parameter $\tilde{\omega}$. Recent approaches for solving nonlinear EVPs of the form (13) are based on contour integration [27, 28] and on rational approximations [29]. In this contribution, the block SS contour integral method [8] is employed, which essentially transforms the nonlinear EVP (13) into a generalized EVP

$$\mathbf{H}_1 \boldsymbol{\psi} = \lambda \mathbf{H}_2 \boldsymbol{\psi}, \tag{14}$$

with the eigenvector $\boldsymbol{\psi}$ and eigenvalue λ. The latter is identical to the complex eigenfrequency $\tilde{\omega}$ of the structural acoustic system in Eq. (13). The block matrices $\mathbf{H}_1, \mathbf{H}_2 \in \mathbb{C}^{KL \times KL}$ are defined as

$$\mathbf{H}_1 = \begin{bmatrix} \mathbf{M}_0 & \mathbf{M}_1 & \cdots & \mathbf{M}_{K-1} \\ \mathbf{M}_1 & & & \vdots \\ \vdots & & & \mathbf{M}_{2K-3} \\ \mathbf{M}_{K-1} & \cdots & \mathbf{M}_{2K-3} & \mathbf{M}_{2K-2} \end{bmatrix}, \quad \mathbf{H}_2 = \begin{bmatrix} \mathbf{M}_1 & \mathbf{M}_2 & \cdots & \mathbf{M}_{K} \\ \mathbf{M}_2 & & & \vdots \\ \vdots & & & \mathbf{M}_{2K-2} \\ \mathbf{M}_{K} & \cdots & \mathbf{M}_{2K-2} & \mathbf{M}_{2K-1} \end{bmatrix}, \tag{15}$$

where K and L are positive integers and need to be set by the user. The product KL determines the subspace dimension. The moments $\mathbf{M}_l \in \mathbb{C}^{L \times L}$ are obtained from

$$\mathbf{M}_l = \frac{1}{2\pi \mathrm{i}} \oint_C z^l \mathbf{U}^{\mathrm{H}} \mathbf{A}^{-1}(z) \mathbf{V} \, \mathrm{d}z, \quad l = 0, \ldots, 2K-1, \tag{16}$$

where the matrices $\mathbf{U}$ and $\mathbf{V}$ contain random source vectors as columns. The original system $\mathbf{A}$ is evaluated at the complex-valued shifts z. The latter is defined along an elliptic contour C that has its major axis aligned with the real axis of the complex plane. The two vertices on the real axis correspond to the lower and upper frequency bounds $(f_{\min}, f_{\max})$ respectively. A suitable definition of an ellipse can be given as

$$z(\theta) = \gamma + \rho \left(\cos\theta + \mathrm{i}\zeta \sin\theta\right), \quad \theta \in [0, 2\pi), \tag{17}$$

where $\gamma = (f_{\max} + f_{\min})/2$ and $\rho = (f_{\max} - f_{\min})/2$. After solving the generalized EVP (14), the fluid-loaded structural modes are obtained from $\boldsymbol{\Psi} = \mathbf{S}\boldsymbol{\psi}$. With the definition in Eq. (17) at hand, approximations of the matrices $\mathbf{M}_l$ and $\mathbf{S} = \left[\mathbf{S}_0, \ldots, \mathbf{S}_{K-1}\right]$ can be computed using the N-point trapezoidal rule, i.e.

$$\hat{\mathbf{S}}_l = \frac{1}{\mathrm{i}N} \sum_{j=1}^{N} \left(\frac{z(\theta_j) - \gamma}{\rho}\right)^l z'(\theta_j) \mathbf{A}^{-1} z(\theta_j) \mathbf{V}, \quad \hat{\mathbf{M}}_l = \mathbf{U}^{\mathrm{H}} \hat{\mathbf{S}}_l, \tag{18}$$

where N denotes the number of integration points on the contour and $\theta_j = 2\pi(j - 1)/N$, $j = 1, \ldots, N$. The block matrices are assembled according to Eq. (15) using the approximated moments $\hat{\mathbf{M}}_l$. Finally, the generalized EVP $\hat{\mathbf{H}}_1 \hat{\boldsymbol{\psi}}_j = \hat{\lambda}_j \hat{\mathbf{H}}_2 \hat{\boldsymbol{\psi}}_j$ can be solved and the eigenfrequencies $\tilde{\omega}_j$ as well as the fluid-loaded modes $\boldsymbol{\Psi}_j$, $j = 1,...,KL$ can be obtained by

$$\tilde{\omega}_j = \gamma + \rho\hat{\lambda}_j, \quad \boldsymbol{\Psi}_j = \hat{\mathbf{S}}\hat{\boldsymbol{\psi}}_j. \tag{19}$$

The modal loss factor corresponding to a structural mode $\boldsymbol{\Psi}_j$ is determined by [5]

$$\eta_j = -2\frac{\mathrm{Im}\left(\tilde{\omega}_j\right)}{\mathrm{Re}\left(\tilde{\omega}_j\right)}. \tag{20}$$

Note that $\mathrm{Im}\left(\tilde{\omega}_j\right)$ is negative since the harmonic time dependency is defined as $e^{-\mathrm{i}\omega t}$. Assuming $\mathbf{D} = \mathbf{0}$, the modal loss factor in Eq. (20) quantifies the extent of radiation damping for the corresponding fluid-loaded structural mode, since the energy dissipation due to sound radiation is the only damping contribution. In that case, the modal loss factor in Eq. (20) is equivalent to the harmonic loss factor given by Eq. (10) at the respective eigenfrequency $f_j = \mathrm{Re}\left(\tilde{\omega}_j\right)/2\pi$, assuming that the respective mode is excited in the response analysis.

2.3 *Hybrid Experimental-Numerical Assessment of Radiation Damping*

While the above described coupled FEM-BEM approach facilitates an accurate representation of the acoustic conditions and associated effects such as scattering and short-circuiting in the low frequency range, it suffers from two disadvantages when compared to an experimental evaluation of sound radiation. First, finite element models usually introduce vast simplifications of boundary conditions such as *clamped* or *simply supported*, which can yield large errors in acoustic quantities when sound radiation is mainly induced by edge and corner motions [23]. Second, vibroacoustic behavior is largely determined by the elastic material properties of the structure at hand. It is clear that accurate predictions of radiation damping therefore demand precise material models. Ironically, radiation damping is particularly relevant in complex material configurations that are designed to achieve a high ratio of bending stiffness to mass. These issues can be addressed by a hybrid experimental-numerical method developed in [15]. It is briefly reproduced below.

The structural finite element model is omitted and instead, the structure is characterized by

$$\mathbf{v}_\mathrm{r} = \mathbf{Y}(\omega)\left[\mathbf{f}_\mathrm{e} + \mathbf{C}_\mathrm{ef}\left(\mathbf{p}^\mathrm{i} - \mathbf{H}^{-1}(\omega)\mathbf{G}(\omega)\mathbf{v}_\mathrm{f}^\mathrm{i}\right)\right], \tag{21}$$

in which $\mathbf{Y}(\omega)$ is a mobility matrix containing experimentally determined transfer functions that relate the force excitation to the surface velocity. The vector $\mathbf{v}_\mathrm{r}$ contains the surface velocities at the nodes of the response grid. The excitation vector in square brackets on the right-hand side of Eq. (21) comprises structural excitation $\mathbf{f}_\mathrm{e}$ as well as excitation by an incident acoustic field. The excitation vector is defined on a (possibly different) excitation grid. The force associated with the incident acoustic field is computed by BEM resulting in a similar expression for the force vector as on the right-hand side of Eq. (9). Having evaluated $\mathbf{v}_r$ for a given excitation, the particle velocities on the nodes of the boundary element model are computed by

$$\mathbf{v}_f = \mathbf{C}_{fr}\mathbf{v}_r, \tag{22}$$

and the pressure field and sound radiation can be obtained by Eqs. (4) and (5). The coupling matrix $\mathbf{C}_{ef}$ in Eq. (21) relates the acoustic quantities on the boundary element mesh to the excitation grid, and similarly, $\mathbf{C}_{fr}$ in Eq. (22) establishes the coupling between the response grid and the boundary element nodes.

In order to assess radiation damping via Eq. (10), the time-averaged total vibrational energy is evaluated by

$$E_{tot} = \frac{1}{2}\mathbf{v}_r^T \left(\mathbf{M}_r - \frac{i}{\omega}\mathbf{C}_{rf}\mathbf{H}^{-1}(\omega)\mathbf{G}(\omega)\mathbf{C}_{fr} \right) \mathbf{v}_r^*, \tag{23}$$

which is similar to the expression in Eq. (12). The mass matrix $\mathbf{M}_r$ of the structure under test is assembled by the structural mass contributions of each element on the response grid. In this work, the transfer functions contained in $\mathbf{Y}(\omega)$ are obtained by scanning laser Doppler vibrometry (LDV) and excitation by an automated modal hammer. In many situations, it is not necessary to have the whole matrix $\mathbf{Y}(\omega)$, e.g. when only a local excitation is of interest. Moreover, symmetry of the structure and boundary conditions can be exploited to reduce the measurement effort. Details on this hybrid experimental-numerical approach are given in [15]. The experimental procedure for determining the mobility matrix $\mathbf{Y}(\omega)$ is reported in Sect. 3.2.

3 Application

This section studies the acoustic radiation damping of a honeycomb sandwich panel using the described numerical as well the hybrid experimental-numerical method. Modal and harmonic loss factors for different types of excitations and acoustic boundary conditions are computed.

The panel consists of two aluminum face sheets enclosing an aluminum honeycomb core. The dimension and the material properties are listed in Table 1. Two different load cases are considered. A point force is located at ($x = 0.062$ m, $y = 0.188$ m) and a monopole sound source is located at ($x = 1.376$ m, $y = 1.25$ m, $z = 0.3$ m). The origin of the corresponding coordinate system coincides with the center of the panel.

Table 1 Geometry and material properties of the aluminum honeycomb sandwich panel provided by the manufacturer. Assumed values are marked with a star*

Aluminum face sheets		
Thickness	t	0.5 mm
Density	ρ_f	2690 kg/m^3
Young's modulus	E	70 GPa
Poisson's ratio	ν_a	0.3
Aluminum honeycomb core		
Thickness	h	4.5 mm
Density	ρ_c	135 kg/m^3
Young's modulus	E_x, E_y	10 MPa*
Young's modulus	E_z	360 MPa
Shear modulus	G_{xy}	1 MPa*
Shear modulus	G_{yz}	280 MPa
Shear modulus	G_{xz}	140 MPa
Poisson's ratio	ν_c	0.01*
Dimensions	$l_x \times l_y$	0.748 × 0.5 m^2

3.1 *Results Obtained by Coupled Finite and Boundary Element Analyses*

In this section, radiation damping of the sandwich panel is studied using the coupled FEM-BEM framework described in Sect. 2.1. The panel itself is modeled using twenty-noded hexahedral solid finite elements for the representation of the core and eight-noded quadrilateral shell finite elements based on the Reissner-Mindlin theory for the two face sheets. A uniform mesh of 36 × 24 elements along the in-plane directions ensures that at least six elements capture one bending wave length in the considered frequency range. A single solid element is used to discretize the core in the out-of-plane direction. Simply supported boundary conditions are applied to the two short edges of the panel. The long edges are unconstrained. Structural damping is neglected in the following. Figure 1 shows the finite element mesh of the panel with the position of the point force excitation.

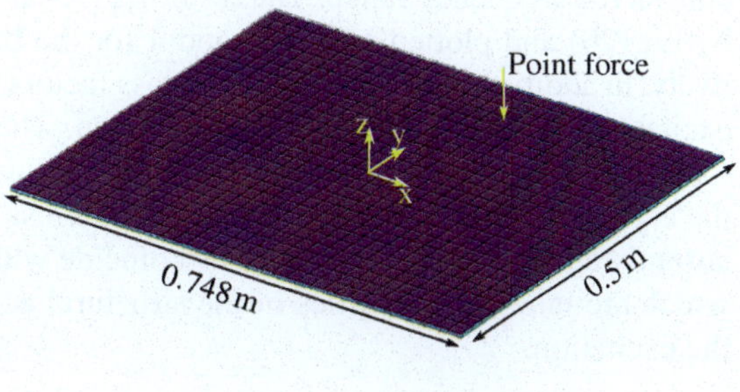

Fig. 1 Finite element model of the sandwich panel

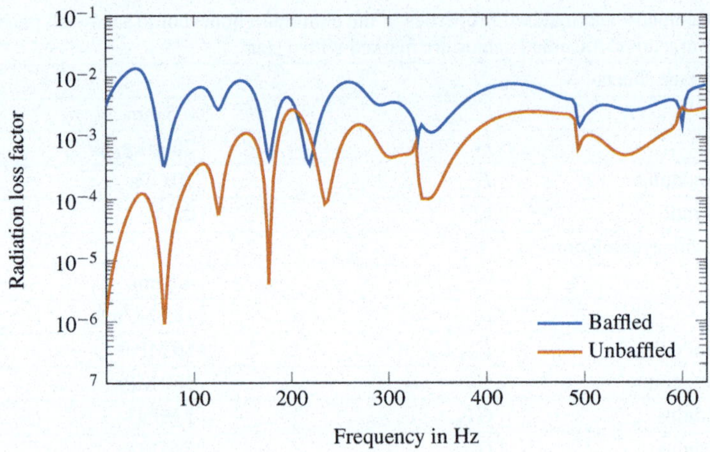

Fig. 2 Radiation loss factor by harmonic response of the honeycomb sandwich panel subject to excitation by a point force. Comparison between baffled and unbaffled acoustic boundary conditions

The structural finite element mesh is coupled to a boundary element model for the representation of the surrounding acoustic field. The boundary elements coincide with the finite elements. Two different acoustic conditions are studied: First, the panel is confined in an infinitely extended baffle, which prohibits acoustic flow between the two sides of the panel. Second, the panel is situated in a free acoustic field, which will be referred to as *unbaffled*.

Figure 2 shows radiation loss factors obtained by harmonic response analyses of the panel subject to excitation by a point force. The frequency range up to 625 Hz is considered. The baffled and the unbaffled cases are compared with each other. The acoustic short circuiting occurring in the unbaffled panel decreases radiation efficiency and thus radiation damping at low frequencies. The two curves converge to each other towards higher frequencies. The apparent dips in the loss factors correspond to structural modes of the panel associated with sound pressure cancellation among neighboring half-cells. This cancellation effect disappears at higher frequencies.

Besides harmonic analyses, modal analyses of the air-loaded panel are performed using the algorithm described in Sect. 2.2.2. Thirteen eigenfrequencies occur in the considered frequency range. The resulting modal radiation loss factors are obtained by Eq. (20) and plotted in Figs. 3 and 4 for the baffled and unbaffled case, respectively. In addition, harmonic radiation loss factors for both point force excitation and excitation by a monopole source are shown. The latter yields significantly higher loss factors at frequencies between resonances due the spatially uniform loading. At the eigenfrequencies, however, the harmonic loss factors of the two load cases consistently agree with each other. They coincide with the modal radiation loss factors, which are inherent properties of the structural acoustic system and independent of the excitation.

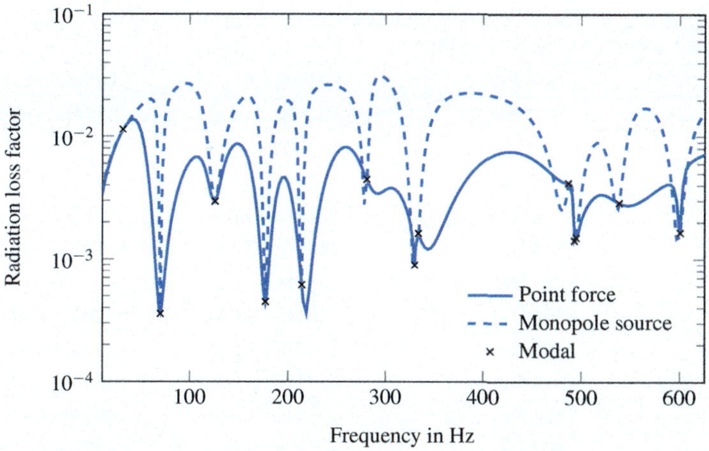

Fig. 3 Radiation loss factor of the honeycomb sandwich panel confined in an acoustically rigid baffle. Comparison between point force excitation and excitation by a monopole source. Additionally, the modal radiation loss factors are plotted at their respective eigenfrequencies

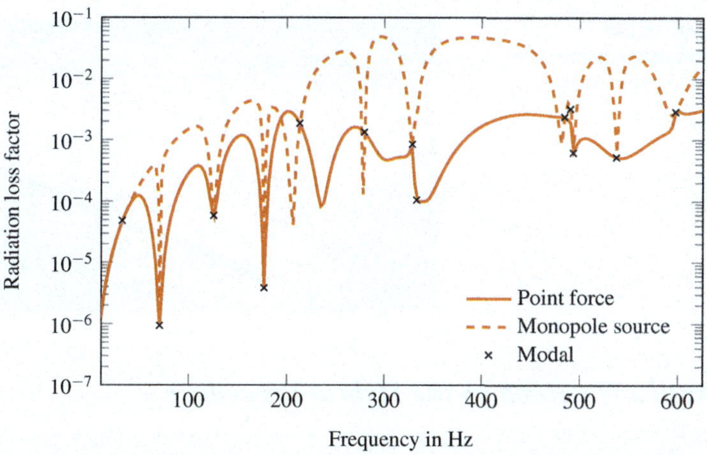

Fig. 4 Radiation loss factor of the unbaffled honeycomb sandwich panel. Comparison between point force excitation and excitation by a monopole source. Additionally, the modal radiation loss factors are plotted at their respective eigenfrequencies

Table 2 lists the eigenfrequencies of the panel in vacuo as well as the eigenfrequencies resulting from a modal analysis including air loading for the baffled and the unbaffled case. Eigenfrequency shifts of up to 5 % for the baffled and the unbaffled panel clearly illustrate the effect of additional mass and damping due to the acoustic field.

Table 2 Eigenfrequencies of the honeycomb sandwich panel in vacuo and in air considering baffled and unbaffled acoustic boundary conditions

No.	In Vacuo	Baffled (%)	Unbaffled (%)
1	32.7 Hz	30.7 Hz (−5.9)	31.0 Hz (−5.0)
2	70.7 Hz	69.9 Hz (−2.6)	69.9 Hz (−2.5)
3	130 Hz	126 Hz (−3.2)	126 Hz (−3.5)
4	181 Hz	177 Hz (−2.1)	177 Hz (−2.2)
5	218 Hz	214 Hz (−1.8)	213 Hz (−2.3)
6	287 Hz	281 Hz (−2.2)	279 Hz (−2.7)
7	336 Hz	330 Hz (−1.7)	329 Hz (−2.1)
8	340 Hz	334 Hz (−1.8)	333 Hz (−2.0)
9	498 Hz	487 Hz (−1.7)	484 Hz (−2.3)
10	501 Hz	493 Hz (−1.6)	490 Hz (−2.1)
11	503 Hz	495 Hz (−1.6)	493 Hz (−2.0)
12	547 Hz	540 Hz (−1.6)	537 Hz (−1.8)
13	610 Hz	539 Hz (−1.6)	598 Hz (−2.0)

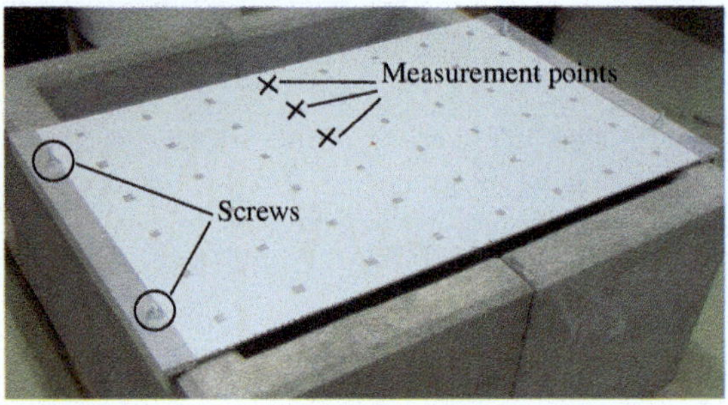

Fig. 5 Set-up of the honeycomb sandwich panel for mobility measurements

3.2 *Results Obtained by the Hybrid Procedure*

The finite element model of the sandwich panel is now replaced by measured mobility data of the panel in order to apply the hybrid procedure described in Sect. 2.3 for the assessment of radiation damping. The sandwich panel is mounted onto a concrete foundation by gluing the short edges of the panel into aluminum F-profiles and fixing them with screws. The set-up is shown in Fig. 5. The excitation is carried out on a uniform grid with 6×4 patches by means of an automated modal hammer (NV Tech SAM1). The force transmitted at the hammer tip is measured in order to derive the respective transfer functions.

On the opposite side of the panel, scanning LDV (Polytec PSV 500) is performed on a uniform response grid of 8×6 patches in order to measure the surface vibrations of the panel. The signal to noise ratio is increased by applying reflective tape at the laser positions. The resulting transfer function matrix $\mathbf{Y}(\omega)$ is of size 48×24.

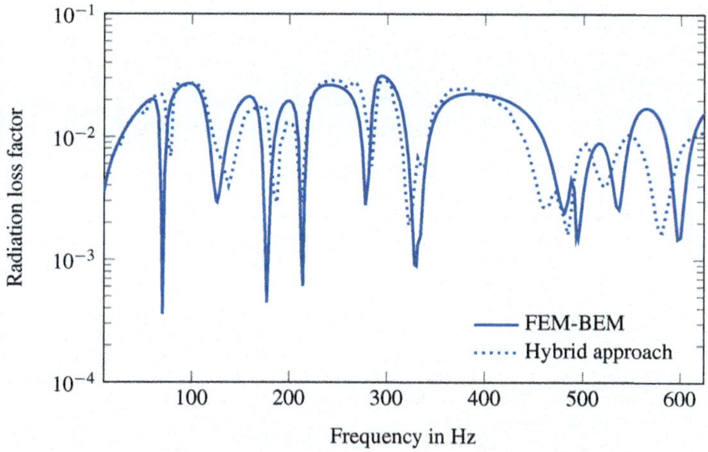

Fig. 6 Radiation loss factor of the baffled honeycomb sandwich panel subject to excitation by a monopole sound source. Comparison between the results obtained by the FEM-BEM approach and the hybrid approach

The transfer functions are obtained by relating the Fourier transforms of the velocity signal to the force signal. The measurement time is long enough so that all signals die out. No window functions are required. Complex frequency domain averaging mitigates the noise that is not phase correlated.

Figures 6 and 7 show the radiation loss factors of the panel in the baffled and unbaffled condition, respectively. An excitation by a monopole sound source is considered. In the hybrid approach, the incident sound field is created using the boundary element model and the resulting force vector is obtained by coupling the boundary element mesh to the experimental excitation grid. The velocity response is then computed by multiplication with the measured matrix of mobilities, c.f. Eq. (21). Finally, the radiated sound field is again computed by BEM.

Figures 6 and 7 and indicate an excellent agreement between the numerical and the hybrid experimental-numerical approach. The numerically predicted eigenfrequencies match the experimental results well, which suggests that the simply supported boundary conditions in the finite element model are a reasonable approximation of the actual mounting condition in the experiment. The material modeling in the structural finite element model also accurately reflects the properties of the sandwich panel. In other cases in which the mounting condition and material properties are subject to uncertainties or prestress occurs during manufacturing, the hybrid approach facilitates an accurate low frequency assessment of radiation damping [15].

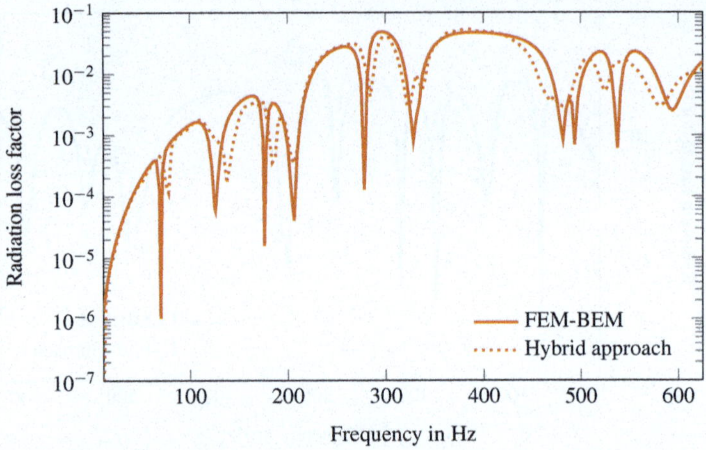

Fig. 7 Radiation loss factor of the unbaffled honeycomb sandwich panel subject to excitation by a monopole sound source. Comparison between the results obtained by the FEM-BEM approach and the hybrid approach

4 Summary and Conclusion

This contribution has reviewed two recent approaches for assessing acoustic radiation damping in the low frequency range. The methodological basis for both approaches is the acoustic BEM, which allows to predict the radiated sound power of structures exhibiting complex geometrical configurations and acoustic boundary conditions.

The first approach is purely numerical and employs a finite element model to characterize the vibrational behavior of the submerged solid structure. Harmonic radiation loss factors are derived based on the forced response of the coupled structural acoustic system. In addition, excitation-independent modal radiation loss factors are obtained by solving the underlying nonlinear structural acoustic eigenvalue problem by a contour integral method.

The second approach is based on an experimentally obtained matrix of mobilities. The mobility matrix, which contains rtransfer functions relating the surface velocity to the force excitation, is coupled to a numerical model of the surrounding acoustic field based on BEM. This approach allows consideration of the actual mechanical boundary conditions and elastic material properties without the need of acoustic measurement facilities.

The two approaches were used to analyze the radiation damping of a honeycomb sandwich panel. The results illustrate the pronounced influence of the type of excitation and acoustic boundary conditions on radiation damping in the low frequency range. At the structural resonance frequencies, however, the harmonic radiation loss factors consistently agree with the excitation-independent modal loss factors. The hybrid experimental-numerical method has achieved excellent agreement with the numerically predicted results. The results in [15, 29] indicate that the methods presented here are also applicable to more complex geometries.

Ongoing and future work includes application of the presented method in the design process of damping treatments and exploitation of radiation damping for deliberate energy dissipation.

Acknowledgements This work was supported by the German Research Foundation (DFG project MA 2395/15-2) in the context of the priority program 1897 "Calm, Smooth and Smart—Novel Approaches for Influencing Vibrations by Means of Deliberately Introduced Dissipation."

References

1. Ehrig, T., Holeczek, K., Modler, N., Kostka, P.: Dynamic behaviour adaptation of lightweight structures by compressible constrained layer damping with embedded polymeric foams and nonwovens. Appl. Sci. **9**(3), 3490 (2019)
2. Gnanasambandham, C., Fleissner, F., Eberhard, P.: Enhancing the dissipative properties of particle dampers using rigid obstacle-grids. J. Sound Vib. **484**, 115522 (2020)
3. Tan, A.S., Belkner, J., Stroschke, A., Sattel, T.: Damping adjustment utilizing digital electrorheological valves with parallely segmented electrodes. Smart Mater. Struct. **28**(7), 075013 (2019)
4. Meyer, C.H., Lerch, C., Rothe, S., Hoffmann, S., Langer, S.C., Lohmann, B., Rixen, D.J.: Study on model reduction for position optimization of acoustic blackholes. Proc. Appl. Math. & Mech. **21**(1), e202100123 (2021)
5. Mead, D.J.: Passive Vibration Control. John Wiley & Sons Ltd, Chichester, England (1999)
6. Maidanik, G.: Response of ribbed panels to reverberant acoustic fields. J. Acoust. Soc. Am. **34**(6), 809–826 (1962)
7. Crocker, M.J., Price, A.J.: Sound transmission using statistical energy analysis. J. Sound Vib. **9**(3), 469–486 (1969)
8. Baydoun, S.K., Marburg, S.: Investigation of radiation damping in sandwich structures using finite and boundary element methods and a nonlinear eigensolver. J. Acoust. Soc. Am. **147**(3), 2020–2034 (2020)
9. Zhou, R., Crocker, M.J.: Sound transmission loss of foam-filled honeycomb sandwich panels using statistical energy analysis and theoretical and measured dynamic properties. J. Sound Vib. **329**(6), 673–686 (2010)
10. Scrosati, C., Scamoni, F., Bassanino, M., Mussin, M., Zambon, G.: Uncertainty analysis by a Round Robin Test of field measurements of sound insulation in buildings: single numbers and low frequency bands evaluation—Airborne sound insulation. Noise Control. Eng. J. **61**(3), 291–306 (2013)
11. Kirkup, S.: The boundary element method in acoustics: a survey. Appl. Sci. **9**(8), 1642 (2019)
12. Kaltenbacher, M.: Computational Acoustics. Springer, Cham, Switzerland (2018)
13. Astley, R.J., Macaulay, G.J., Coyette, J.-P., Cremers, L.: Three-dimensional wave-envelope elements of variable order for acoustic radiation and scattering. part I. formulation in the frequency domain. J. Acoust. Soc. Am. **103**(1), 49–63 (1998)
14. Roozen, N.B., Leclère, Q., Urbán, D., Méndez Echenagucia, T., Block, P., Rychtáriková, M., Glorieux, C.: Assessment of the airborne sound insulation from mobility vibration measurements; a hybrid experimental numerical approach. J. Sound Vib. **432**, 680–698 (2018)
15. Baydoun, S.K., Roozen, N.B., Marburg, S.: Hybrid assessment of acoustic radiation damping combining in-situ mobility measurements and the boundary element method. Acta Acust. **6**, 44 (2022)

16. Marburg, S., Nolte, B.: A unified approach to finite and boundary element discretization in linear time-harmonic acoustics. In: Marburg, S., Nolte, B. (eds.) Computational Acoustics of Noise Propagation in Fluids. Finite and Boundary Element Methods, pp. 1–34. Springer (2008)
17. Ochmann, M., Brick, H.: Acoustical radiation and scattering above an impedance plane. In: Marburg, S., Nolte, B. (eds.) Computational Acoustics of Noise Propagation in Fluids. Finite and Boundary Element Methods, pp. 459–494. Springer (2008)
18. Bathe, K.-J.: Finite Element Procedures. Englewood Cliffs, New Jersey: Prentice Hall (1996)
19. Peters, H., Marburg, S., Kessissoglou, N.: Structural-acoustic coupling on non-conforming meshes with quadratic shape functions. Int. J. Numer. Methods Eng. **91**(1), 27–38 (2012)
20. Baydoun, S.K., Voigt, M., Jelich, C., Marburg, S.: A greedy reduced basis scheme for multifrequency solution of structural acoustic systems. Int. J. Numer. Methods Eng. **121**(2), 187–200 (2020)
21. Jelich, C., Baydoun, S.K., Voigt, M., Marburg, S.: A greedy reduced basis algorithm for structural acoustic systems with parameter and implicit frequency dependence. Int. J. Numer. Methods Eng. **122**(24), 7409–7430 (2021)
22. Baydoun, S.K., Voigt, M., Marburg, S.: Low-rank iteration schemes for the multi-frequency solution of acoustic boundary element equations. J. Theor. Comput. Acoust. **29**(3), 2150004 (2021)
23. Fahy, F., Gardonio, P.: Sound and Structural Vibration. Academic Press, Oxford, United Kingdom (2007)
24. Petyt, M., Lea, J., Koopmann, G.: A finite element method for determining the acoustic modes of irregular shaped cavities. J. Sound Vib. **45**(4), 495–502 (1976)
25. Banerjee, P.K., Ahmad, S., Wang, H.C.: A new BEM formulation for the acoustic eigenfrequency analysis. Int. J. Numer. Methods Eng. **26**(6), 1299–1309 (1988)
26. Roman, J.E., Campos, C., Romero, E., Tomàs, A.: Slepc users manual. Technical report, Universitat Politècnica de València (2020)
27. Kimeswenger, A., Steinbach, O., Unger, G.: Coupled finite and boundary element methods for fluid-solid interaction eigenvalue problems. SIAM J. Numer. Anal. **52**(5), 2400–2414 (2014)
28. Zheng, C.-J., Bi, C.-X., Zhang, C., Gao, H.-F., Chen, H.-B.: Free vibration analysis of elastic structures submerged in an infinite or semi-infinite fluid domain by means of a coupled FE-BE solver. J. Comput. Phys. **359**, 183–198 (2018)
29. Baydoun, S.K., Voigt, M., Goderbauer, B., Jelich, C., Marburg, S.: A subspace iteration eigensolver based on Cauchy integrals for vibroacoustic problems in unbounded domains. Int. J. Numer. Methods Eng. **122**(16), 4250–4269 (2021)

Suppressing Brake Vibrations by Deliberately Introduced Damping

Dominik Schmid, Nils Gräbner, Utz von Wagner, and Volker Mehrmann

1 Introduction

Brake vibrations are a typical example of Noise, Vibration, Harshness (NVH) problems in the automotive industry, dominating the development costs in several fields like brakes. The high-frequency brake vibration phenomena investigated here are caused by self-excitation based on the friction forces between pad and disk or drum, see e.g. [50]. Particularly brake squeal represents this kind of noise phenomenon in the audible frequency range between 1 and 15 kHz, see e.g. [23]. Results regarding the influence of damping on high frequency brake vibrations have already been published in [16]. In general, models for brake squeal are nonlinear. For describing this phenomenon these models have to contain the self-excitation mechanism caused by the non-conservative friction forces and nonlinearities for restricting the vibrations to a limit cycle. The experimentally observed stationary vibration often possesses a dominant frequency during squeal, e.g. [17].

In practice, it is still state of the art to use linearized models around a stationary solution for describing the potential for brake squeal. Thereby the instability of the desired quiet solution is interpreted as squeal. If brake squeal models are linearized

D. Schmid (✉) · N. Gräbner · U. Wagner
Technische Universität Berlin, Einsteinufer 5, 10587 Berlin, Germany
e-mail: dominik.schmid@tu-berlin.de

N. Gräbner
e-mail: nils.graebner@tu-berlin.de

U. Wagner
e-mail: utz.vonwagner@tu-berlin.de

V. Mehrmann
Technische Universität Berlin, Straße des 17. Juni 136, 10623 Berlin, Germany
e-mail: mehrmann@math.tu-berlin.de

P. Eberhard (ed.), *Calm, Smooth and Smart*, Lecture Notes in Applied and Computational Mechanics 102, https://doi.org/10.1007/978-3-031-36143-2_13

with respect to stationary operation states, the resulting equations of motion can be written as

$$\mathbf{M}\ddot{\mathbf{y}} + (\mathbf{D} + \mathbf{G})\dot{\mathbf{y}} + (\mathbf{K} + \mathbf{N})\mathbf{y} = \mathbf{0}, \quad (1)$$

with a symmetric and positive definite mass matrix **M**, symmetric and positive semi-definite damping matrix **D** and stiffness matrix **K** as well as a skew-symmetric gyroscopic matrix **G** and a circulatory matrix **N**, while **y** is the vector of displacement or rotation angles which can either result from FE models oder models with a low number of degrees of freedom, e.g. Multibody systems (e.g. [17]).

Fundamental properties of self-excited systems like stability and bifurcation behavior are strongly influenced by dissipation which therefore in general might have a decisive influence on noise and vibrations. More precisely, damping is capable to suppress vibrations but also to excite vibrations in self-excited systems, see e.g. [19]. The general influence of damping on brake models is investigated in [16]. Despite its general importance, damping mechanisms are often omitted or only very fundamentally considered in state of the art simulations both in industry and academia. A quantitative modeling is difficult due to scattering or a lack of parameters. The objective of the current investigations is to optimize the vibration behavior of brakes—following directly the title of the priority program—with respect of being calm and smooth by using deliberately introduced damping. This contribution shows results achieved in both phases of the project. In the first period the focus was on the investigation of damping influence of shims in disk brakes while the second period investigates more general the influence of related damping devices on drum brakes. Results regarding disk brake investigations are taken from the corresponding doctoral thesis [37]. Essential analytical, numerical as well as experimental outcome is shown subsequently, see more detailed results in the corresponding Sects. 4, 5 and 6 in [37]. Furthermore, results for a model of a duplex drum brake have already been presented in [45]. A modified FE model of a simplex drum brake, where one brake shoe acts as leading and the other as trailing shoe is shown below describing the experimentally investigated drum design in a more realistic manner.

The paper is organized as follows. First, shims, thin composite structures consisting of elastic and viscoelastic layers applied on brake pad back plates, are modeled and investigated experimentally. These investigations especially consider methods for Finite Element (FE) models regarding a homogenization process for multilayer shim compounds. Furthermore, different shim set-ups are tested for achieving maximum damping and studying the influence on the investigated brake system. In next steps, shims are applied to back plates as well as modified disks to assess the squealing behavior using a dynamometer test rig. The second part deals with the investigation of possibilities to include damping mechanisms in drum brakes. The focus is on damping devices capable to be fixed on the free surfaces of shoes (inner surface) and drum (outer surface). Especially the drum with its large free outer surface offers new possibilities compared to disk brakes. Experimental modal analyses of drum brake components are carried out. Furthermore, a Finite-Element drum brake model with respect to squeal is built up focusing on essential brake parts and damping measures.

2 Shims: Principle, Application and Modeling

In disk brakes thin shim structures are often applied on pad back plates as shown in Fig. 1a in order to avoid brake noise. Their layer thicknesses usually are in the range of a few tenths of a millimeter and they are intended for increasing damping or changing other dynamic properties. Figure 1b depicts the shim configuration used for the entire investigations. Here, a steel core is wrapped with two elastomer layers.

Currently in industrial applications, selecting appropriate shims requires a variety of tests. In the present investigations, the influence of essential factors like geometrical dimensions and rheological properties is considered systematically based on corresponding shim models. In FE-systems the modeling of thin shim composites in general is not easy. Often a multi-layer approach is used considering each layer individually [29]. This approach can cause large element distortions and reduce the quality of results in simulations strongly. Furthermore, modeling of damping—the key feature of shims—is often neglected completely leading to inaccurate computations. Besides sufficient damping often low weight and high static stiffness characteristics are required as well [20]. So, the aim of the actual investigations is to get a manageable modeling of the shims including all relevant effects for the considered brake vibrations.

In general shims possess a multi-layer structure. The damping increase of brake pads using shims is mainly based on the properties of the thin elastomer layers. A deformation of the pad compound due to bending shown in Fig. 2 results in a shearing behavior of the soft viscoelastic core. This shear deformation contributes to a much greater dissipation of brake pads. The principle applied here is called *Constrained-layer-damping* (CLD), where a thin damping core, often an elastomer, is placed between a stiff constraining layer and a thick carrier structure (beam).

Usually the beam possesses a much larger thickness than all additional layers affecting the stiffness of the entire CLD structure. Note that the main effect of shims may not consist in providing additional damping but rather pursues the increase of the pad stiffness [15]. Additional elastomer top (constraining) layers of shims can be described by the Free-layer-damping (FLD) concept [32]. FLD is a simple measure for damping of flexural vibrations. Therefore, a damping layer is applied to the structural surface to be damped. Similar to the just mentioned CLD compound, energy is dissipated by cyclic deformation of the structure, but in this context primarily by

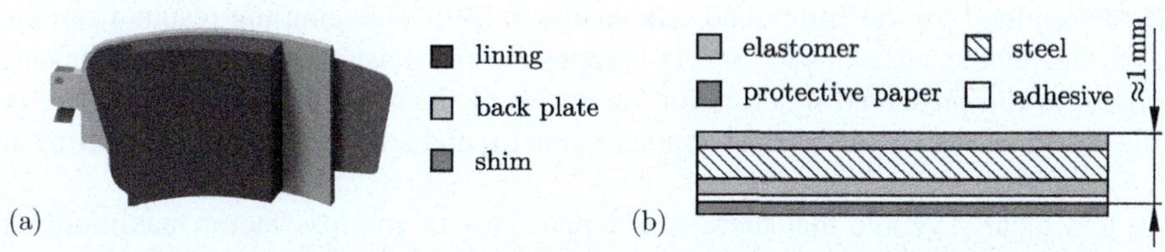

Fig. 1 **a** Typical brake pad composition [36] and **b** shim structure used for the investigations according to [37, 52][Fig. 2.2]

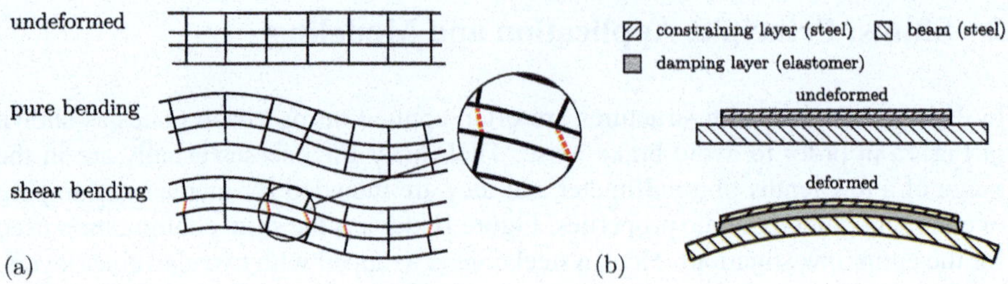

Fig. 2 **a** Shear deformation of viscoelastic core [43] according to [41, Fig. 5], [25, pp. 31] and **b** CLD mechanism according to [35],[pp. 62], see [37],[Fig. 4.8]

elongation and contraction of the viscoelastic damping material [3, 22], not by shear. This damping principle has shown to be less effective on component level since there only has been a limited increase of damping. These elastomer top layers are intended primarily for affecting the contact between pad and piston respectively carrier during braking and contribute to an improved noise behavior.

Subsequently, analytical approaches are presented allowing the consideration of damping in CLD structures. A homogenization approach is used combining essential layers of the shim and providing equivalent system properties. The homogenized shim structure is implemented into an existing FE model of the overall brake system and validated using measurement data. The objective is to achieve equivalent system properties of the multi-layer CLD structure by just using one layer in the FE model.

In the following a homogenization method for a CLD assembly is presented based on the approaches from NASHIF et al. [31] and ROSS et al. [35]. Using the theory from ROSS the result is a homogenized stiffness parameter used for subsequent FE-simulations. The theory was originally developed for a rectangular cross-sectional composite while in general brake pad shims do not have a rectangular shape. Results of the present investigations for rectangular plates as well as brake pad back plates bonded with shims have already been published in [42] containing also experimental investigations. Due to a more complex geometry of pad back plates more iterations are needed to achieve a comparable match with experimental results as for rectangular plates. An additional analytical approach based on [24, 34] is used for considering essential parameters like shim length and layer thickness particularly affecting damping of shim composites. This theory describes lateral vibrations of a three layer CLD structure. Natural frequencies and loss factors [10, 48] of each bending mode can be determined for the compound, see results in [39]. The damping results from the shearing of the viscoelastic core is mapped by the elastomer loss factor. Results based on the presented theories for variations of the shim coverage are shown in Fig. 3 comparing a symmetrical application in (a) and an edge application starting at $x = 0$ in (b).

It is clearly visible that there is a dependency of the loss factor maximum on the bending mode as well as the degree of coverage. Considering a symmetrical application the first mode shape shows the highest loss factor at approximately 50% coverage whereas the maximum at higher modes moves more and more towards a

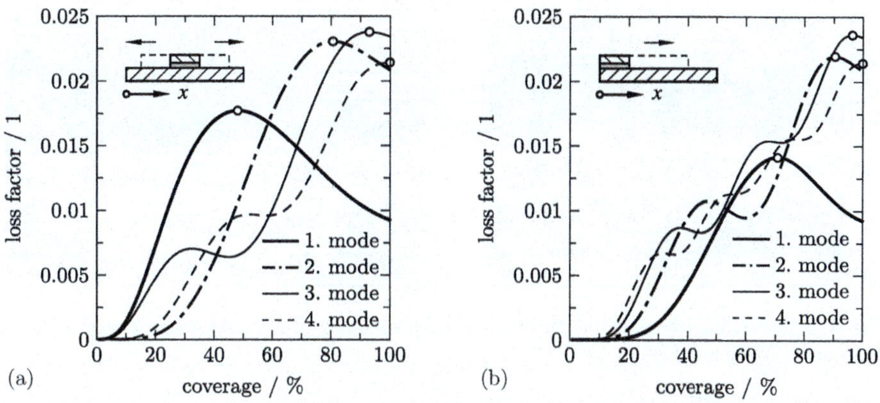

Fig. 3 Influence of shim length considering a rectangular CLD structure (180x50x5 mm^3) on **a** symmetrical application [37][Fig. 4.16 a] and **b** application at $x = 0$ [37],[Fig. 4.17 a]

fully covered carrier structure. Considering the configuration in (b) a much higher coverage tends to be needed for achieving an equivalent damping potential.

Summarizing, the following aspects can be noted. First of all, it is of relevance to have knowledge about what mode shape shall be damped preferably because there is a strong dependency on system properties.

- A maximum loss factor of the shim is not always the primary design goal. Stiffness and mass distribution often play a decisive role, see mode decoupling scenarios [27].
- The selection of the damping material is essential. Good dissipation properties over a wide temperature and frequency range are desirable, see dynamic-mechanical analysis and modal analysis carried out in [38, 40].
- Increasing the stiffness and thickness of the constraining layer as well as using elastomers with a high loss factor have proven to be effective [13, 22]. Partial coverage can lead to higher as well as lower loss factors and should be taken into account when selecting shims. Therefore it can not be assumed that an increase of the shim length contributes to larger loss factors.
- An additional fixation of shims with the back plate by riveting should be avoided to achieve the highest possible damping.

Experimental tests on component level considering temperature influence as well as coverage variation and delamination scenarios have been carried out in [37, 38, 40]. Support has been provided by student workers [4, 7, 11, 12, 26, 47] and [9] on specific modeling and experimental aspects during the project, see [37]. Furthermore, the homogenization approach used has been validated in [42]. The results showed a very good agreement with experimentally determined natural frequencies by simultaneously maintaining all mode shape as well as the order of appearance. The reference to damping is considered later in Sect. 4.

(a) (b)
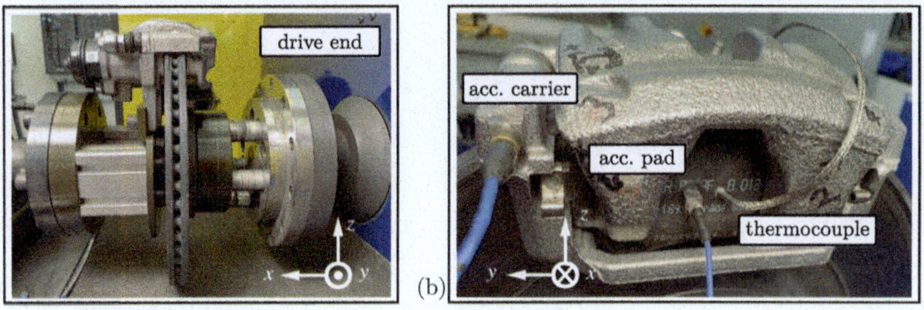

Fig. 4 **a** Investigated floating caliper brake and **b** positioning of triaxial accelerometers and thermocouple [37],[Fig. 5.17]

3 Brake Set-up

The industrial floating caliper brake shown in Fig. 4 is used for the entire experimental investigations. To measure time series during squealing two triaxial accelerometers are applied, placed on the caliper and shim as well as an external thermocouple for recording temperature. All test campaigns were carried out on a LINK dynamometer test rig D 1500 at Chair of Mechatronics and Machine Dynamics at Technische Universität Berlin.

To determine the influence of shims on brake noise, identical test campaigns were carried out for brake pads with shims and after removing shims. One test cycle included 180 brakings considering different pressure and speed levels. Test bench trials were repeated ten times to achieve a better quality of squeal information. Figure 5 shows the squealing frequencies determined from the time data.

Comparing both test scenarios it is obvious that applying shims contributes to a more silent brake system. Figure 5 clarifies that a lower number of squealing events, lower acceleration amplitude levels correlated with the volume of the perceived squealing noise and a lower number of squeal frequencies occurred using shims. However, there has been no complete elimination of squealing. It is noticeable that even a new squeal frequency with a dominating y-direction at 1.4 kHz appears, which was not present before. A damping increase due to shearing of the elastomer core requires an appreciable deformation of the pad. A reduction of noise by using shims is therefore often only noticeable for higher squealing frequencies.

The question therefore is, if damping devices can be applied more efficiently with respect to the avoidance of squeal. This has led to tests of further measures. We hereby focused on the damping of the gray cast brake disk. The intention by these investigations was not primarily to find a countermeasure directly applicable in real brakes, but to get basic insights into the effectiveness of damping of components on the behavior of the overall system. In WEHNER et al. [51] it has been proved theoretically using a FE model, that damping of the disk is very effective (and much more effective than damping pads) with respect to the avoidance of squeal. According to [8], the friction rings of the disk (i.e. the part in contact with the pad) account for the

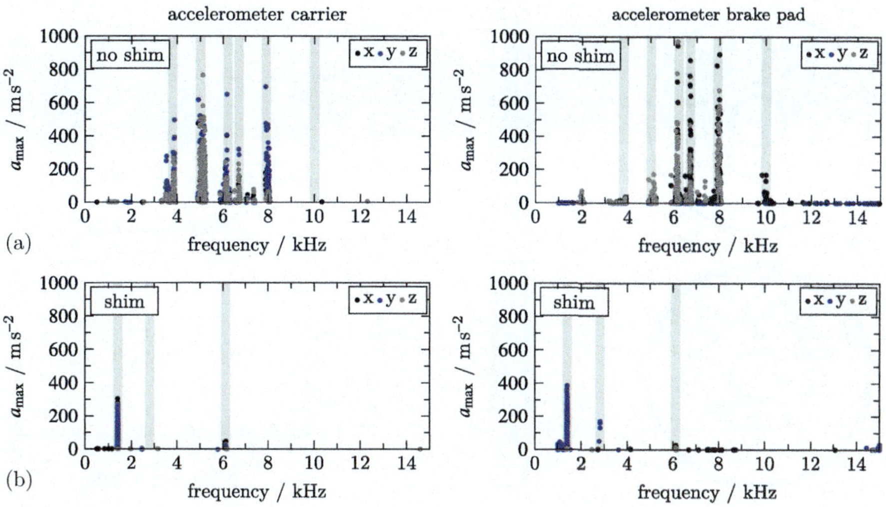

Fig. 5 Maximum accceleration $a_{\max}$ measured during squealing: brake pads **a** without shims and **b** using shims [37],[Fig. 5.21]

largest part of noise radiation, approximately about 70%. Unfortunately modifying disks is often a demanding task for thermal and safety reasons. In our tests the outer cooling fins were removed as depicted in Fig. 6 and shims were positioned inside the groove and along the circumferential surface. As already mentioned these measures are not directly applicable to series and have exclusively been used for our academic trial purposes.

On component level experimental modal analysis were carried out for both test set-ups, brake disk with and without shims. All measurement objects were placed on soft foam to realize a free-free equivalent support. For excitation an automatic impulse hammer and for measuring the output signal a laser vibrometer have been applied. The frequency response functions are shown in Fig. 6.

The shims used ensure that there is a reduction of peak amplitudes particularly in the frequency range above 3 kHz. Table 1 provides information about the first eigen shapes considering the response behavior. Comparing damping ratios determined from transfer functions of the modified disk and shims used by default on back plates show that there is a lower damping characteristic for the modified disk. However, a closer look clarifies a significant increase in damping compared to conventional grey cast iron brake disks possessing damping ratios in the order of 0.001 and below. Additionally as argued before, a larger influence of disk damping on the noise problems is expected, especially in the lower frequency range.

In fact, dynamometer testing emphasizes how effective the measure is, as shown in Fig. 7. The detected squeal distribution is compared subsequently. Each marker correlates with a squealing event measured. Shims placed on disk brakes increase the damping capacity significantly and show a clear benefit on the noise behavior.

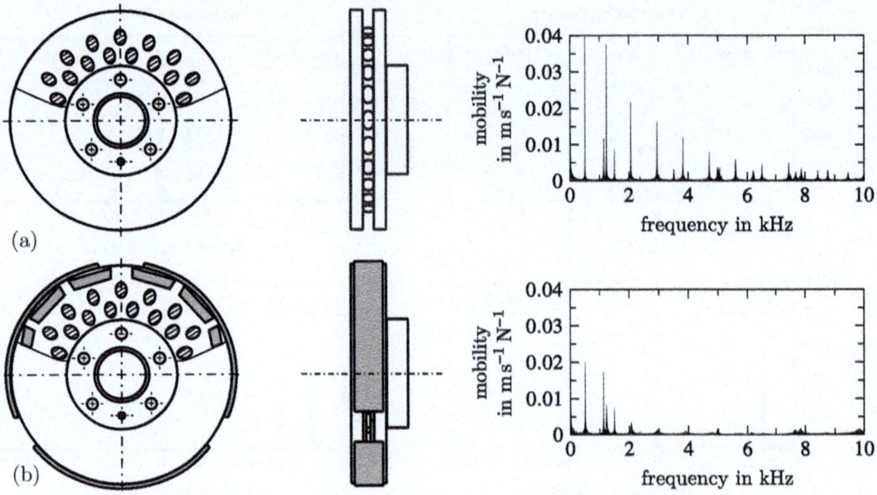

Fig. 6 Modifications of brake disk in order to investigate measures to increase damping. Corresponding mobility (free-free support): **a** disk with removed outer cooling fins and **b** disk with removed outer cooling fins and added shims in groove and along circumference [37],[Fig. 5.14]

Table 1 Influence of shims on system properties of brake components: free-free supported modified brake disk and back plate applied with shims [37],[Table 12], [37][Table 18]

	f_1 Hz	ϑ_1 –	f_2 Hz	ϑ_2 –	f_3 Hz	ϑ_3 –	f_4 Hz	ϑ_4 –	f_5 Hz	ϑ_5 –
Disk	518	0.005	1130	0.001	1231	0.005	1497	0.002	2078	0.008
Back plate	2475	0.009	3548	0.008	5817	0.007	7622	0.006	9755	0.005

f—natural frequency, ϑ—damping ratio

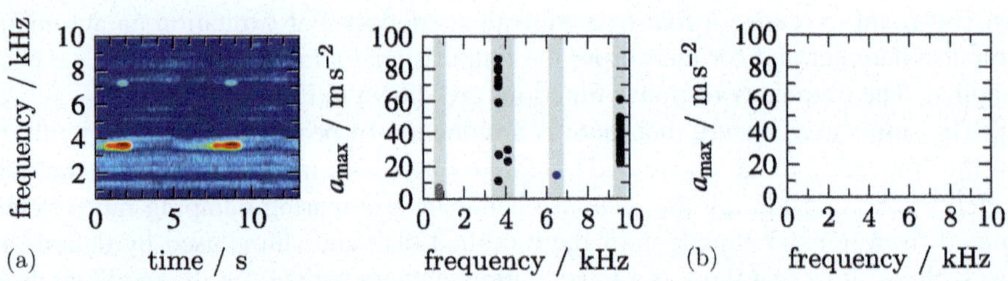

Fig. 7 Spectrogram of acceleration signals and overview of occurring squealing frequencies: **a** squeal events detected for brake disk without shims and **b** no squeal events occured for brake disk using shims [37],[Fig. 5.27]

The squeal affinity could be reduced entirely by the tested measures. As a result, additional damping measures should focus on increasing the damping of the disk which is not an easy task to realize in serial mass produced specimen.

4 Finite-Element Modeling

Shims are inevitably a problem when modeling via Finite-Element programs due to very thin layers. A finer meshing and thus an increase of degrees of freedom is not desired for complex brake models and contributes to significantly longer computation times. Experimental modal analyses of identical brake pads have shown that there is a non-negligible variation especially in damping characteristics [37]. Therefore, the mapping of damping has to be considered within suitable tolerances in simulations. While implementing damping, the choice of Rayleigh-damping [5] seems to be the most appropriate one. This kind of damping gives a sufficiently precise characterization of the dissipation behavior in the squeal relevant range considering the variation captured in test set-ups. FE computations at component level of homogenized shim structures in [42] showed a very good agreement with experimental tests. The damping therein has been mapped realistically as well as the natural frequencies. The shim homogenization process has shown to obtain equivalent results and has been used subsequently for implementation in an already existing industrial FE brake model of the experimentally examined brake.

The state of the art in squeal simulations in industry is to perform the so-called Complex Eigenvalue Analysis (CEA). For CEA the eigenvalues of the brake system are computed. Due to the friction forces between pad and disk asymmetries in the displacement proportional terms, i.e. a stiffness matrix and a circulatory matrix, and additionally gyroscopic terms, eigenvalues with positive real parts $\mathrm{Re}(\lambda)$ are possible, see e.g. [17, 33]. This means that the trivial solution of the (with respect to an equilibrium position linearized) equations of motion becomes unstable and the corresponding vibration mode and frequency are considered to belong to potential corresponding squeal. A comparison is made between a conventional multi-layer shim approach in (b) and a homogenized shim compound in (c), see Fig. 8. For validating the simulation results squeal events measured have been considered in (a).

Comparing the two approaches in (b) and (c) a clear difference can be seen. The conventional multi-layer approach in (b) shows a poorer agreement with the experimentally determined squealing events in the frequency range up to 10 kHz, while also (c) only determines some of the experimentally found squealing frequencies. When examining the deformation behavior of mode shapes in (b) large element distortions occurred locally due to the thin element thicknesses, see results in [37]. These (unphysical) distortions [30] may lead to (also unphysical) positive real parts and therefore worsen the quality of the results. However, the homogenized shim approach in (c) constitutes a significant improvement in modeling [37]. As a result, realistic mode shapes can be observed as well as squeal frequencies are mapped in a much better way especially at 4 kHz and 6 kHz.

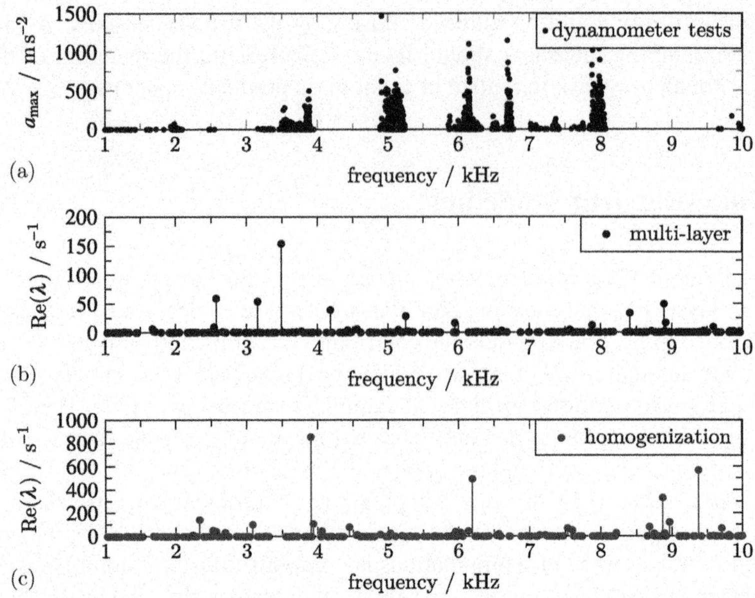

Fig. 8 Validation of squeal events: **a** dynamometer tests and results of CEA: **b** multi-layer and **c** homogenized shim approach [37],[Fig. 6.8]

To reduce computation time in multi-parameter simulations and complex eigenvalue analysis, model reduction methods for the associated FE models have been developed and implemented as Python and Matlab codes in the first project phase [18]. These methods were employed in [6] to develop homotopy methods. By first reformulating the second order system (1) as a first order perturbed dissipative Hamiltonian system $E\dot{z} = (J - R_D)z - R_N z$, with

$$J := \begin{bmatrix} -G & -(K + \frac{1}{2}N) \\ (K + \frac{1}{2}N^T) & 0 \end{bmatrix}, \tag{2}$$

$$E := \begin{bmatrix} M & 0 \\ 0 & K \end{bmatrix}, \quad R := R_D + R_N = \begin{bmatrix} D & 0 \\ 0 & 0 \end{bmatrix} + \begin{bmatrix} 0 & -\frac{1}{2}N \\ -\frac{1}{2}N^T & 0 \end{bmatrix}, \tag{3}$$

one sees that the circulatory term N associated with the FE nodes on the contact surface is solely responsible for the eigenvalues with positive real part, since the system with $N = 0$ always is Lyapunov stable, i.e. it has a spectrum in the closed left half plane [28] with all eigenvalues on the imaginary axis being semisimple.

Using a full scale industrial FE model and bringing in the perturbation term $R_N Q$ via a homotopy $E\dot{z} = (J - R_D)z - \alpha R_N z, \alpha \in [0, 1]$. it was observed that for $\alpha = 0$ the maximum real part is $-5.0462e - 06$ and for $\alpha = .1$ it is already $2.0336e - 05$, see Fig. 9, where the largest real part of an eigenvalue is plotted vs. the parameter α, i.e. the unperturbed problem is already close to instability under unstructured perturbations.

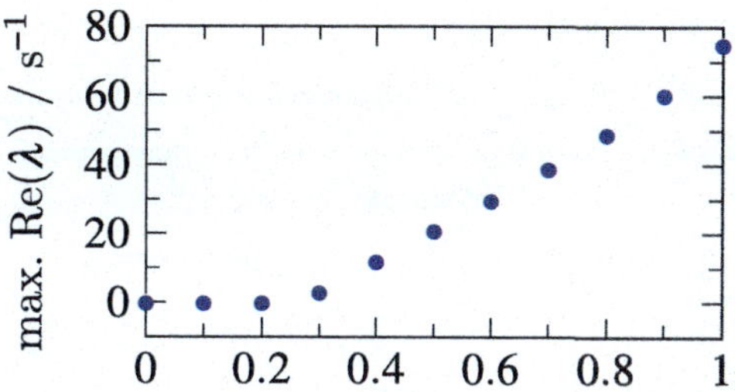

Fig. 9 Eigenvalue plot for $\alpha \in [0, 1]$ [6] [Fig. 4.1]

In order to decide already in a design phase whether the norm of the matrix N is tolerable to preserve the asymptotic stability in [2] a new efficient method was developed to determine the structured distance to instability for large scale problems and to determine the imaginary part of the eigenvalue, where the eigenvalue is crossing the imaginary axis. The method combines nonlinear eigenvalue optimization methods with the described model reduction methods.

5 Investigations on Drum Brakes

Drum brakes have largely disappeared in passenger cars in the last years. However, this type of friction brake is returning in terms of the emergence of electrically powered vehicles and particle emissions. When used, drum brakes are often applied within the rear-axels in cars [1] and in buses or trailers [21]. Generally, electric vehicles are in wide operational states decelerated using the regenerative brake of the electric motor additionally to friction brakes. The consequence is that conventional brake systems are no longer permanently exposed to wear. For this reason, drum brakes are being used more and more frequently at least for rear-axles. Lower costs in general [14] and the reduction of abrasion emissions due to the encapsulated design [49] are two main advantages compared to disk brake systems. Nevertheless, noise in the audible frequency range remains during braking, particularly before standstill.

Usually, industrial drums consist of cast material possessing a high damping potential among other suitable mechanical properties [14]. These industrial brakes are often optimized in such way that there are almost no squealing events occuring in academic test series. To investigate the influence of damping a steel drum, manufactured at the Chair of Mechatronics and Machine Dynamics, instead of gray cast iron is used for all further investigations presented here. This steel drum possesses a much lower material damping than cast materials. The simpler structure neglecting cooling fins also simplifies the Finite-Element modeling. A steel drum with an outer diameter of 214 mm has been turned from a hollow steel cylinder and has been used entirely for research purposes.

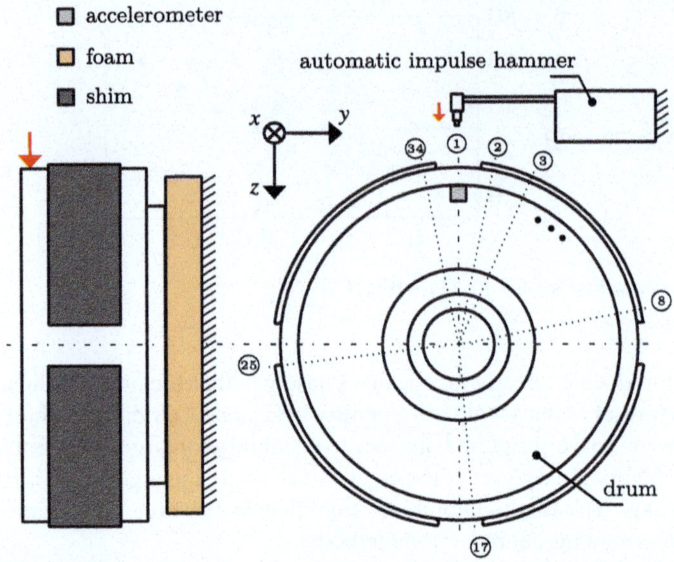

Fig. 10 Modal analysis set-up of non-industrial steel drum considering 34 collocation points [44] and [45] [Fig. 3]

Table 2 Damping ratios of non-industrial steel drum, [45, Table 2]

	ϑ_1 %	ϑ_2 %	ϑ_3 %	ϑ_4 %	ϑ_5 %	ϑ_6 %	ϑ_7 %	ϑ_8 %	ϑ_9 %
No shim	0.057	0.070	0.079	0.061	0.030	0.210	0.019	0.232	0.135
Shim	0.252	0.424	0.703	0.632	0.351	0.841	–	0.831	0.568
Increase	342	506	790	936	1070	300	–	258	320

ϑ—damping ratio

Results showing CEA of a modified duplex drum brake have already been described in [45]. The modeling and outcome presented herein correspond to the real simplex drum brake also experimentally investigated at the Chair of Mechatronics and Machine Dynamics. For increasing the dissipation of the drum, four rectangular shim plates (width length 53 mm, length 139 mm) are applied along the circumference equidistantly. For determining system properties like natural frequencies and damping ratios of brake components the modal analysis set-up shown in Fig. 10 is used.

Damping ratios ϑ for both drum set-ups—with and without shims - are shown subsequently in Table 2. Adding shims results in an increase in damping caused by the CLD mechanism as described priorly. Damping ratios are about 3 to 10 times larger than in the turned steel drum. Missing values in Table 2 and 3 could not be determined in the applied evaluation procedure.

Fig. 11 Industrial brake shoe attached with shims investigated in [9]

Table 3 Damping ratios of industrial brake shoes, data from [9, Table 4.4]

	ϑ_1 %	ϑ_2 %	ϑ_3 %	ϑ_4 %	ϑ_5 %	ϑ_6 %	ϑ_7 %	ϑ_8 %
No shim	–	0.6	0.6	0.6	0.4	–	0.6	0.8
4 shims	0.9	1.4	1.6	2.1	1.4	–	–	1.5

ϑ—damping ratio

Besides the drum surface passive damping structures like shims can be attached to brake shoes as well which has been investigated in the student thesis [9]. One version tested therein is visible in Fig. 11.

The damping ratios of the variant that have shown in [9, Table 4.4] the biggest impact on damping is summarized in Table 3. Damping ratios of the measure are compared with the industrial brake shoe. The lining itself contributes to a relatively high damping potential of approximately 0.6 %. Adding shims increases the damping of the brake shoes once again significantly, especially in the tested partial coverage execution.

The tests yield that these two placements (drum and shoes) show a considerable potential for increasing damping. Future tasks might deal with e.g. how to apply them in vehicles.

The FE modeling of the drum brake in Fig. 12 concentrates on fundamental components like drum and brake shoes including lining. All components are modeled with a simplified geometry. The focus is rather how damping can influence the noise behavior of drum brakes in general. Hexahedral solid elements with a quadratical approach (C3D20) are used for meshing all brake parts. Reference points positioned near the brake shoes and the center of the drum allow the application of concentrated forces and bearings (pinned-pinned) as well as pivot of the brake parts. The results presented below relate to the simplex drum brake configuration in Fig. 12, where one brake shoe acts as leading shoe and the other as trailing shoe. The contact between lining and drum is based on a surface to surface formulation. Steel components like drum and back plate are modeled using isotropic properties. For modeling shims a homogenization approach is used. The lining consisting of several components [46] shows a nonlinear and transversal isotropic material behavior [53]. Direction-dependent parameters for the friction material are implemented using engineering constants.

The approach is rather to show where damping is particularly effective. For these positions, damping is implemented in a realistical manner using Rayleigh parameters

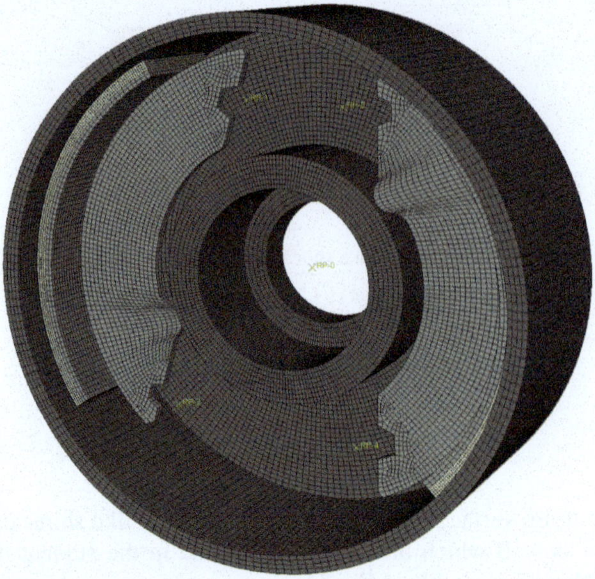

Fig. 12 FE drum brake model

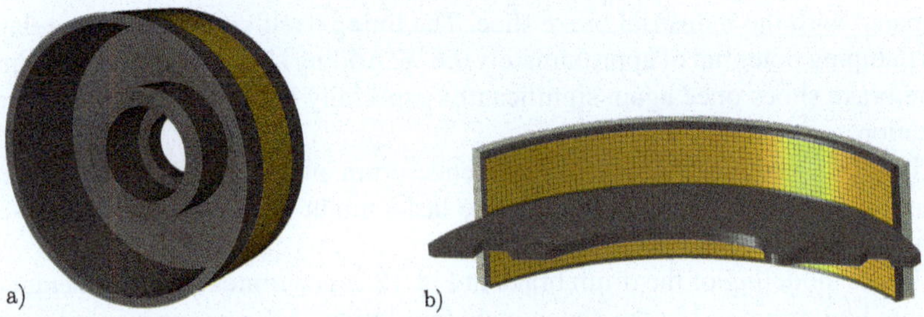

Fig. 13 Shims applied to: **a** drum surface and **b** brake shoes

determined from experimental damping ratios, see Table 3. A discussion on a higher term approach like using a Caughey series has been investigated in [45]. The Rayleigh parameters are implemented in all shim components as well as the drum and brake shoes. A shim ring with a width of 40 mm and a thickness of 1 mm is attached to the outer surface of the drum via tie-constraint. Each brake shoe is equipped with two shims filling the given space. A detailed view of the brake components including shims is shown in Fig. 13.

The results of CEA are summarized in Fig. 14. The eigenvalues show that damping the outer surface drum contributes to an improvement in noise behavior. Much fewer eigenvalues show a positive real part in the investigated frequency range. Shims applied on brake shoes show a high effectiveness in the frequency range above 5 kHz.

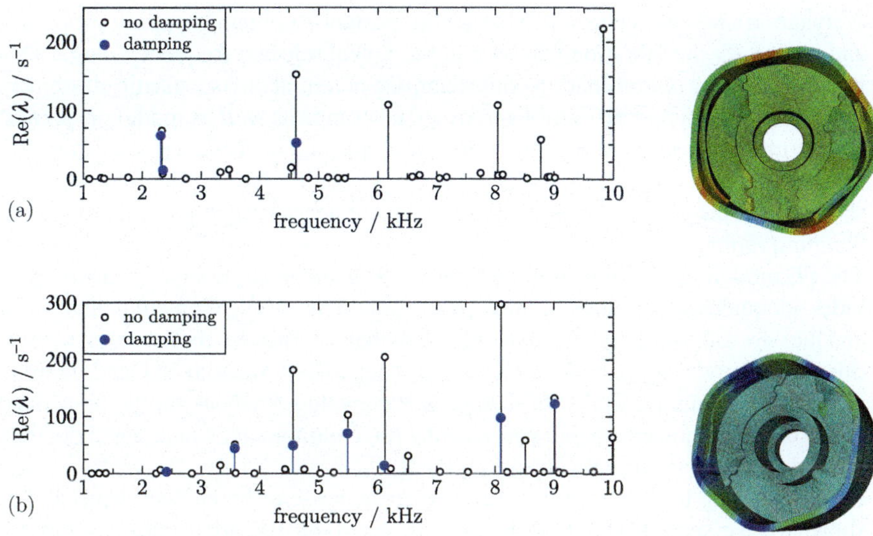

Fig. 14 Results of CEA considering damping in: **a** brake shoes (mode shape at 4616 Hz) and **b** drum (mode shape at 6121 Hz)

All positive real parts have been shifted successfully in the negative half-plane in this frequency range.

Nevertheless, there are still three positive real parts in the lower frequency range remaining. Therefore, it can be stated that damping the drum and brake shoes are one of the key aspects for reducing or even preventing squeal in the future. A combination of both measures should be aimed to achieve silent drum brakes. Note that additional masses and the overall stiffness influence the entire brake system.

A comparison with dynamometer tests considering the academic steel drum are intended to be focused in near future in a separate paper.

6 Conclusions

Within the actual DFG priority program "Calm, Smooth and Smart" countermeasures against high frequency brake vibrations were investigated. These vibrations are caused by self-excited vibrations, and this problem is a typical example of NVH issues in automotive industry. Countermeasures for disk as well as drum brakes were investigated in detail focusing on damping. Damping is well-known to be a powerful countermeasure against self-excited vibrations, if applied appropriately. In the first project period thin composite structures, called shims, were examined. As a result an improved approach for the FE-modeling of shims based on homogenization was developed. Further key aspects that have emerged can be summarized as follows:

- A dynamic characterization of shims is important to obtain damping properties and to classify the vibration behavior in the squeal relevant frequency range. This includes both the determination of the damping potential of viscoelastic elastomers within the essential temperature and frequency range as well as modal properties. Furthermore, the shim length and positioning plays a decisive role for achieving a maximum damping effect for the considered configuration. The investigations showed that increasing the shim length does not necessarily contribute to greater damping ratios.
- The FE modeling of shims as homogenized equivalent single-layer structure provides adequate aspect ratios of the elements, prevents strong elemental distortions and therefore shows a realistic deformation behavior. Besides these central aspects this type of modeling provides a reduction of degrees of freedom and leads to more efficient computation times. In Complex Eigenvalue Analysis the homogenized shim structure showed a better agreement with dynamometer tests carried out than the conventional used multi-layer approach.
- For eliminating low-frequency squealing the increase of brake disk damping has shown to be essential. The friction ring surfaces of the disk are usually responsible for a large part of sound radiation. In test set-ups a completely silent disk brake has been achieved by positioning shims instead of the outer cooling fin ring and towards the circumferential direction.
- Damping the drum as well as the brake shoes is one of the key measures for achieving silent drum brakes. Experimental modal analysis showed a significant increase in damping when applying shims on both surfaces. The Complex Eigenvalue Analysis for the developed drum brake model considering the essential brake parts confirmed this.

Acknowledgements This work is funded by the Deutsche Forschungsgemeinschaft (DFG, German Research Foundation) WA1427/27-1,2, project number 314964071 and ME790/36-1,2 "Suppressing brake vibrations by deliberately introduced damping" within the PP 1897 "Calm Smooth and Smart - Novel Approaches for Influencing Vibrations by Means of Deliberately Introduced Dissipation." We would also like to thank Audi AG, Wolverine Advanced Materials and AL-KO Kober SE for their support.

References

1. Ahmed, I.: Modeling of vehicle drum brake for contact analysis using ansys. SAE Technical Paper 2012-01-1810 (2012)
2. Aliyev, N., Mehrmann, V., Mengi, E.: Computation of stability radii for large-scale dissipative Hamiltonian systems. Adv. Comput. Math. **46**, 6 (2020)
3. ASTM E 756-05: Standard test method for measuring vibration-damping properties of materials, 10-2005, pp. 1–14 (2005)
4. Bartholomäus, T.: Erstellung einer MATLAB-Routine zur automatisierten Dämpfungsbestimmung. Bachelorarbeit, MMD, TU Berlin (2017)
5. Bathe, K.J., Wilson, E.L.: Numerical Methods in Finite Element Analysis. Prentice-Hall, New Jersey (1976)

6. Beckesch, A.: Pfadverfolgung für Finite-Elemente-Modelle parametrischer mechanischer Systeme, Master Thesis, TU Berlin, 2018
7. Bökemeier, L.: Untersuchungen des Einflusses von Nichtlinearitäten und Temperatur auf Systemeigenschaften von Bremsbelägen. Bachelorarbeit, MMD, TU Berlin (2017)
8. Buck, A.: Simulation von Bremsenquietschen (Brake Squeal), Dissertation TU München (2008). Shaker Verlag, Aachen
9. Conrad, J.: Bestimmung modaler Eigenschaften von Trommelbremskomponenten und Erprobung von Maßnahmen zur Dämpfungserhöhung, Bachelor thesis, MMD, TU Berlin, 2019
10. Cremer, L., Heckl, M.: Körperschall. Springer, Berlin Heidelberg New York (1967)
11. Dannwolff, H.: Analytische Ermittlung von Shimverlustfaktoren unter Verwendung von Constrained-layer-damping Theorien. Masterarbeit, MMD, TU Berlin (2018)
12. Exposito, S.C.: Untersuchung zum Quietschen von Trommelbremsen. Bachelorarbeit, MMD, TU Berlin (2018)
13. Flint, J.: Disc Brake Squeal, Dissertation University of Southern Denmark Engineering College of Odense. Vester Kopi (2002)
14. Ganguly, S., Tong, H., Dudley, G., Connolly, F., Hoff, S.: Eliminating drum brake squeal by a damped iron drum assembly. SAE Technical Paper 2007-01-0592, pp. 1–10 (2007)
15. Glisovic, G., Miloradović, D.: Eliminating brake noise problem. Int. J. Mobil. & Veh. Mech. **36**(3), 37–51 (2010)
16. Gräbner, N., Gödecker, H., von Wagner U.: On the Influence of Damping on Brake Vibrations. International Conference on Engineering Vibration, pp. 1–11 (2015)
17. Gräbner N: Analyse und Verbesserung der Simulationsmethode des Bremsenquietschens, Dissertation TU Berlin, 2016
18. Gräbner, N., Mehrmann, V., Quraishi, S., Schröder, S.: von Wagner, U: Numerical methods for parametric model reduction in the simulation of disc brake squeal. Zeitschrift für Angewandte Mathematik und Mechanik. **96**, 1388–1405 (2016)
19. Hagedorn, P., Heffel, E., Lancaster, P., Müller, P.C., Kapuria, S.: Some recent results on MDGKN-systems. ZAMM **95**(7), 695–702 (2015)
20. Hajela, P., Lin, C.-Y.: Optimal design of viscoelastically damped beam structures. Appl. Mech. Rev. **44**(11) Part 2, 96–106 (1991)
21. Hamid, M.N.A., Teoh, C.-Y., Ripin, Z.M.: The operational deflection shapes and transient analysis of the brake shoes in drum brake squeal. Proc. Inst. Mech. Eng. Part D J. Automob. Eng. **227**(6), 866–884 (2013)
22. Kerwin, Jr. E.M., Ungar, E.E.: Requirements imposed on polymeric materials by structural damping applications. In: Corsaro, R.D., Sperling, L.H. (eds.) Sound and Vibration Damping with Polymers. ACS Symposium Series, vol. 424, pp. 317–345 (1990)
23. Kinkaid, N.M., O'Reilly, O.M., Papadopoulos, P.: Automotive disc brake squeal. J. Sound Vib. **267**(1), 105–166 (2003)
24. Lall, A.K., Asnani, N.T., Nakra, B.C.: Damping analysis of partially covered sandwich beams. J. Sound Vib. **123**(2), 247–259 (1988)
25. Lazan, B.J., Metherfell, A.F., Sokol G.: Multiple-band surface treatments for high damping. Technical Report AFML-TR-65-269, pp. 1–45 (1965)
26. Maaß, C.: Einfluss von Klebstoffschichten auf Verlustfaktoren eines Constrained-layer-damping Aufbaus und Ermittlung von Streuungskenngrößen bei Bremsbelägen. Bachelorarbeit, MMD, TU Berlin (2019)
27. Mead, D.J.: The practical problems of assessing damping treatments. J. Sound Vib. **1**(3), 270–291 (1964)
28. Mehl, C., Mehrmann, V., Wojtylak, M.: Linear algebra properties of dissipative Hamiltonian descriptor systems. SIAM J. Matrix Anal. Appl. **39**, 1489–1519 (2018)
29. McDaniel, J.G., Li, X., Elvenkemper, A., Wegman, E., Wang, A., Chen, S.E., Flint, J.: Simulating the effect of insulators in reducing disc brake squeal. SAE Technical Paper 2005-01-3944, pp. 1–7 (2005)

30. Morris, A.: A Practical Guide to Reliable Finite Element Modelling. John Wiley & Sons Ltd., Chichester (2008)
31. Nashif, A.D., Jones, D.I.G., Henderson, J.P.: Vibration Damping. John Wiley & Sons, Chichester (1985)
32. Oberst, H., Frankenfeld, K.: Über die Dämpfung der Biegeschwingungen dünner Bleche durch fest haftende Beläge. Acta Acustica united with Acustica **2**(4), 181–194 (1952)
33. Popp, K., Rudolph, M.: Brake Squeal, Detection, Utilization and Avoidance of Nonlinear Dynamical Effects in Engineering Applications Final Report, pp. 197–225. Shaker Verlag, Aachen (2001)
34. Rao, D.K.: Frequency and loss factors of sandwich beams under various boundary conditions. J. Mech. Eng. Sci. **20**(5), 271–282 (1978)
35. Ross, D., Ungar, E., Kerwin, Jr. E.M.: Damping of plate flexural vibrations by means of viscoelastic laminae. In: Ruzicka, J.E. (eds.) Structural Damping, pp. 49–87. Pergamon Press, New York (1959)
36. Schmid, D.: Brake pad. figshare. Dataset., Lizenz CC BY4.0 (2018). https://doi.org/10.6084/m9.figshare.7046663.v1. https://figshare.com/articles/Brakepad/7046663
37. Schmid, D.: Zum Einfluss von Dämpfung auf Bremsenschwingungen, Dissertation TU Berlin, Lizenz CC BY4.0, 2020
38. Schmid, D., Gräbner, N., von Wagner, U.: Experimental Investigations of Brake Pad Shim Properties. Proceedings in Applied Mathematics and Mechanics, vol. 17, No. 1, pp. 41–44. Wiley (2017)
39. Schmid, D., Gräbner, N., von Wagner, U.: Characterization of Brake Shims Using Analytical Constrained Layer Damping Theories. Proceedings in Applied Mathematics and Mechanics, vol. 18, No. 1, pp. 1–2. Wiley (2018)
40. Schmid, D., Sessner, V., Gräbner, N., von Wagner, U., Weidenmann, K.A.: Parameter Identification of Brake Pad Shims for Complex Eigenvalue Analysis. Proceedings in Applied Mathematics and Mechanics, vol. 19, No. 1, pp. 1–4. Wiley (2019)
41. Parfitt, G.G., Lambetz, D.: The damping of structural vibrations, C. P. No. 596 Aeronautical Research Council. Her Majesty's Stationery Office, London (1962)
42. Schmid, D., Gräbner, N., von Wagner, U.: On brake pad shim characterization: a homogenization approach and finite element analysis. In: Advanced Structured Materials—New Achievements in Continuum Mechanics and Thermodynamics, vol. 108, pp. 447–464. Springer (2019)
43. Schmid, D., Constrained-layer-damping shear bending. figshare. Figure, Lizenz CC BY4.0 (2019). https://doi.org/10.6084/m9.figshare.8143808.v1. https://figshare.com/articles/Constrained-layer-damping_shear_bending/8143808
44. Schmid, D.: Modal analysis set-up of non-industrial steel drum. figshare. Figure, Lizenz CC BY4.0 (2022). https://doi.org/10.6084/m9.figshare.17715890.v1. https://figshare.com/articles/figure/Modal_analysis_set-up_of_non-industrial_steel_drum/17715890
45. Schmid, D., Gräbner, N., von Wagner U.: Friction induced noise in drum brakes: finite-element modeling and experiments with special focus on damping. Arch. Appl. Mech. (2022)
46. Selamat, M.S.B.: Friction materials for brakes application. J. Ind. Technol. **14**(2), 9–25 (2005)
47. Seltitz, S.: Experimentelle Dämpfungsuntersuchungen an Bremsbelägen. Bachelorarbeit, MMD, TU Berlin (2016)
48. Ungar, E.E., Kerwin, E.M., Jr.: Loss factors of viscoelastical systems in terms of energy concepts. J. Acoust. Soc. Am. **34**(7), 954–957 (1962)
49. Vey, C., Pfeffer, P.E.: Brake systems 2025—Future Trends. In: 9th International Munich Chassis Symposium 2018. Springer, Heidelberg, Dordrecht, London, New York (2019)
50. Wallaschek, J., Hach, K.-H., Stolz, U., Mody, P.: A survey of the present state of friction modelling in the analytical and numerical investigation of brake noise generation. In: Proceedings of the ASME Vibration Conference DETC99/VIB-8357, pp. 1–12 (1999)
51. Wehner, J.H., Jekel, D., Sampaio, R., Hagedorn, P.: Damping Optimization in Simplified and Realistic Disc Brakes. Springer Briefs in Applied Sciences and Technology, pp. 1–50 (2018)
52. Wolverine Advanced Materials, Shim datasheets (2016)
53. Yuhas, D.E., Ding, J., Venkatesan, S.: Non-linear aspects of friction material elastic constants. SAE Technical Paper 2006-01-3193, pp. 1–10 (2006)

Vibration Reduction by Energy Transfer Using Shape Adaption

Alexander Nowak, Kai Willner, and Alexander Hasse

1 General Approach

The approach is illustrated in Fig. 1 with an exemplary structure. The slender, beam-like structure consists of specially designed compliant ribs with selective compliance, interconnected by a hull structure. By actuating the compliant ribs, the structure can be modified from the configuration with shape I to shape II. This shape adaption of the cross-sections – which in turn modifies the beam's second moment of area – enables a dynamic adaption of the beam's bending stiffness.

In the bottom part of Fig. 1, a possible time law for the two stiffness states is shown. The stiffness change is determined by the *vibration mode* – which is typically a critical, low-frequency bending mode (see Fig. 1, black curve). At zero-crossings of this mode's amplitude, the stiffness is increased by adapting the cross-sections to shape II and returned to the original stiffness with shape I at the extremal points (see Fig. 1, gray curve).

The reduction of vibrations is – besides the present damping – based on two mechanisms. In the quarter cycles between zero points and maxima of the vibration mode, the higher stiffness causes a certain amplitude reduction. This effect – not directly visible in Fig. 1 – can be understood as a counterforce and is called *active effect* in the following. The second effect is a transfer of part of the kinetic energy subtracted from the critical vibration mode into a specifically designed, higher frequency *absorber*

A. Nowak · A. Hasse (✉)
Professorship Machine Elements and Product Development, Chemnitz University of Technology, Reichenhainer Str. 70, 09126 Chemnitz, Germany
e-mail: alexander.hasse@mb.tu-chemnitz.de

A. Nowak
e-mail: alexander.nowak@mb.tu-chemnitz.de

K. Willner
Institute of Applied Mechanics, Friedrich Alexander University Erlangen Nuremberg, Egerlandstr. 5, 91058 Erlangen, Germany
e-mail: kai.willner@fau.de

P. Eberhard (ed.), *Calm, Smooth and Smart*, Lecture Notes in Applied and Computational Mechanics 102, https://doi.org/10.1007/978-3-031-36143-2_14

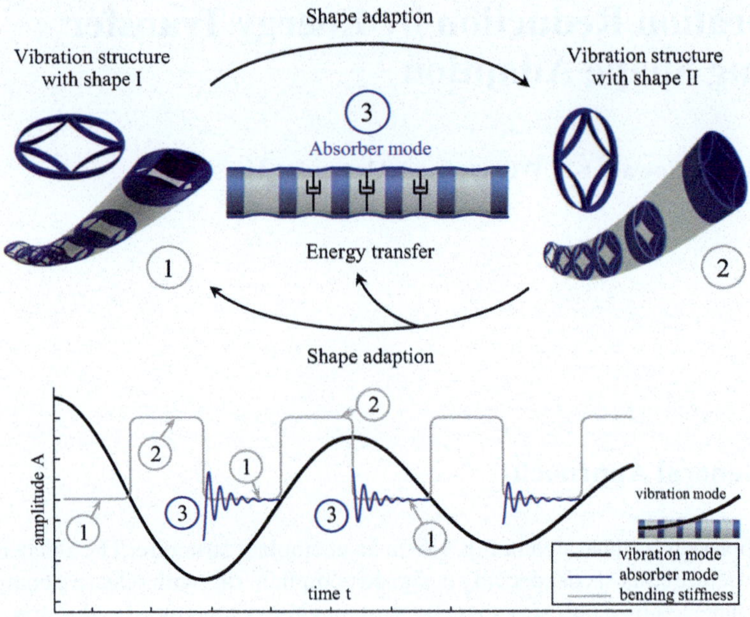

Fig. 1 Approach for vibration reduction by stiffness variation using shape adaption

mode. This effect, called *semi-active effect* in the following, is illustrated by the blue curve. The high frequency of the absorber mode enables a faster amplitude decay through structural damping, which can potentially be enhanced by dampers particularly designed for this range.

In Sect. 2, a proof-of-concept is presented by an experiment with a shape-adaptive structure. The influence of shape adaption on the structural stiffness and the dynamics of the system is shown. In Sect. 3, the analysis of the two effects introduced above (active and semi-active) is discussed in detail. In Sect. 4, a method is presented for determining a spatially reasonable distribution of stiffness changes by actuation of the ribs and to perform parameter studies with respect to different types of cross-sectional modifications and absorber modes.

2 Experimental Study with a Shape-Adaptive Structure

In contrast to active methods, where actuators act directly on the degrees of freedom of the vibrating system, and passive methods, in which the system's inherent damping is supplemented by additional dissipative elements without further energy requirements, semi-active approaches are defined as methods that influence the system parameters [12].

The literature offers numerous semi-active approaches for vibration reduction. Many of these relate to adaptive dampers, e.g. for wind and earthquake protection of buildings [26] or for car suspension systems [30]. Variable stiffness is often employed in tunable absorbers, e.g. with the use of shape memory alloys [28], piezo actuators [18], magnetostrictive [10] or magnetorheological materials [8]. Cyclical variations in stiffness are found less frequently. In most of the contributions, as for instance in [21] and [27], discrete spring-mass systems are considered. With shape adaption – a research field that is driven primarily by the aerospace sector [1] – there are some obvious benefits, such as the suitability for stiffness adjustment of continuous structures and the possibility of intermediate values in the stiffness change. Cyclical stiffness variations for vibration reduction based on shape adaption have, however, hardly been investigated. For this reason, the experimental setup [22] presented below was developed at the beginning of the project to evaluate the general feasibility of this approach.

2.1 Experimental Setup

The test object is shown in Fig. 2. It is a cantilever steel plate with elongating piezo patch actuators attached to it. When actuated, the piezo patches cause the plate's cross-section to bend (see Fig. 2, bottom right) and thus the second moment of area with respect to the x-axis is changed.

Between the fixation and the first piezo patch, a shaker can be attached for forced vibrations in y-direction. The response of the structure is measured with a laser Doppler vibrometer capturing displacement and velocity at the end of the plate.

The fixation is designed in a way that the curvature of the plate is free to change. Under the subsequent assumption of a constant cross-sectional curvature along the beam, the second moment of area I_x of the deformed configuration can be calculated with the outer and inner radius (r_1, r_2) and the angle of curvature ϑ by

$$I_x = \int_{-\vartheta/2}^{\vartheta/2} \int_{r_1}^{r_2} (r \cdot \cos \vartheta - s)^2 \cdot r \, dr \, d\vartheta \tag{1}$$

The stiffening factor $\mu(\vartheta)$, i.e. the ratio between I_x of the initial and the deformed geometry, is given by

$$\begin{aligned}
\mu(\vartheta) &= \frac{12}{wh^3}\left(A(\vartheta)w^3h + B(\vartheta)\frac{wh^3}{12}\right) \\
A(\vartheta) &= \frac{1}{\vartheta^2}\left(\frac{1}{2}(1 + \frac{\sin\vartheta}{\vartheta}) - 4\frac{\sin^2(\vartheta/2)}{\vartheta^2}\right) \\
B(\vartheta) &= \frac{3}{2}(1 + \frac{\sin\vartheta}{\vartheta}) - 8\frac{\sin^2(\vartheta/2)}{\vartheta^2}
\end{aligned} \tag{2}$$

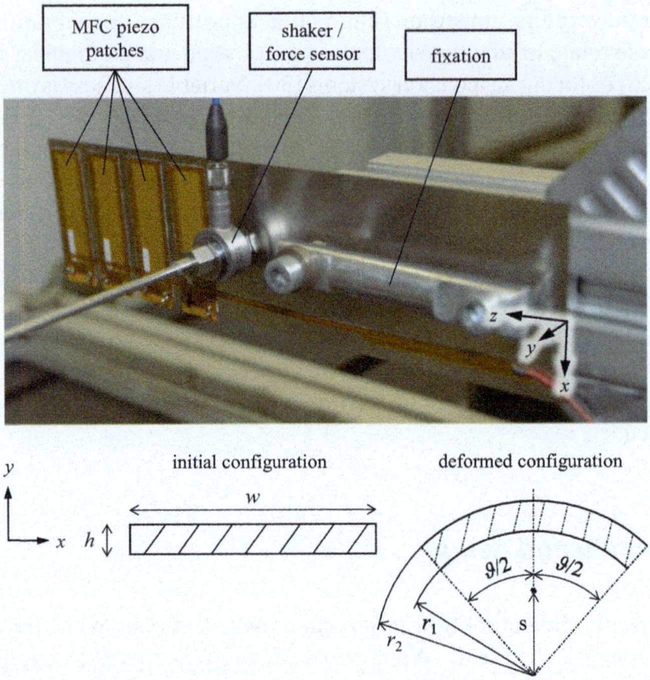

Fig. 2 Setup with a shape-adaptive structure

At maximum measured curvature, a value of approx. 2 is obtained for μ. The experimental bending measurements also yield a doubling of the bending stiffness at small deflection. As already mentioned above, it is naturally possible to realize intermediate values for μ depending on the voltage on the piezos.

2.2 *Wavelet-based Analysis*

The stiffness variation obviously implies a non-linear, time-variant system. A classical frequency response function offers only limited suitability for the analysis of such systems.

To reveal the time-variant dynamics of the structure, continuous wavelet transform is therefore applied. The wavelet-based spectra of the input excitation E and output response R are given by

$$E(a,b) = \frac{1}{\sqrt{a}} \int_{-\infty}^{\infty} e(t)\psi^*\left(\frac{t-b}{a}\right) dt, \quad R(a,b) = \frac{1}{\sqrt{a}} \int_{-\infty}^{\infty} r(t)\psi^*\left(\frac{t-b}{a}\right) dt \tag{3}$$

with b and a as operators for locality in time and frequency, respectively, and $\psi(t)$ as the analyzing wavelet function. The time-variant frequency response, defined as the H_1 estimator is

$$H_1(a,b) = \frac{G_{RE}(a,b)}{G_{EE}(a,b)} \tag{4}$$

with the wavelet-based cross-power spectra G_{RE} and auto-power spectra G_{EE}:

$$\begin{aligned} G_{RE}(a,b) &= R(a,b)E^*(a,b) \\ G_{EE}(a,b) &= E(a,b)E^*(a,b) \end{aligned} \tag{5}$$

Equation 4 is a two-dimensional function representing a set of frequency response functions for all specific values of time. For a more detailed consideration of this topic and an extension to wavelet-based coherence, please refer to [9].

With the presented method, the time variant characteristics of the structure shown in Fig. 2 were investigated. The structure was excited with band-limited white noise and simultaneously, the piezo patches were actuated with square and sine signals. For clarification, it should be mentioned here that the piezos were actuated very slowly to achieve a good resolution of the results. This investigation here is therefore not to be confused with a stiffness variation time law based on a modal amplitude as shown in Fig. 1.

The wavelet-based input-output analysis is shown in Fig. 3. The results clearly show the varying natural frequency of the system. The stiffness variation provides two limits at the natural frequency, specified in Fig. 3 with *deformed conf.* at maximum voltage and *initial conf.* with switched off piezos. It can also be seen that a square-wave signal allows a relatively sharp transition between those frequencies, while the sinusoidal actuation leads to a continuous adaption of the natural frequencies with

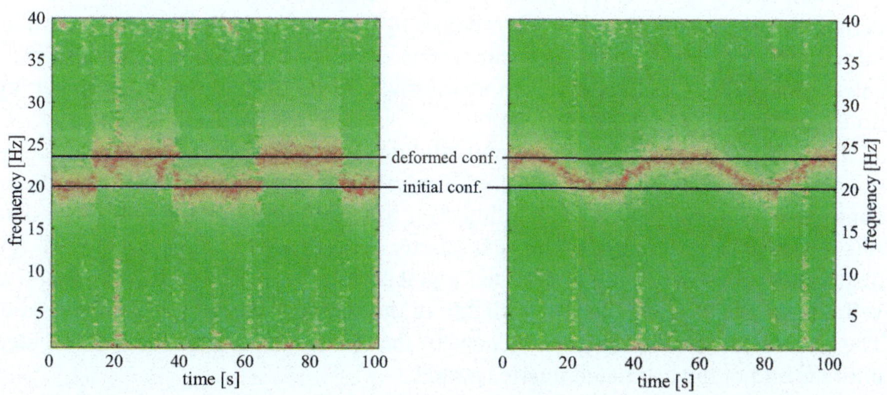

Fig. 3 Time-variant frequency response function due to actuation with a square-wave signal (left) and a sinusoidal signal (right)

all intermediate values. This offers a considerable advantage over variable systems as in [21, 27], which, as mentioned above, allow only two stiffness states.

2.3 Vibration Reduction Analysis

The analysis carried out in the previous section demonstrates that a time-variant structural behavior with a defined change in stiffness is possible via shape adaption. This section describes the application to vibration reduction.

First, this requires the definition of a suitable time law for the stiffness variation. Preliminary studies on simplified systems (see also Sect. 3) have revealed that a cyclic stiffness increase at the zero-crossing and recovery of the original stiffness value at the maximum of the modal amplitude (see Fig. 1, gray curve) is best suited. In the experiment, this time law is realized by a feedback control based on the y-displacement at the beam end. Thereby, several signal shapes are conceivable. Figure 4 shows three of these variants, which all have their starting point at the zero-crossing and end point at the maximum, but with increasingly smoother shapes (in the following called *square*, *rounded square* and *sine* actuation). For the square signal, the voltage V on the piezos can be stated in simplified form as

$$V = \begin{cases} \text{max}, & \text{if } |u_k| > |u_{k-1}| \\ \text{off}, & \text{else} \end{cases} \tag{6}$$

with u_k as the measured displacement at a timestep k. This requires an ideal (damped) sinusoidal displacement curve like the blue one in Fig. 4. However, superposition with higher-frequency components occurs due to the (desired) internal energy transfer (see Fig. 1 blue curve), resulting in additional zero-crossings and maxima in the displacement curve. With application of (6), this leads to frequent and abrupt stiffness increases mostly not related to the modal amplitude of the vibration mode, which turned out to be highly counterproductive in the experiment and in the numerical calculations. Filter techniques have also proven to be unsuitable, as the resulting time delay severely impairs the results.

The following procedure was eventually implemented: the high-frequency components are filtered out of the measured displacement and velocity, but not for direct control of the piezo voltage according to a time law like (6). Instead, these data sets serve for an estimation of the next quarter period based on previous cycles. If a displacement zero-crossing is detected together with an exceedance of an adaptive velocity threshold, the actuation with one of the signal shapes in Fig. 4 is triggered. The actuation is then no longer influenced by the current measurement but is executed until the end of the estimated quarter period.

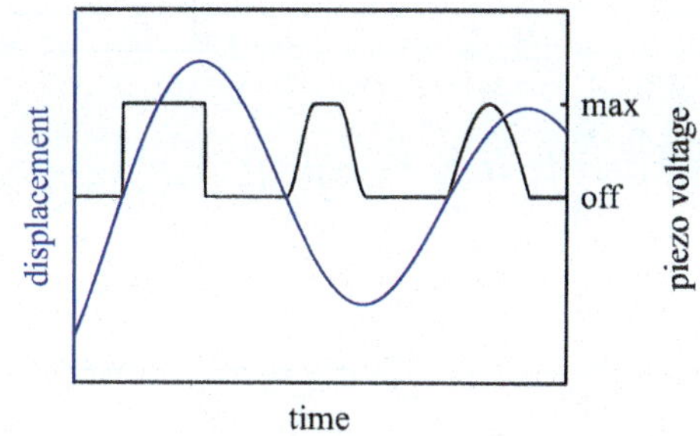

Fig. 4 Possible time laws for actuation of the piezos

Alternatively, the possibility for multiple sensors and real-time transformation into the relevant modal amplitudes as a basis for the actuation signals should be mentioned, which is planned for upcoming experimental studies.

The presented method to modify the stiffness was applied in different test scenarios, including free vibration after initial deflection, sweep over the range of the first natural frequency and with random excitation forces. For comparison, the structure in the initial configuration (low stiffness) and with permanently activated piezos (high stiffness) was considered. Additional damping elements were not employed.

A first interesting result was observed in the comparison of the three investigated signal shapes of Fig. 4. The best results are achieved with sinusoidal actuation, followed by rounded square and square. Particularly in contrast to stiffness switching with a square signal, the sinusoidal adaption of the structure does not act as an impulse on the structure and thus excitation by the actuators is reduced. Besides, a long dwell time on the maximum stiffness is not that significant, but rather the difference between maximum and minimum stiffness – which is identical for all shapes. This observation, however, is very dependent on the structure, type and position of the actuators and should therefore not be understood in a generalized way.

In all tested cases and also for all signal types, improved performance was achieved by applying the variable stiffness concept compared to the time-invariant structure. Figure 5 shows two exemplary results. For free vibration (see Fig. 5a, comparing constant low stiffness k_l to stiffness variation with a sine signal k_s), the amplitude decay is considerably faster. For harmonic loads (see Fig. 5b, comparing k_l, k_s and constant high stiffness k_h under swept sine excitation), the resonance of k_s is clearly less intense and it is evident that there is no typical resonance peak. The response curve is rather flat, which is due to the fact that the energy is transferred to higher-frequency modes.

3 Energetic Considerations

An important aspect in the evaluation of the presented method are energetic considerations. On the one hand, the internal energy transfer to a higher mode (see Fig. 1, blue

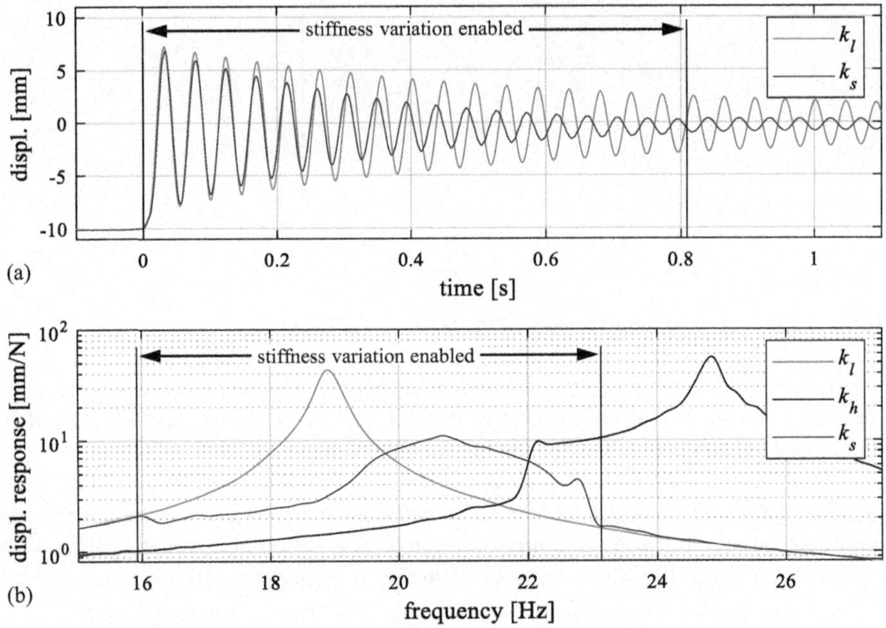

Fig. 5 Experimental results for **a** free vibration and **b** sweep excitation

curve) needs to be investigated. On the other hand, efficiency must be evaluated. In the approach discussed here, actuators are involved and, as already mentioned, there is definitely an active component which is generally associated with high energy requirements.

The realization of the stiffness variation evidently has a significant influence on this consideration. Several approaches can be found in the literature on vibration reduction by cyclically modified stiffness, including piezoceramic actuators switched from open-circuit to short-circuit state [4], controllable magnetorheological dampers [17], tension-controlled strings [25], connection of the main structure with an elastic brace whose stiffness is controlled by a control valve [29, 31] or employment of a Voigt element with an adaptive damper [16]. Adaption of spring stiffness values is also frequently found, e.g. by manipulating the effective length of a spring [27], connecting and detaching a secondary system [15, 21] or changing the shape in which the springs are arranged [20].

In all mentioned publications, the respective approaches are considered semi-active. A decisive advantage of semi-active methods often emphasized is their low energy consumption compared to active methods [16, 17, 21, 29, 31]. However, justifications based on energetic calculations or quantitative comparisons with other methods are not provided. Almost all approaches are either based on adaptive dampers or include devices that dissipate energy through e.g. friction or electrical resistance.

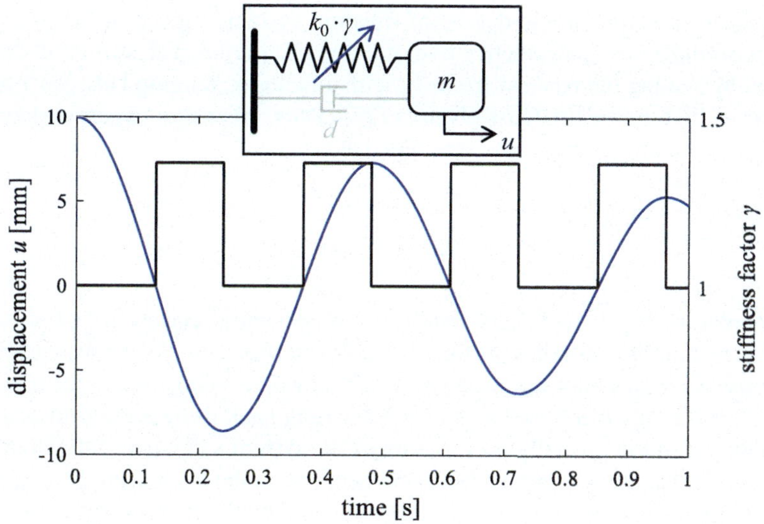

Fig. 6 Spring-mass oscillator with variable stiffness

While this is certainly reasonable in terms of vibration reduction, it makes it more difficult to separate the different effects.

The experiment presented in the last section could also hardly be considered for such energetic observations. Although the excitation of higher modes can be shown in a numerical simulation, a separation of active and semi-active components is practically impossible. In addition, the actuators cover almost the entire free area and cannot be controlled individually. A parameter variation and determination of the origin of the internal energy transfer is therefore not feasible.

For this reason, discretized systems similar to [21, 27] were chosen for fundamental energetic considerations. A detailed study in this regard is given in [24]. The following section will provide a brief overview of this topic.

3.1 Single-Degree-of-Freedom System with Variable Stiffness

The system considered in this section is inspired by [27]. In this contribution, the stiffness of a spring can be increased by a motor-controlled arm blocking some of the spring coils. Equivalent to the time law based on a modal amplitude shown in Sect. 1, the time law in [27] is defined as increase of stiffness between zero points and maxima and vice versa – in this case related to the measured displacement of the mass. In the above-mentioned publication, the system is treated as a single spring-mass oscillator, which is also considered in this section. For the sake of explanation, the system is further assumed to be undamped.

The corresponding single-degree-of-freedom system is given in Fig. 6, together with an exemplary displacement curve of a free oscillation. The stiffening device is realized by scaling the basic stiffness k_0 with a factor γ, ranging between 1 and the quotient of high to low stiffness given in [27]. Thus, no energy expenses due to e.g. deformation of the structure are considered at this point.

Under these assumptions, the potential energy change at the points of the stiffness switches results in

$$\Delta U = \frac{1}{2}\Delta k u^2 \tag{7}$$

Considering Eq. 7, two facts stand out: the potential energy remains constant when increasing the stiffness, as the displacement u is zero. As a consequence, the amplitude decreases after the stiffness switch; however – considering the undamped system – no energy has been lost so far. Switching back to k_0 at the next zero point would thus result in virtually no variation of the system's dynamics afterwards.

The second point concerns the stiffness decrease at the maximum of the displacement. In this case, there is a drop of energy, defined by the stiffness difference of the two states $\Delta k = k_0(1-\gamma)$. The energy is extracted from the system by the stiffening device performing negative work on the system. By definition, it is still a semi-active procedure, since a system parameter is varied. A corresponding actuator, however, has to perform the same work as given in (7), which de facto does not distinguish it from active vibration reduction. In such cases where an amplitude reduction is based on negative work instead of damping – as shown in Fig. 6 – the corresponding energy loss is called *active component* here and in the following.

3.2 Serial System with Variable Stiffness

The preliminary study of the last section is now extended by considering damping and dividing the spring into two sections. As mentioned, the system's spring in [27] can be separated by an arm, which is represented here by two springs with basis stiffness values k_{01} and k_{02}. The (comparatively small) mass of the physical spring is represented by m_1.

The corresponding system is illustrated in Fig. 7. As displayed, both springs can now be scaled with a factor γ_1 or γ_2, respectively. Thus the high stiffness of the total system corresponds to

$$k_h = \frac{\gamma_1\gamma_2 k_{01}k_{02}}{\gamma_1 k_{01} + \gamma_2 k_{02}} \tag{8}$$

First, the case of [27] is considered in which an increase in stiffness is performed by completely blocking the left spring. Therefore, the corresponding scaling factor γ_1 tends towards infinity. With Eq. 8 and $\gamma_1 \to \infty$, the high stiffness is consequently equal to k_{02}. Alternatively, both springs can be scaled with the same factor $\gamma_1 = \gamma_2 = \gamma = k_h/k_l$, which is in accordance with the single-degree-of-freedom system

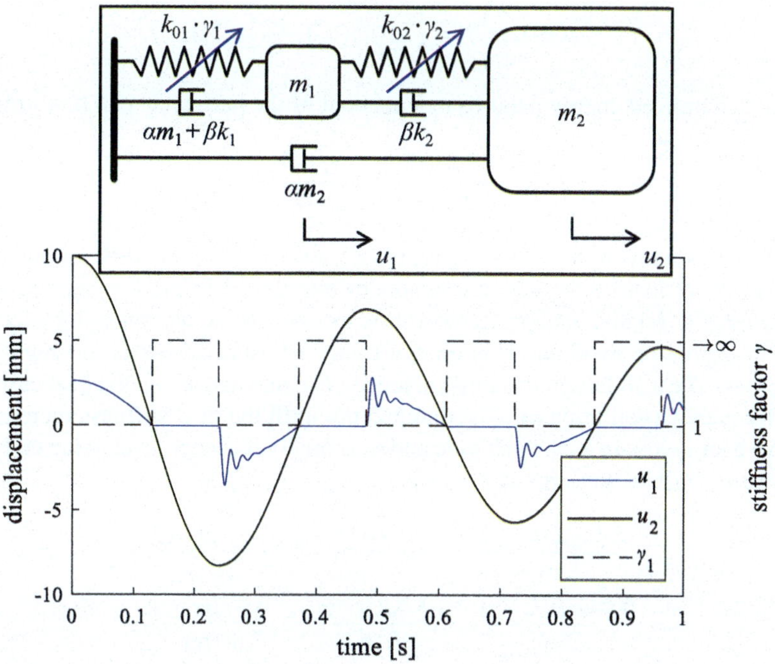

Fig. 7 Serial system with variable stiffness

in Sect. 3.1. The case of scaling one of the spring stiffness values is called *local case* here; scaling the whole system is called *global case*.

The curves in Fig. 7 show a free oscillation with the local case. After releasing the blocked spring, a noticeable high-frequency oscillation of u_1 can be observed, which already resembles the modal amplitude of the absorber mode in Fig. 1. In this quarter cycle, the proportionality of u_1 to u_2

$$u_1 = \frac{k_2}{k_1 + k_2} u_2 \tag{9}$$

which otherwise results from the low inertia of m_1, can obviously no longer apply. However, Eq. 9 is valid for any high stiffness quarter cycle and the equation of motion derived for the second degree of freedom is identical in the local and global case before and after the stiffness increase:

$$m_2 \ddot{u}_2 + \frac{k_1 k_2}{k_1 + k_2} u_2 = 0 \tag{10}$$

Consequently, the loss of vibration energy (without damping) over two extremal points t_1 and t_2 is for both cases

$$\Delta E(t_1, t_2) = \frac{1}{2} k_h (u_2^2(t_2^-) - u_2^2(t_1^-)) \tag{11}$$

The generalized energy balance independent of the half-cycle can be formulated for the global case as

$$\frac{\Delta E_a(t_1, t_2)}{E(t_1)} = \frac{k_l - k_h}{k_h} \tag{12}$$

The same relative value of energy is periodically extracted from the second degree of freedom, reflecting the active component introduced in the last section. In the local case, the relative energy extraction is exactly the same, but is transferred to the first degree of freedom. This energy transfer is stated here as the *semi-active component* ΔE_s. If damping is now included (E_p) and hence – in the best case – the high-frequency oscillation of u_1 is damped out until the next stiffness increase, the relative energy loss in each half cycle and thus the total energy level at the end of the observation will be the same in both cases:

$$\begin{aligned} \frac{\Delta E_{\text{local}}(t_1, t_2)}{E(t_1)} &= \frac{\Delta E_s(t_1, t_2)}{E(t_1)} + \frac{\Delta E_p(t_1, t_2)}{E(t_1)} = \\ \frac{\Delta E_{\text{global}}(t_1, t_2)}{E(t_1)} &= \frac{\Delta E_a(t_1, t_2)}{E(t_1)} + \frac{\Delta E_p(t_1, t_2)}{E(t_1)} \end{aligned} \tag{13}$$

The specific situation of a completely blocked degree of freedom is of course not the general case. Hence, the example of modifying the second spring with finite values for γ_2 is discussed below. Due to this adjustment, it is no longer possible to reduce the vibration without an active component. The stiffness change now implies negative work and the energy loss is determined by all three components ΔE_a, ΔE_p and ΔE_s. To compare the two variants – local and global stiffness variation – Fig. 8 shows the energy lost through damping for three pairs with increasing intensity of the stiffness variation. Each pair of same color results in the same total energy at the end of the observation, with the dotted lines representing the global case and the solid lines representing the corresponding local case.

With increasing value for γ_2, the damped energy also increases significantly in the local case. In the global case, the dissipated energy even decreases with increasing γ. This phenomenon can be best explained by considering the system in modal space. With the matrix of eigenvectors $\mathbf{\Phi}$ of the generalized eigenvalue problem with the stiffness matrix $\mathbf{K}$ and the mass matrix $\mathbf{M}$

$$\mathbf{K\Phi} = \mathbf{M\Phi\Lambda} \tag{14}$$

applied for modal transformation

$$\tilde{\mathbf{K}} = \mathbf{\Phi}^T \mathbf{K\Phi} = \mathbf{\Lambda} \tag{15}$$

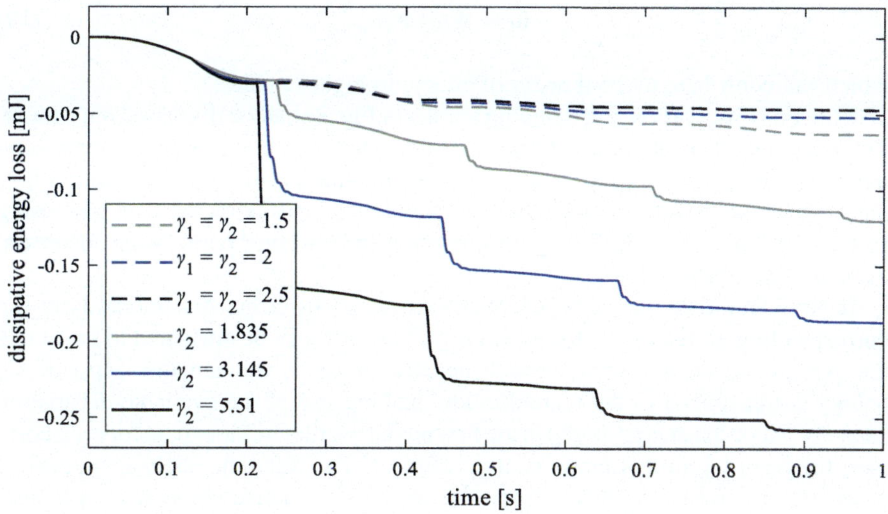

Fig. 8 Energy loss by damping, compared for global (dashed curves) and local (solid curves) case

it is evident that globally scaling each spring stiffness with γ leads to exactly the same set of eigenvectors:

$$\gamma \mathbf{K}\boldsymbol{\Phi} = \mathbf{M}\boldsymbol{\Phi}\bar{\boldsymbol{\Lambda}} \tag{16}$$

The eigenvalues λ_i as the diagonal entries of $\boldsymbol{\Lambda}$ are all scaled according to γ

$$\bar{\lambda}_i = \gamma \lambda_i \tag{17}$$

and thus also the total modal stiffness matrix $\tilde{\mathbf{K}}$ is scaled with the same factor as the physical one. The modal basis $\boldsymbol{\Phi}$ however remains constant and likewise the relation of the entries in $\tilde{\mathbf{K}}$. The mutual relationship of the modal amplitudes is not affected and the modal oscillators do not show any additional response. Without further external excitation, the first modal amplitude remains dominant in the present case and again, the system can be considered as a single-degree-of-freedom system. The active component can be expressed with the modal amplitudes $\tilde{\mathbf{u}}$ at a point t of stiffness decrease

$$\Delta E_a = \frac{1}{2}(1 - \gamma)\tilde{\mathbf{u}}^T(t)\tilde{\mathbf{K}}\tilde{\mathbf{u}}(t) \tag{18}$$

which basically corresponds to Eq. 7.

This consideration no longer applies in the case presented here, nor in general, when performing local changes in stiffness. The entries in the stiffness matrix are changed individually and consequently, there are two modal bases (and possibly corresponding intermediate states). When switching from high to low stiffness, the vector of the modal amplitudes $\tilde{\mathbf{u}}_\mathbf{h}$ must be transformed into the new space

$$\tilde{\mathbf{u}}_{\mathrm{l}} = \boldsymbol{\Phi}_l^{-1} \boldsymbol{\Phi}_h \tilde{\mathbf{u}}_{\mathrm{h}} \tag{19}$$

which can contribute to the transfer of energy to higher modes.

With this observation, the damping shown in Fig. 8 shall be discussed once more. The second mode is not or hardly excited in the global case. Accordingly, there is no damping with respect to the second mode and the dissipated energy shown in Fig. 8 takes place exclusively in mode 1. Since more energy is extracted with increasing factor γ, the energy dissipated by damping over the observed period even decreases, as mentioned above.

In the local case, the same energy is extracted from the first mode as in the corresponding global case. However, only a part of it can be attributed to the active component, which is associated with negative work by the actuator. Most of the energy is transferred to the second mode, making use of the available dissipation capacity. Since the energy stored in the first modal oscillator is identical for each color pair, the damping in this mode is also identical. The difference between a dashed curve and a solid curve thus corresponds to the semi-active component, which is achieved by local stiffness variation.

Finally, two points remain to be named for continuous structures with stiffness changes performed by shape adaption. On the one hand, the actuator must be explicitly considered in the calculation of the active component. The extracted energy is not directly reflected in the system energy, but via the force and stroke applied by the actuator. On the other hand, it should be noted that shape adaption always leads to a change of the modal basis. Accordingly, there will always be an internal energy transfer and thus a semi-active component. This also applies to the test structure presented in Sect. 2. With the observations outlined in this section – and those of the next section – this transfer can be enhanced allowing for a more efficient implementation.

4 Synthesis of Shape-Adaptive Beams

In the last two sections, the focus was on the analysis of the dynamics associated with variable stiffness. For practical relevance, a method is presented in this section to design structures for the required stiffness change.

As described in the introduction section, beam-like structures are considered in the context of this project with the stiffness change being achieved by shape adaption. Shape adaption is generally associated with compliant mechanisms [6]. Thereby, the deformability of the structure is exploited in order to achieve predefined kinematics without requiring sliding or rolling components of conventional mechanisms. The design of such mechanisms is typically approached by structural optimization procedures [7]. Most of those approaches to synthesize compliant mechanisms for shape adaption are however limited to planar structures and do not take into account their dynamic behavior.

A versatile approach for shape adaption of three-dimensional structures is the belt-rib concept presented in [3] and extended by a synthesis method in [2]. Compliant inner rib structures are employed here to deform the surrounding outer envelope. Related concepts are still being developed to this day, mostly in the field of wing structures for e.g. the modification of aerodynamic characteristics [19]. As it is well suited for the adaption of beam-like exterior structures, a design method based on compliant ribs was also developed within this project with special emphasis on structural dynamics and realization of stiffness variations.

4.1 Surrogate Model

The energetic considerations in Sect. 3.2 have demonstrated that a homogeneous increase in the stiffness of a structure is not optimal or efficient. For three-dimensional, continuous structures, the dynamic behavior is much more complex and difficult to predict. The initial focus of the design is therefore not on maximizing stiffness changes, but on the pre-definition of an efficient model that can represent the dynamics of the 3D structure under stiffness variations and allows for parameter studies. This surrogate model [23] is presented in the following.

An exemplary target structure is shown in Fig. 9. It consists of the outer hull structure, the ribs to be defined and possibly additional stiffening elements. The corresponding surrogate model of the ribs, on which the design of a stiffness matrix is based, is shown on the right-hand side. The surrogate model is a reduced order model defined at master nodes that are connected to the hull structure. The structural target behavior is realized by an inverse modal transformation performed on the reduced model. The general objective is to modify the critical bending mode and, if necessary, to embed a particular absorber mode. The model should also provide

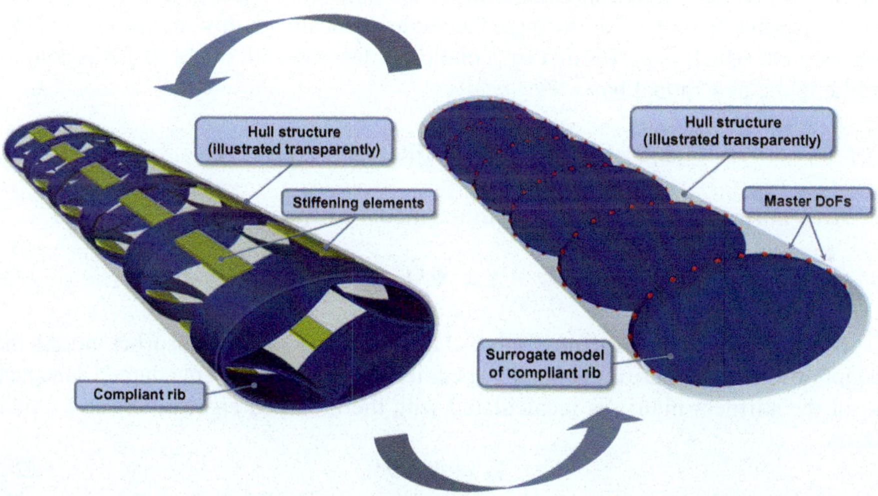

Fig. 9 Shape-adaptive beam with surrogate model

the possibility to parameterize the cross-sectional deformation, absorber mode and placement of the ribs in an efficient way.

The first step is the definition of the master (m) and slave (s) degrees of freedom (DoFs). All DoFs at which loads are applied and the coupling points with the ribs are placed in the set of the m-DoFs, the others in the set of the s-DoFs. The global stiffness matrix is rearranged on basis of these two sets and a transformation matrix **T** is calculated

$$\mathbf{K} = \begin{bmatrix} \mathbf{K}_{mm} & \mathbf{K}_{ms} \\ \mathbf{K}_{sm} & \mathbf{K}_{ss} \end{bmatrix}, \quad \mathbf{T} = \begin{bmatrix} \mathbf{I} \\ -\mathbf{K}_{ss}{}^{-1}\mathbf{K}_{sm} \end{bmatrix} \tag{20}$$

which is the basis of the well-known Guyan reduction [13]. Since the dynamic behavior is relevant, **T** can be extended by eigenvectors derived from the generalized eigenvalue problem of the slave stiffness and mass matrix [5].

The following calculation only takes place on the partition $\mathbf{K}_{mm}$ and is therefore independent of the expensive inversion $\mathbf{K}_{ss}^{-1}$ in the transformation matrix **T**.

The criterion here is not the specification of a desired displacement, but the manipulation of a certain eigenvalue of the stiffness matrix. Thus, it is a matter of finding a stiffness matrix for each rib i which is essentially ruled by the corresponding desired deformation mode φ_d. This desired mode (or modes, if the absorber mode is included) has to be specified and may be considered as several linear combinations of the cross-sectional modes in the parameter study. The matrix of eigenvectors $\mathbf{\Phi}$ and eigenvalues $\mathbf{\Lambda}$ of a rib's master DoF subspace $\mathbf{k_i}$ are then determined and compared with the desired mode:

$$\mathbf{a} = \mathbf{\Lambda}^{-1}(\mathbf{\Phi}^T \mathbf{k_i} \varphi_d) \tag{21}$$

Via **a**, the least influence on the matrix due to a modal adjustment can be determined. In the corresponding column of $\mathbf{\Phi}$, the eigenvector is replaced by φ_d. Since the orthogonality of $\mathbf{\Phi}$ is lost through the mode replacement, this condition must first be restored, which is performed by a cholesky decomposition and qr-factorization, yielding the new modal basis $\mathbf{\Psi}$

$$\mathbf{k_i} = \mathbf{C}\mathbf{C}^*, \quad \mathbf{C}^T\mathbf{\Phi} = \mathbf{Q}\mathbf{R}, \quad \mathbf{\Psi} = (\mathbf{C}^T)^{-1}\mathbf{Q} \tag{22}$$

After normalization of $\mathbf{\Psi}$, the modified modal stiffness values are obtained

$$\bar{\mathbf{\Lambda}} = \mathbf{\Psi}^T \mathbf{k_i} \mathbf{\Psi} \tag{23}$$

By scaling the modal stiffness value of the desired mode (and absorber mode), the adaption of the structure is eventually specified. Finally, the current adapted subspace $\bar{\mathbf{k}}_\mathbf{i}$ of the stiffness matrix is recalculated with the modified eigenvalues and modal basis

$$\bar{\mathbf{k}}_\mathbf{i} = \mathbf{\Psi}\bar{\mathbf{\Lambda}}\mathbf{\Psi}^T \tag{24}$$

and is returned to $\mathbf{K}_{mm}$.

4.2 Application of the Surrogate Model

As stated in the last section, the presented calculation is independent of the pre-solved part in (20). As only the master DoFs change, the transformation matrix **T** remains valid and does not have to be recalculated. Thus, parameter studies can be carried out with little expense by the reduced matrices

$$\tilde{\mathbf{K}} = \mathbf{T}^T\mathbf{K}\mathbf{T}, \quad \tilde{\mathbf{M}} = \mathbf{T}^T\mathbf{M}\mathbf{T} \tag{25}$$

The main parameters to be considered are the form of the actuation, the form of the absorber mode and the number and placement of the ribs. A detailed parameter study would be too extensive at this point, but referring to the considerations in Sects. 2 and 3, one example shall be discussed here. Figure 10 shows three results for a free oscillation case. In the three instances, all parameters such as the form of actuation were identical, except for the distribution of 5 possible ribs along the beam. The fastest amplitude reduction can be observed in the configuration with maximum number of ribs (blue curve). However, it is outperformed by the black configuration – only having two ribs – after a little over half of the observation time.

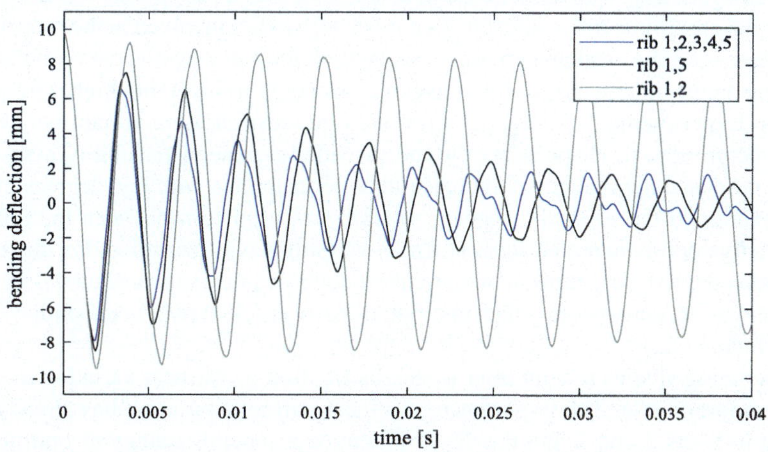

Fig. 10 Parameter assessment based on the surrogate model

Both parameter settings reveal a performance decline after an initially strong amplitude reduction, which reminds of the experimental observations in Sect. 2.3. The impulse excitation by the stiffness variation has a negative effect and should therefore be terminated below a certain velocity limit.

With reference to Sect. 3.2, the following can further be deduced: with only two ribs (black) as opposed to five (blue), less actuators and thus significantly less energy is required for a comparable result. The ribs in the blue case are distributed over the entire length of the beam, providing an almost global increase in stiffness. In the black case, the ribs are locally positioned at two points, leaving space for vibrations within the structure and enhancing the highlighted semi-active component by internal energy transfer. However, considering the gray curve with also two ribs, it is evident that a local application of few ribs is no assurance of good performance, which justifies the employment of such preliminary investigations.

Subsequently to such a study and the identification of suitable parameters, these serve as a basis for the design of the rib structures, which will not be covered here. The relevant literature offers a variety of approaches, and the interested reader is referred to e.g. [14] or [11].

5 Conclusions and Outlook

This report has summarized the subjects that were addressed within an approach for vibration reduction by stiffness variation. Shape adaption was employed to vary the structures' stiffness, which proved to be a powerful approach since it allows to modify the stiffness of continuous beam structures and there is no limitation to pure on-off switches, as there is with classical stiffness switching concepts. The stiffness is varied cyclically, with an increase in stiffness between zero points and maxima of the modal amplitude to be reduced. Two mechanisms are involved in the reduction of vibrations – on the one hand an active component through negative work of the actuator and on the other hand a semi-active one by energy transfer into higher frequency modes. Experimental investigations revealed the desired, time-variant behavior of the structure due to shape adaption and provided a general feasibility assessment of the proposed approach. Analytical studies of structures with few degrees of freedom examined the energetic aspects, whereby a correlation between the efficient, semi-active component and local stiffness variations was identified. For the design of shape-adaptive structures, a numerical model based on an inverse modal transformation was presented with which computationally efficient parameter studies can be performed.

Upcoming studies mainly refer to the distribution of stiffness variation in spatial and temporal sense. The importance of a local stiffness variation has already been shown in Sects. 3 and 4, but the development of a general strategy or optimization approaches for the spatial distribution of stiffness changes over a structure is still pending. Temporal optimization of the stiffness variations has not yet been operated at all (except for signal shapes). When several areas for stiffness modification are available, as is the case with the belt-rib concept, they can be actuated independently of each other, which is particularly promising for the reduction of modes other than the first one.

References

1. Barbarino, S., Bilgen, O., Ajaj, R.M., Friswell, M.I., Inman, D.J.: A review of morphing aircraft. J. Intell. Mater. Syst. Struct. **22**(9), 823–877 (2011)
2. Campanile, L.F.: Modal synthesis of flexible mechanisms for airfoil shape control. J. Intell. Mater. Syst. Struct. **19**(7), 779–789 (2008)
3. Campanile, L.F., Sachau, D.: The belt-rib concept: a structronic approach to variable camber. J. Intell. Mater. Syst. Struct. **11**(3), 215–224 (2000)
4. Corr, L.R., Clark, W.W.: Energy dissipation analysis of piezoceramic semi-active vibration control. J. Intell. Mater. Syst. Struct. **12**(11), 729–736 (2001)
5. Craig, R.R., Bampton, M.C.C.: Coupling of substructures for dynamic analyses. AIAA J. **6**(7), 1313–1319 (1968)
6. Daynes, S., Weaver, P.M.: Review of shape-morphing automobile structures: concepts and outlook. Proc. Inst. Mech. Eng. Part D J. Automob. Eng. **227**(11), 1603–1622 (2013)
7. Deepak, S.R., Dinesh, M., Sahu, D.K., Ananthasuresh, G.K.: A comparative study of the formulations and benchmark problems for the topology optimization of compliant mechanisms. J. Mech. Robot. **1**(1), (2009)
8. Deng, H.X., Gong, X.L., Wang, L.H.: Development of an adaptive tuned vibration absorber with magnetorheological elastomer. Smart Mater. Struct. **15**(5), (2006)
9. Dziedziech, K., Nowak, A., Hasse, A., Uhl, T., Staszewski, W.J.: Wavelet-based analysis of time-variant adaptive structures. Philos. Trans. R. Soc. Math. Phys. Eng. Sci. **376**(2126), (2018)
10. Flatau, A.B., Dapino, M.J., Calkins, F.T.: High bandwidth tunability in a smart vibration absorber. J. Intell. Mater. Syst. Struct. **11**(12), 923–929 (2001)
11. Frecker, M.I., Ananthasuresh, G.K., Nishiwaki, S., Kikuchi, N., Kota, S.: Topological synthesis of compliant mechanisms using multi-criteria optimization. J. Mech. Des. **119**(2), 238–245 (1997)
12. Garrido, H., Curadelli, O., Ambrosini, D.: Semi-active friction tendons for vibration control of space structures. J. Sound Vib. **333**(22), 5657–5679 (2014)
13. Guyan, R.J.: Reduction of stiffness and mass matrices. AIAA J. **3**(2), 380 (1965)
14. Hasse, A., Campanile, L.F.: Design of compliant mechanisms with selective compliance. Smart Mater. Struct. **18**(11), (2009)
15. Ledezma-Ramirez, D.F., Ferguson, N., Zamarripa, A.S.: Mathematical modeling of a transient vibration control strategy using a switchable mass stiffness compound system. Shock. Vib. **2014**, (2014)
16. Liu, Y., Matsuhisa, H., Utsuno, H.: Semi-active vibration isolation system with variable stiffness and damping control. J. Sound Vib. **313**(1–2), 16–28 (2008)
17. Liu, Y., Matsuhisa, H., Utsuno, H., Park, J.G.: Variable damping and stiffness vibration control with magnetorheological fluid dampers for two degree-of-freedom system. JSME Int. J. Ser. C Mech. Syst. Mach. Elem. Manuf. **49**(1), 156–162 (2006)
18. Mayer, D., Herold, S.: Passive, adaptive, active vibration control, and integrated approaches. Vibration Analysis and Control in Mechanical Structures and Wind Energy Conversion Systems, pp. 1–22 (2018)
19. Meguid, S.A., Su, Y., Wang, Y.: Complete morphing wing design using flexible-rib system. Int. J. Mech. Mater. Des. **13**(1), 159–171 (2017)
20. Nagarajaiah, S., Sahasrabudhe, S.: Seismic response control of smart sliding isolated buildings using variable stiffness systems: an experimental and numerical study. Earthq. Eng. Struct. Dyn. **35**(2), 177–197 (2006)
21. Nitzsche, F.: The use of smart structures in the realization of effective semi-active control systems for vibration reduction. J. Braz. Soc. Mech. Sci. **34**, 371–377 (2012)
22. Nowak, A., Willner, K., Campanile, L.F., Hasse, A.: Active vibration control of excited structures by means of shape adaption. In: Proceedings of the 28th International Conference on Adaptive Structures and Technologies, Cracow, Poland (2017)
23. Nowak, A., Willner, K., Hasse, A.: Model and parameter study of a shape-adaptable beam for vibration control. Proc. Appl. Math. Mech. **18**(1), (2018)

24. Nowak, A., Campanile, L.F., Hasse, A.: Vibration reduction by stiffness modulation—a theoretical study. J. Sound Vib. **501**, 116040 (2021)
25. Onoda, J., Sano, T., Kamiyama, K.: Active, passive, and semiactive vibration suppression by stiffness variation. AIAA J. **30**(12), 2922–2929 (1992)
26. Ou, J., Li, H.: Analysis of capability for semi-active or passive damping systems to achieve the performance of active control systems. Struct. Control. Health Monit. **17**(7), 778–794 (2010)
27. Ramaratnam, A., Jalili, N.: A switched stiffness approach for structural vibration control: theory and real-time implementation. J. Sound Vib. **291**(1–2), 258–274 (2006)
28. Savi, M.A., De Paula, A.S., Lagoudas, D.C.: Numerical investigation of an adaptive vibration absorber using shape memory alloys. J. Intell. Mater. Syst. Struct. **22**(1), 67–80 (2011)
29. Tan, P., Zhou, F., Yan, W.: A Semi-active Variable Stiffness and Damping System for Vibration Control of Civil Engineering Structures. ANCEER Annual Meeting, Honolulu, Hawaii (2004)
30. Tseng, H.E., Hrovat, D.: State of the art survey: active and semi-active suspension control. Veh. Syst. Dyn. **53**(7), 1034–1062 (2015)
31. Xinghua, Y.: Model and analysis of variable stiffness semiactive control system. In: Proceedings of the Twelfth World Conference on Earthquake Engineering, Auckland, New Zealand (2000)

A Combined Numerical-Experimental Approach for the Damping Evaluation of Non-Linear Dissipative Vibration Systems

Gleb Kleyman, Martin Jahn, Sebastian Tatzko, and Lars Panning-von Scheidt

1 Introduction

Almost all established methods for experimental modal analysis rely on linear system theory [5], that is a linear relationship between deformation and restoring force. If that is not the case, for example due to geometric properties of the structure, sliding joints between single parts of the assembly, or deliberately introduced nonlinearities, the obtained modal parameters will most likely be erroneous [30]. The nonlinear behavior can further cause bifurcations which imply the splitting of solution curves and may lead to multiple stable and unstable oscillation states at the same harmonic excitation. Conventional experimental methods fail to excite those unstable states, which is often observed as the so-called jump phenomenon in frequency sweeps [16]. Within the first phase of SPP1897 the focus of our project was set to numerical methods for solving nonlinear systems of any size by harmonic balance and continuation [10]. Therein, nonlinear normal modes have been investigated, where resonance frequencies and damping ratios depend on the vibration amplitudes. However, to date there are only few experimental methods that can account for this fact. One promising approach is experimental continuation. Experimental continuation exploits numerical methods to trace solution curves (frequency response, resonance frequency and damping, or S-curves) of real physical systems. Therefore, experimental continuation enables accurate measures of resonance frequency and damping ratio even for strongly nonlinear systems. The experiences regarding continuation in the first phase of SPP1897 provide the foundation for the present study. This is how a comprehensive tool for experimental continuation of almost arbitrary nonlinear systems was developed.

Experimental continuation was originally introduced to overcome limitations of the time-delayed feedback control, a method known from chaos control [25].

G. Kleyman (✉) · M. Jahn · S. Tatzko · L. P. Scheidt
Institute of Dynamics and Vibration Research, An der Universität 1, 30823 Garbsen, Germany
e-mail: kleyman@ids.uni-hannover.de

P. Eberhard (ed.), *Calm, Smooth and Smart*, Lecture Notes in Applied and Computational Mechanics 102, https://doi.org/10.1007/978-3-031-36143-2_15

Later it was found that experimental continuation is also perfectly suited for the non-parametric identification of nonlinear systems [26]. Different work-groups have applied the method in its original formulation, exploiting a quasi-Newton method [3], as well as a simplified form [21]. Also, different approaches have been published, that address stability analysis [2], acceleration of measurement time [1] and increased robustness against noise [23]. Recently, the authors of the present study have successfully applied a simplified form of experimental continuation for estimation of forced response, resonance frequency, and damping curves of two nonlinear systems [12]. These results lead to an extensive framework for experimental continuation. In the current implementation, as presented in this paper, a pseudo-arclength continuation algorithm based on Newton's method is exploited. The core is a real-time processor with integrated AD/DA board, which allows communication between sensors resp. actuators and the continuation algorithm. This approach is demonstrated by the example of a geometrically nonlinear beam with additional strongly nonlinear damping.

2 Methodology of Experimental Continuation

Experimental continuation is inspired by its numerical counter part [24]. In numerical studies, continuation methods have been applied to nonlinear problems for several decades, not specifically addressing vibration problems. In recent years, the continuation principal along with harmonic balance and shooting methods has become a powerful tool for steady state vibration analysis in nonlinear dynamics. This is because it enables to efficiently calculate frequency response curves [6] and nonlinear normal modes of large-scale nonlinear conservative [17] and non conservative systems [13]. The main idea of continuation is to iteratively find solutions of a parameter dependent nonlinear problem. Starting from one solution point, an initial guess for a neighboring solution point is generated. A root finding algorithm is then applied to minimize an error measure, the so called residual function, in the vicinity of the initial guess. Of course, for physical structures there is no explicit expression for the equation of motion available, this is why input and output data of the system under test is used instead, as illustrated in Fig. 1.

In experimental continuation frequency domain data is processed. First, time domain data is collected in a buffer and transferred to the frequency domain by Fourier-transform. Next, the Fourier coefficients are processed by the continuation algorithm and a new excitation signal u_e is calculated. The excitation signal is transformed back into the time domain. As there may exist unstable oscillation states, which can not be excited in open-loop, an additional feedback control is applied. The feedback control works in a sample-by-sample manner. That means at each cycle of the AD-converter the system output x is compared to a reference signal x_R and a control signal u_c is calculated from their difference. If parts of the solution curve are unstable, the continuation algorithm fails, therefore proper stabilization is essential. But, the feedback control may introduce higher harmonic distortion to the

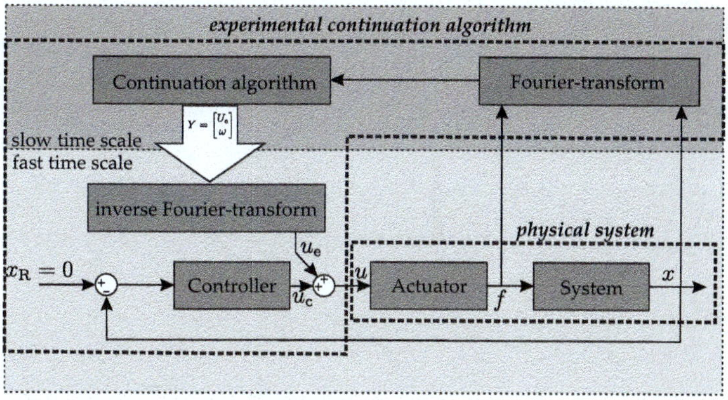

Fig. 1 Schematic representation of experimental continuation

control signal. These higher harmonic components have to be dealt with actively by elimination. Therefore, the entire experimental continuation process can be divided into three basic steps: stability analysis and stabilization (Sect. 2.1), elimination of harmonic distortion (Sect. 2.2) and the continuation algorithm (Sect. 2.3).

2.1 Stability Analysis and Stabilization

As already stated, nonlinear systems can exhibit unstable steady states. While in a purely numerical problem it is still possible to calculate those states and analyze their stability regarding small perturbations [14], in experiments unstable steady states may become problematic. Here, an analytical example shall explain why unstable steady states exist and how they can be dealt with. As an illustrative example we consider a nonlinear, harmonically excited single degree of freedom system with linear damping and nonlinear stiffness:

$$m\ddot{x} + c\dot{x} + kx + k_{\mathrm{nl}}x^3 = f \tag{1}$$

An analytical approximate solution for this system can be found by harmonic linearization [8], where the excitation $f(t)$ and the response $x(t)$ are considered harmonic:

$$\begin{aligned} x(t) &\approx \tilde{x}(t) = \hat{x}\sin(\Omega t) \\ f(t) &= \hat{f}_{\sin}\sin(\Omega t) + \hat{f}_{\cos}\cos(\Omega t). \end{aligned} \tag{2}$$

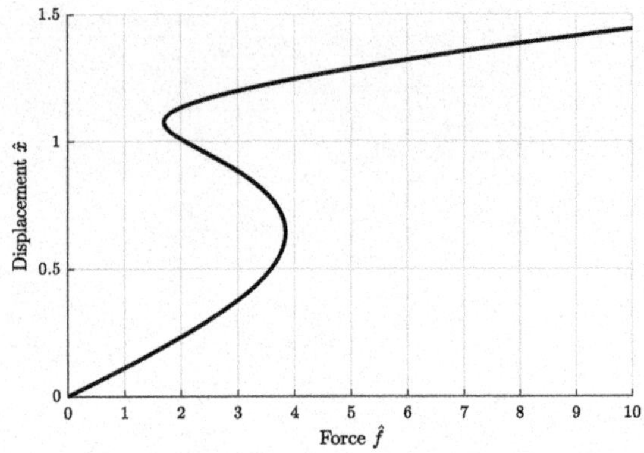

Fig. 2 S-curve of a duffing oscillator with $m = 1$ kg, $k = 1$ N/m, $c = 0.5$ Ns/m, $k_{nl} = 10$ N/m^3 and $\Omega = \pi$ 1/s

The nonlinear part $\tilde{x}^3$ can be expressed in terms of its Fourier series. If only the fundamental harmonic is considered and its amplitude $\hat{f}$ is calculated from the sine and cosine components $\hat{f} = \sqrt{\hat{f}_{\sin}^2 + \hat{f}_{\cos}^2}$, the linearization leads to:

$$\hat{f} = \hat{x}\sqrt{c^2\Omega^2 + \left(-m\Omega^2 + k + k_{nl}\frac{3}{4}\hat{x}^2\right)^2}. \tag{3}$$

Equation 3 describes a so called S-curve. In the simplified experimental continuation, as for example described in [1], the S-curves play a major role. They can be exploited to characterize the dynamics of nonlinear systems near a resonance frequency. This is because from several S-curves, all measured at different frequencies, the frequency response curves can be reconstructed, as Fig. 9 shows. Compared to the simplified version, the original approach followed in this work enables the direct estimation of nonlinear frequency response curves, the implementation is however more complex. The derivation of S-curves is still useful to analytically assess stability of periodic solutions, though. For a Duffing system a S-curve is shown in Fig. 2.

As it can be seen, the S-curve is multivariate with respect to the excitation amplitude $\hat{f}$. The solutions where the gradient $\frac{\mathrm{d}\hat{x}}{\mathrm{d}\hat{f}}$ becomes negative are unstable. A negative value for $\frac{\mathrm{d}\hat{x}}{\mathrm{d}\hat{f}}$ means that inevitable stochastic disturbances cause the system immediately to drop out of this periodic motion and settle at a stable solution with $\frac{\mathrm{d}\hat{x}}{\mathrm{d}\hat{f}} > 0$ instead. Stabilization of those unstable states can be achieved by closed-loop feedback control. If control parameters are adjusted properly, the control loop can impose a periodic motion, even if it is originally unstable. In control theory, a state is unstable if it can not be maintained stationary, when a harmonic input signal $u_e(t) = \hat{u}_{\sin}\sin(\Omega t) + \hat{u}_{\cos}\cos(\Omega t)$ is applied. For an experimental setup, where the force acting between vibration exciter and structure is proportional to the actuators input voltage, applying a differential controller with gain c_c and reference signal $x_R = 0$ the vibrating system is described by:

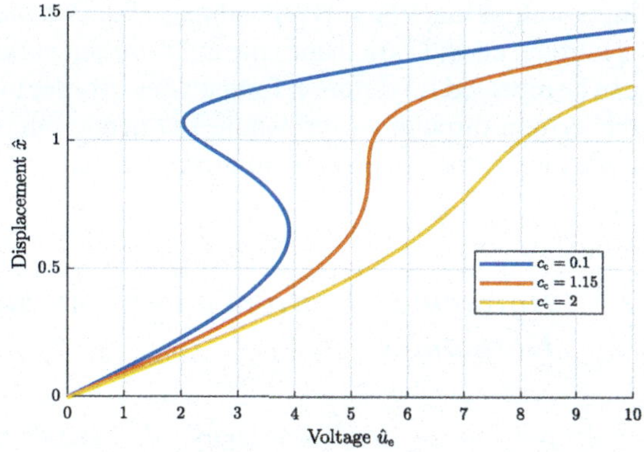

Fig. 3 Input-output function of a feedback controlled duffing oscillator with $m = 1\,\text{kg}$, $k = 1\,\text{N/m}$, $c = 0.5\,\text{Ns/m}$, $k_{\text{nl}} = 10\,\text{N/m}^3$ and $\Omega = \pi\ 1/\text{s}$

$$m\ddot{x} + c\dot{x} + kx + k_{\text{nl}}x^3 = \gamma(u_{\text{e}} - c_{\text{c}}\dot{x}). \tag{4}$$

Following the Ansatz from Eq. 2 the solution reads:

$$\hat{u}_{\text{e}} = \frac{\hat{x}}{\gamma}\sqrt{(c + \gamma c_{\text{c}})^2\Omega^2 + \left(-m\Omega^2 + k + k_{\text{nl}}\frac{3}{4}\hat{x}^2\right)^2}. \tag{5}$$

Equation 5 describes the input-output relationship of the closed-loop system. For the trivial case, where $c_{\text{c}} = 0\,\text{Vs/m}$ and $\gamma = 1\,\text{N/V}$ this curve equals the S-curve in Fig. 2. For the non-trivial case, three curves for different values of the feedback gain are shown in Fig. 3. Note that these curves represent displacement versus voltage now. It can be seen, that by feedback control the input-output function is unfold to the point where the curve becomes unique (yellow curve). That means, compared to the lowest gain setting (blue curve), there is only one response state for each harmonic excitation with amplitude $\hat{u}_{\text{e}}$. With respect to the stability of the closed-loop system, this implies that all states are stable. It is essential to point out that although the input-output function is altered by the feedback loop, the S-curve remains the same for arbitrary values of c_{c} and γ. In other words, the feedback loop serves to stabilize without changing the dynamics of the system regarding the force-displacement relationship. This proof can be performed in the same way for a proportional or proportional-differential controller.

The given example includes a simplified model of a vibration exciter with an ideal transfer function (proportionality factor γ). Often, vibration exciters have more complex transfer functions, which additionally may affect the stability of the closed-loop system. Therefore, in real life experiments, the feedback gain has to be determined experimentally. Best practice is to start with frequency sweeps in order to provoke the jumping phenomenon if present. This will reveal if the system behaves hardening or softening near a certain mode. Next a constant excitation frequency is set, which lies well above or below the linearized resonance frequency of the investigated mode. At this specific frequency an amplitude up-and down-sweep is performed and the input-output function corresponding to Fig. 3 is measured. Finally, the feedback gain is introduced and varied until the input-output curve becomes unique. This procedure can be repeated for different frequencies in order to find settings that reliably stabilize the system throughout the whole frequency range of interest while preserving the dynamics of the underlying nonlinear system.

2.2 Harmonic Distortion and Elimination of Higher Harmonics

In the previous section it was analyzed, that a closed-loop feedback control is capable of stabilizing steady states of strongly nonlinear systems. However, only the fundamental harmonic of oscillation was considered. In a real life scenario when a nonlinear system is excited by a perfectly harmonic input, it may still respond non-harmonic. As the system response is fed back to the input those higher harmonic components become part of the excitation signal. These frequency components in turn excite higher harmonics, leading to an excitation force which has a significantly high harmonic distortion. Furthermore, it has been shown that even in linear structures without feedback, significant harmonic distortion can occur at high deflection amplitudes due to the nonlinear behavior of the vibration exciter itself [28]. This occurs particularly strongly in lightly damped structures near resonance. Typically, modal testing is intended to investigate the response of a structure to harmonic excitation. Thus, experimental continuation attempts to keep the excitation signal as harmonic as possible. For this purpose, higher harmonics are added to the input signal and adjusted so that all frequency components except the fundamental component vanish:

$$u_{\mathrm{e}}(t) = \sum_{n=1}^{N} u_{\sin,n} \sin(n\Omega t) + u_{\cos,n} \cos(n\Omega t). \tag{6}$$

Here N is the number of controlled harmonics. The corresponding Fourier coefficients $u_{\sin,N}$ and $u_{\cos,N}$ are iteratively found by Newton's algorithm as described in the following section.

2.3 Continuation Algorithm

Assume $\mathbf{y}$ is the solution vector of a nonlinear problem and λ is a free parameter (e.g. the excitation frequency). In the particular case of experimental continuation, where a harmonic excitation signal is desired, $\mathbf{y}$ is expressed by:

$$\mathbf{y} = \left[u_{\sin,1},\, u_{\cos,1},\, \ldots,\, u_{\sin,N},\, u_{\cos,N}\right]^{\mathsf{T}}. \tag{7}$$

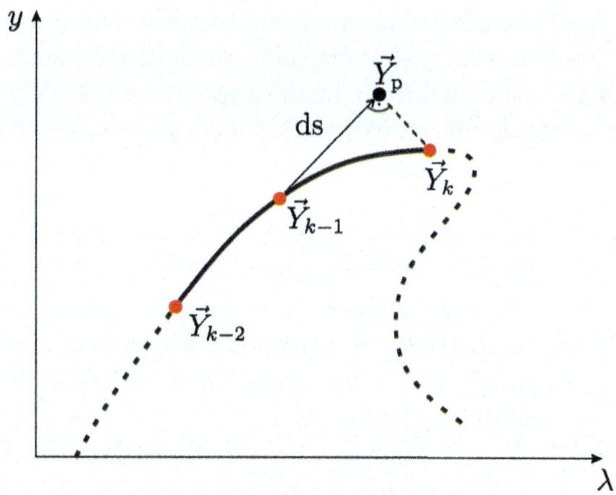

Fig. 4 Generalized pseudo-arclength continuation procedure

A parametrization of λ describes a solution curve in a hyper dimensional space ($2N+1$ dimensions). Single points of the curve can be approximated by solutions $\mathbf{Y_k}$, compare Fig. 4:

$$\mathbf{Y}_k = \begin{bmatrix} \mathbf{y} \\ \lambda \end{bmatrix}. \tag{8}$$

Given an adequate initial guess each solution point can be estimated by a root finding algorithm, like the Newton's method:

$$\mathbf{Y}_{k,j+1} = \mathbf{Y}_{k,j} - \boldsymbol{J}^{-1}\mathbf{R}(Y_{k,j}). \tag{9}$$

Here, $\mathbf{R}$ is the residual vector and $\boldsymbol{J}$ its Jacobian matrix. The Newton algorithm iteratively approaches the solution curve until a user defined threshold value of the residual function is reached. Thereby the Jacobian defines the direction of the new guess in each iteration step. The residual vector can be divided into two parts, a vector $\mathbf{g(Y)}$, which is the vector of function values corresponding to a guess $\mathbf{Y}$ and r_{par}, which describes an additional constraint:

$$\mathbf{R} = \begin{bmatrix} \mathbf{g(Y)} \\ r_{\text{par}} \end{bmatrix} \stackrel{!}{=} 0. \tag{10}$$

The search region is constrained by r_{par}. That means only solutions are accepted that satisfy the condition. A common approach, the so called pseudo-arclength parametrization, is visualized in Fig. 4. As it can be seen, an orthogonality between the solution point (as measured from the initial guess $\mathbf{Y}_\text{p}$) and the secant vector $\mathbf{Y}_\text{p} - \mathbf{Y}_{k-1}$ is imposed:

$$r_{\text{par}} = (\mathbf{Y}_\text{p} - \mathbf{Y}_{k-1})^\top(\mathbf{Y}_\text{p} - \mathbf{Y}_k). \tag{11}$$

This enables the algorithm to trace solution curves even through folds, where multiple solutions at the same value of the parameter λ emerge. Each initial guess $\mathbf{Y}_\mathrm{p}$ is calculated from the most recent solution $\mathbf{Y}_{k-1}$, the tangent vector $\mathbf{t}_{k-1}$ in $\mathbf{Y}_{k-1}$ (which can be approximated by the secant as well) and an arclength parameter ds:

$$\mathbf{Y}_\mathrm{p} = \mathbf{Y}_{k-1} + \mathrm{ds}\mathbf{t}_{k-1}. \tag{12}$$

Depending on whether nonlinear frequency response (NLFR) curves or resonance frequency and damping curves (also referred to as backbone-curves) are to be measured, the first part of the residual vector is composed in different ways. In the case of frequency response curves, $2N$ Fourier coefficients of the excitation force are included in the residual vector:

$$\mathbf{g}_\mathrm{NLFR}(\mathbf{Y}) = \left[f_{\sin,1} - f_0,\ f_{\cos,1},\ \ldots,\ f_{\sin,N},\ f_{\cos,N}\right]^\mathrm{T} \tag{13}$$

Thus, the last $N-1$ harmonics will be eliminated and only the fundamental harmonic remains, which then equals the desired amplitude f_0.
If a backbone-curve (BBC) measurement is performed, the first two entries of the residual vector are replaced by the phase angle $\varphi_{f,\dot{x}}$ between excitation force and velocity at the driving point and $\dot{x}_{\cos,1}$:

$$\mathbf{g}_\mathrm{BBC}(\mathbf{Y}) = \left[\varphi_{f,\dot{x}},\ \dot{x}_{\cos,1},\ \ldots,\ f_{\sin,N},\ f_{\cos,N}\right]^\mathrm{T} \tag{14}$$

The first term constrains phase resonance of the fundamental harmonic components, which is similar to phase resonance testing by Phase-Locked-Loop experiments [22]. The goal of this force appropriation is to compensate the natural dissipation of the structure, which enables investigation of the underlying conservative system. Although single-point excitation with the fundamental harmonic is only an approximation of real phase resonance, several publications have shown that it is accurate enough to extract nonlinear modal parameters [20, 22]. The second entry ($\dot{x}_{\cos,1}$) is necessary to establish a phase relationship between the control and the feedback signal. As the phase resonance is defined by excitation frequency and amplitude, different combinations of $u_{\sin,1}$ and $u_{\cos,1}$ lead to $\varphi_{f,\dot{x}} = 0$. Hence, with $\dot{x}_{\cos,1} = 0$ a certain starting point on the solution curve is arbitrarily chosen.

3 Experimental Setup

In the present study a slender clamped-clamped beam is investigated. An additional nonlinear damping element is later attached to the beam. Such academic systems are often used as basis structures for the development of novel damping concepts. However, even without deliberately introduced nonlinearity, they can exhibit severely nonlinear behavior at moderately large deformation amplitudes in the range of their thickness [4]. This is due to geometric nonlinear effects where bending and stretching

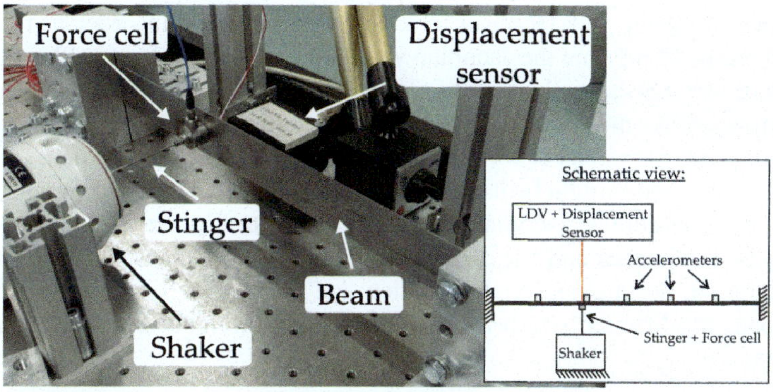

Fig. 5 Experimental setup of a double-clamped beam (left) and schematic representation (right)

of the beam are coupled. For the examined clamped-clamped beam, the equation of motion can be derived analytically, exploiting the nonlinear Bernoulli beam theory considering midplane stretching [27]. An approximate solution considering only the lowest mode of vibration reveals that its dynamics are described by a Duffing oscillator with stiffening character.

A picture of the experimental setup is shown in Fig. 5. The steel beam has dimensions of 370x40x1.5 mm. Both ends are clamped between massive aluminum blocks, which are fixed to a vibration isolation table. The beam is excited by a TIRA S50018 modal shaker driven by a BEAK BAA 120 power amplifier. A Dytran 1053V1 piezoelectric force cell is used to measure the applied force. The beam is excited 251.6 mm apart from one end, which corresponds to 68% of the overall beam length. This driving point position is selected to allow for the excitation of at least the first five bending modes. The velocity of the driving point is measured by a Polytec OFV-552 fiber optic laser-doppler-vibrometer. Additionally, an Allsens AM401 laser displacement sensor is used as a backup option. Five PCB M353B17 accelerometers are positioned equidistantly along the longitudinal axis of the beam. Instead of recording time data, which would require excessive disk storage, all signals were directly stored as Fourier-coefficients up to the 7th harmonic.

The continuation algorithm was implemented exploiting a dSpace MicroLabBox (MLB). The MLB is a hardware device that allows to conveniently implement user-specific code on a real-time processor. The MLB is programmable via Matlab/Simulink and features multiple analog inputs and outputs to control the experiment. In all experiments the cycle time of the MLB was set to 1/8000 s, which means the signals were sampled at 8 kHz. To avoid aliasing effects the input signals are filtered by analog first order low-pass filters with a cut-off frequency of 4 kHz. The block length for the Fourier-transform was set to 4000 samples. To reduce the impact of random disturbances, the results were averaged over 3 blocks. An additional block was added to allow transients to die out after parameter variations. Therefore, for a continuation problem with 7 parameters (3 harmonics + excitation frequency) the determination of the full Jacobian took about 18 seconds. In order to reduce the overall measurement time the Jacobian was calculated only every 3rd Newton step. Between those steps a rank-1 update of the Jacobian, known as Broyden's method, was performed. As the Jacobian is estimated by a finite difference (FD) procedure, the FD-step size parameter **h** is a crucial factor. If the step size is to small, the change of the residual vector might be in the order or even below measurement noise, which gives a poor conditioned Jacobian. If the step size is to big, the assumption of finite differences is violated and the algorithm might predict values far away from the actual solution. For that reason the FD-step size was declared as an adaptive vector. That means the FD-step size **h** follows always the difference between the last two solutions:

$$\mathbf{h} = |\mathbf{Y}_{k-1} - \mathbf{Y}_{k-2}|/2. \tag{15}$$

Minimum and maximum values were assigned to h to prevent it from becoming zero or excessively large. Since the individual entries of the solution vector typically have different orders of magnitude, the solution vector was scaled by so-called preconditioning [14]. A scaling of the Fourier coefficients by a factor of 10^{-2} has proven to be optimal in the conducted experiment. In general, the preconditioning factor depends on the structure, the vibration exciter and its amplifier. The decision for the acceptance of a point as a solution point was set as follows: if the maximum value of the residual vector is less than 1% of the desired force amplitude *OR* only the first entry of the residual vector is less then 1% *AND* the mean value over all entries is less than 1%, the point will be accepted. Same condition applies to the measurement of backbone-curves, with 1° phase difference instead of amplitude deviation. If after 15 Newton iterations no solution has been found, these values were increased by a factor of five, but the corresponding solution was discarded in the post processing. If this happened three times in a row, the algorithm stopped. The arclength parameter ds was set to values between 0.5 and 3, depending on the type of measurement and excitation amplitude. Same as the scaling factor, an optimal arclength parameter must be found by the user in preliminary tests, because as described in [23] a traditional step size control fails in experimental continuation. The controller gain was set to $c_c = 1.5\,\mathrm{Vs/m}$. A differential controller was used and the reference signal was set to zero. For details about stabilization and a heuristic method for estimation of controller gains see Sect. 2.1.

4 Experimental Results

Various experiments were conducted to investigate the potential of experimental continuation. The clamped beam was studied additionally with a particle damper in the form of a locally attached container partially filled with ceramic granular material. First, experiments are discussed in detail for the beam without attachment. The results of the particle damped system are presented below.

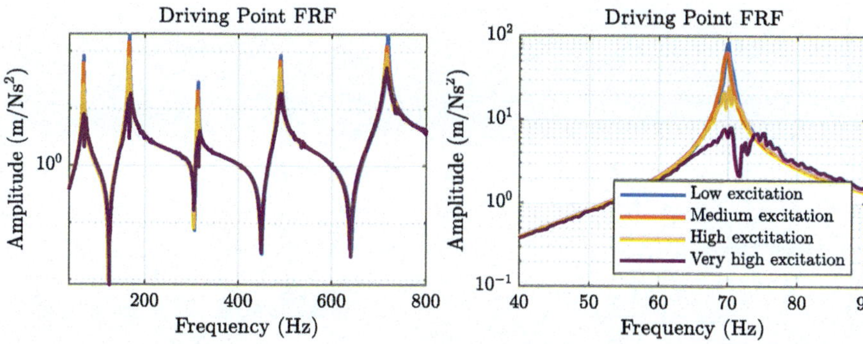

Fig. 6 Frequency response (H_1-estimate) of the double-clamped beam at driving point position excited with a random signal at different levels. First five modes (left) and detailed view of the first mode (right)

4.1 *Double-Clamped Beam*

Results from Linear Analysis

First, a conventional (linear) experimental modal analysis (EMA) was performed. EMA is useful to obtain an overview of the essential characteristics of the system, such as natural frequencies and mode shapes. Furthermore, the EMA has revealed that the identified parameters were not accurate at high excitation levels.

Figure 6 shows the frequency response function (FRF) at the driving point position (DP) estimated from a random excitation at different levels. For each level the linear average of 30 blocks was taken, where each block had 32768 samples, while the sampling frequency was set to 16.834 kHz. A Hanning window and a H_1-estimator were used to calculate the FRFs. At a first glance the results show that the FRFs have only minor differences between the excitation levels. However, the closer view near the first natural frequency at about 70 Hz (right graph) reveals that the resonance peak becomes more and more distorted with increasing excitation level. Note that the distortion is not caused by poor measurement quality as might be assumed, but is a typical behavior of nonlinear systems in experiments [11]. Consistent with this observation the EMA method predicted wrongly damping ratios that increase with the excitation level, compare Fig. 7. The colors correspond to the excitation levels in Fig. 6. Although linear damping was expected from the theory, the estimated values for the first mode span from 0.3 to more than 3%, which is an increase by factor 10. For EMA the polyreference time domain (PTD) method (also known as p-LSCE) from the commercial DAQ-software m+p Analyzer was used. Compared to the deviations in damping ratio the differences in resonance frequency and mode shapes are negligible and are therefore not recalled in detail. The first five mode shapes are plotted in Fig. 7 on the left hand side. All mode shapes are mass normalized, which is important for the nonlinear damping estimation in the following.

Continuation of S-Curves and NLFRs

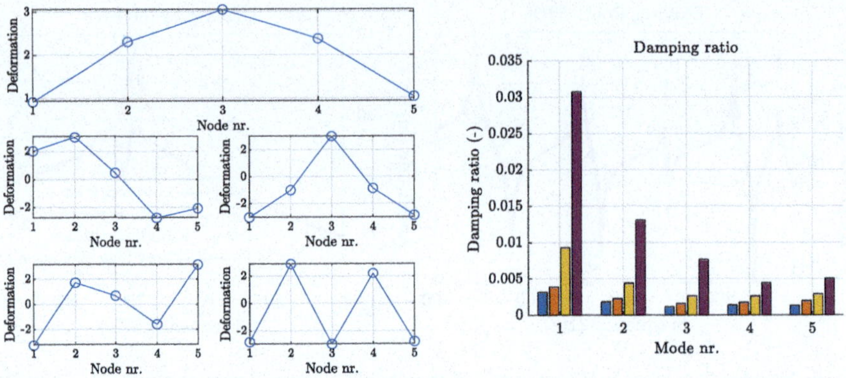

Fig. 7 First five mass normalized mode shapes estimated from linear modal analysis at lowest excitation level (left) and corresponding damping factors (right)

Compared with theory, the linear analysis yields contradictory results. A large shift of resonance frequency but a rather small change of damping was expected prior to the experiment. For that reason we decided to investigate the structure by experimental continuation. After an adequate feedback gain was found, several S-curves of the system were measured in the vicinity of the first linear natural frequency. Figure 8 shows two S-curves at 74 Hz in comparison. In the red curve the second and third harmonic were removed from the excitation force, as described in Sect. 2.2. In the blue curve the same harmonics have been removed from the superimposed driving signal $u = u_e + u_c$ (compare Fig. 1). The elimination of higher harmonics from the driving signal is what usually is performed in the simplified experimental continuation, as described in an earlier publication [12]. Both curves show the maximum displacement per oscillation period over the fundamental harmonic of excitation force. As the Fourier decomposition of the signal and the time domain data on the right side of the figure reveal, the harmonic distortion of the blue curve is significantly higher. That means even if the higher harmonics are eliminated from the driving signal, the excitation force has still a high contribution of 2nd and 3rd harmonic. Note that those graphs refer to the measuring point marked by a red circle near the fold. Surprisingly, the harmonic distortion has hardly any influence on the S-curve in this case. This is due to the fact that there is no other resonance frequency near the 2nd or 3rd harmonic (2nd and 3rd natural frequency are around 167 and 313 Hz, respectively). In other systems higher harmonics may lead to a coupling of modes, however.

Next, the S-curves were investigated in comparison to the nonlinear frequency response curves. Figure 9 shows NLFRs at different excitation amplitudes between 0.02 and 0.28 N in blue and S-curves in red. All curves were measured at the driving point position and the 2nd and 3rd harmonic of the force signal were eliminated. It can be seen, that NLFRs and S-curve span the same three dimensional manifold between force, excitation frequency and displacement. While frequency response curves are two dimensional projections along constant force amplitudes, the S-curves are pro-

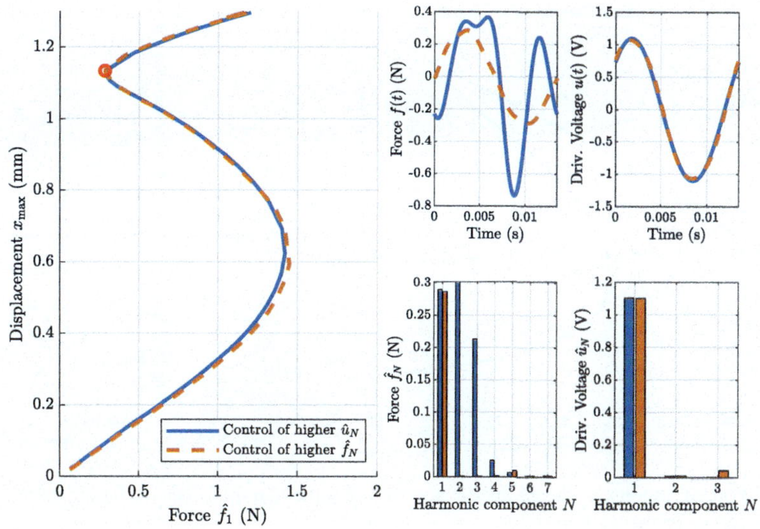

Fig. 8 Comparison between the elimination of higher harmonic components in the control signal (blue) versus in the excitation force (red) at 74 Hz

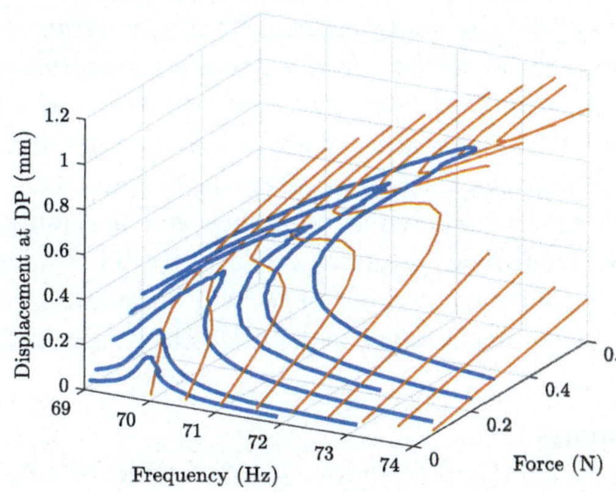

Fig. 9 Comparison between S-curves (red) and nonlinear frequency response curves (blue). Force levels: 0.02, 0.04, 0.1, 0.12, 0.2 and 0.28 N

jections along constant excitation frequencies. Therefore, NLFRs can be calculated from multiple S-curves by interpolation and vice versa. Although, the results clearly indicate that both approaches are equivalent (provided the higher harmonics of force are eliminated in both cases), the present study has been focusing on direct continuation of nonlinear frequency response curves. This is mainly for two reasons: First of all, a NLFR measurement is faster in terms of measurement time, especially if the major interest lies only in a few excitation levels. Second, the NLRF, as it is a direct extension of linear frequency response curves, provides an intuitive characterization of the system's dynamics.

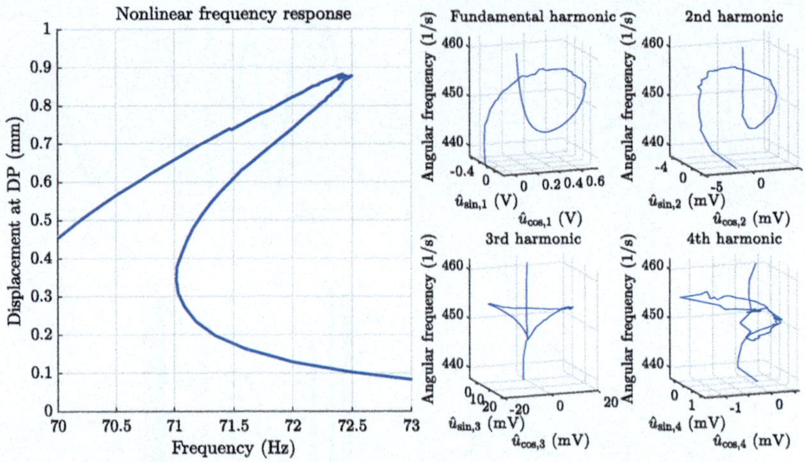

Fig. 10 Forced response of the geometric nonlinear double-clamped beam (left) and corresponding solution curves (right) at $\hat{f} = 0.2\,\mathrm{N}$

As described in Eq. 7, the actual solution space consists of the Fourier coefficients of the driving signal and the angular frequency. Nonlinear frequency response curves are obtained by plotting the maximum displacement for a single solution point versus the corresponding excitation frequency. The hyper-dimensional solution curve itself can only be represented by sub-spaces. In Fig. 10 on the right side, a solution curve is divided into four sub-spaces. Each graph gives the Fourier coefficients (in sine and cosine components) versus the angular frequency. For this measurement, the first four harmonics were controlled. As it can be seen, for the 4th harmonic the curve is unsteady. This is because the content of the 4th harmonic in the force signal is already very low so noise has a larger impact. However, all other curves show the typical helix shape.

Estimation of Damping

The estimation of a modal damping ratio using the results of a Phase-Locked-Loop based phase resonance method, is well described in [18]. This method can be applied in exactly the same manner for results from experimental continuation of backbone-curves. Let $\Phi_r(z_0)$ be the mass normalized mode shape vector of the rth mode at the driving point position and $\hat{x}_r(z_0)$ the displacement amplitude of the fundamental harmonic at the same sensor location. The mode shape follows from an experimental modal analysis, as described at the beginning of the section. The displacement is simply measured during experimental continuation. If the mode shape does not change much with increasing oscillation amplitude, which can be easily checked by EMA at different excitation levels, a modal amplitude $\hat{q}_r$ can be calculated:

$$\hat{q}_r = \Phi_r(z_0)\hat{x}_r(z_0). \tag{16}$$

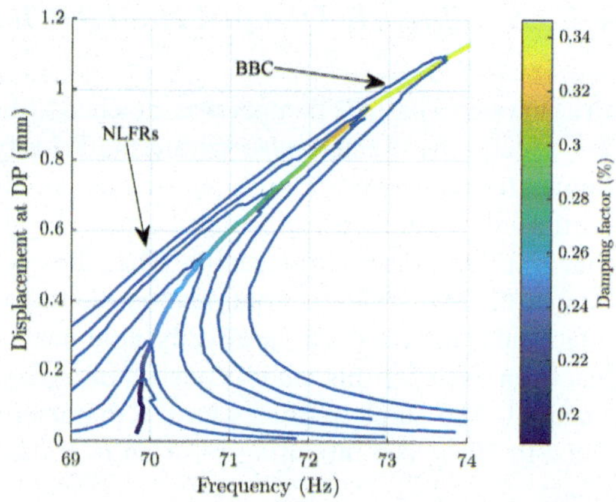

Fig. 11 Forced response curves at different force levels (blue) and nonlinear mode (BBC) of the geometric nonlinear double-clamped beam excited near first resonance frequency (Force levels: 0.02, 0.04, 0.1, 0.12, 0.2 and 0.28 N)

The modal damping follows from a power balance method. Here, it is assumed that for one oscillation period the active power of excitation equals the dissipative power, if the structure oscillates at a resonance angular frequency $\omega_{0,r}$. The power of excitation can be calculated from the excitation force and displacement at the driving point:

$$P_{\mathrm{exc},1} = 0.5 \hat{f}_1 \hat{x}_1 \omega_{0,r} \sin(\varphi_{f,x}). \tag{17}$$

The phase angle between force and displacement is described by $\varphi_{f,x}$. Neglecting higher harmonics, Eq. 17 describes the active power of excitation [19]. Therefore, the nonlinear modal damping ratio is a function of the active power of excitation, the modal amplitude and the natural angular frequency:

$$D_{\mathrm{nl}}(\hat{q}_r) = \frac{P_{\mathrm{exc},1}}{\hat{q}_r^2 \omega_{0,r}^3}. \tag{18}$$

The backbone-curve along with some nonlinear frequency response curves is shown in Fig. 11. Here, the damping follows the color-bar on the right side of the figure. It can be seen, that the BBC almost perfectly matches the peaks of the NLFRs. Indeed, the damping increases with the oscillation level. However, in the measured interval the damping ratio reaches a maximum of only 0.3% which is ten times lower than predicted by EMA. Also, there is a significant shift of resonance frequency (almost 5% at 0.28 N), which was not predicted by EMA. This is because EMA methods are based on linear theory and are thus generally inappropriate for nonlinear systems.

Concluding, it can be stated, that for the investigated system the experimental continuation method gives accurate results in terms of nonlinear frequency response and backbone-curves. The estimated nonlinear damping ratio converges to the results from EMA at low level excitation. Also, the overhanging frequency curves, which indicate a stiffening behavior, match the nonlinear theory.

4.2 Nonlinearly Damped Double-Clamped Beam

In a next step a particle damper was attached to the beam, see Fig. 12. Particle dampers are widely investigated, also within the Priority Programme SPP1897, as they are known for their outstanding performance in vibration mitigation [7, 15]. However, particle filled dampers behave strongly nonlinear and modelling is a challenging task that is still the subject of research. Thus the question that motivated this experiment is: can the experimental continuation method deal with a strongly nonlinear local attachment mounted on an already nonlinear structure? The particle damper, that has been used for this study, consists of a plastic box filled with ceramic granular material. The single particles have a diameter between 0.1 - 1 mm and the box is filled to 50%. The mounting position is at the maximum displacement of the first mode.

Analog to the previous experiment an EMA was performed first, because the mode shapes are needed to calculate the nonlinear damping ratio as explained above. However, the results are not discussed here in detail, as they are close to the case without particle damper. Next an experimental continuation of the BBC was performed. The very first initial guess was adjusted by manual tuning. Again, 2nd and 3rd harmonic were controlled. Almost all settings, such as sampling frequency, feedback gain, etc. retained from the previous experiment. Only the maximum residual error had to be increased from 1% to 2.5%, as the algorithm sometimes had trouble finding solutions at high oscillation amplitudes. Figure 13 shows the backbone-curve obtained from the experiment. It can clearly be seen, that the resonance frequency increases over the entire interval. However, two sections can be distinguished. From 0.02-0.1 mm$\sqrt{\text{kg}}$ the resonance frequency increases strongly and almost linearly. At around 0.1 mm$\sqrt{\text{kg}}$ the slope decreases and the shape of the curve becomes very similar to the beam without particle damper. The damping ratio is strongly nonlinear. It first increases to a maximum value about 4.2% at 0.07 mm$\sqrt{\text{kg}}$ and then decreases asymptotically. Observing the behavior of the particles inside the transparent box one can assign a physical meaning to the curves. Apparently, the resonance frequency is lower at low excitation levels, compared to high excitation levels, as the particles do not move relatively to the beam. With increasing oscillation amplitude the particles begin to move inside the box. On one hand this increases the dissipation of kinetic

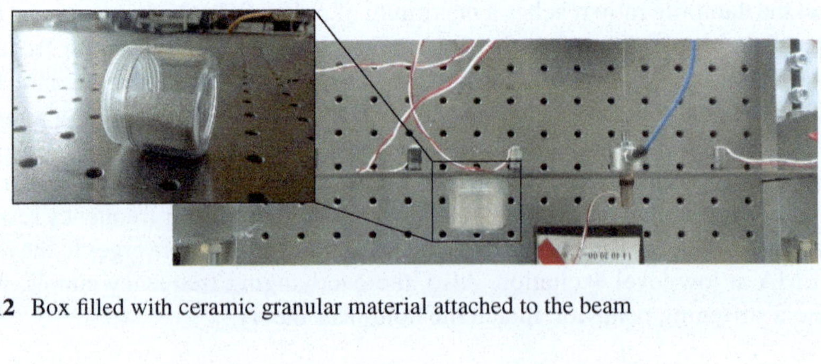

Fig. 12 Box filled with ceramic granular material attached to the beam

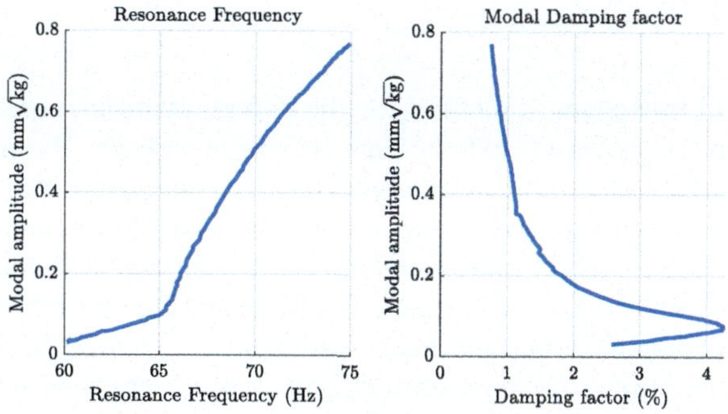

Fig. 13 Resonance frequency and modal damping factor of the first mode of particle damped beam

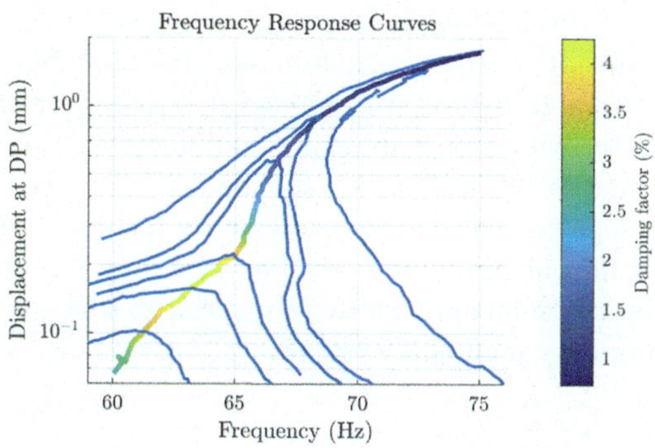

Fig. 14 Forced response curves (blue) and nonlinear mode of the particle damped geometric nonlinear beam excited near first resonance frequency. (Force levels: 0.2, 0.4, 0.5, 0.6, 0.7 and 1.2 N)

energy due to friction and impacts, on the other hand the effective mass is getting reduced, leading to the shift in resonance frequency. When reaching 0.1 mm$\sqrt{\text{kg}}$ it indicated a transition. The effective mass becomes minimal and the geometric nonlinearity begins to dominate the dynamics. At the same time the dissipation decreases. It is remarkable that the maximum damping ratio can be determined quite well. Due to the narrow interval, a linear analysis performed at different excitation levels would probably not be able to determine accurate values.

Finally, NLFRs at different excitation levels have been directly measured. Figure 14 shows the NLFRs along with the resonance frequency and damping ratio in the familiar representation. Again the BBC follows the peaks of the NLFRs. However, it is to mention that the high level frequency response curve (1.2 N) had to be divided into three single measurements, as the algorithm has stopped. This is due to a strong distortion of the force signal presumably caused by chaotic particle motion. Those random impacts introduce high frequency components that would not be eliminated but lead to erroneous Fourier-transform. Filtering and averaging the signals over longer periods could minimize such errors, but has not been tested within this study. Instead, the continuation was resumed starting with the last solution. It is however encouraging to see that the individually measured curves fit well together underlining the reproducibility of experimental continuation.

5 Conclusion

In this paper, a method for the experimental identification of nonlinear frequency response-and backbone-curves, was presented. The method is based on a pseudo-arclength continuation algorithm. To this end, a Newton algorithm is used to find stationary solutions, which minimize a residual vector. Since nonlinear systems can have unstable steady states, an additional feedback-controller was introduced. It was explained how stabilization is achieved and a method for empirical determination of suitable controller gains was presented. As feedback introduces undesired higher harmonics to the excitation force, those harmonics were eliminated iteratively with the Newton's algorithm. The number of controlled harmonics can be easily adjusted with this method. Further, it allows to eliminate any undesired frequency components not only from the excitation force but any other signal as well. The experimental continuation method was applied to a geometrically nonlinear beam clamped on both sides with and without an additional local nonlinearity in the form of a particle damper. For both cases, the method provided comprehensible and plausible results which converge to linear behavior for low vibration levels. Prior knowledge of the nonlinear behavior as well as modeling was not necessary. The experimental continuation has proven to be a versatile tool for the experimental characterization of arbitrary nonlinear systems.

References

1. Abeloos, G., Renson, L., Collette, C., Kerschen, G.: Stepped and swept control-based continuation using adaptive filtering. Nonlinear Dyn. **104**, 3793–3808 (2021)
2. Barton, D.A.W.: Control-based continuation: bifurcation and stability analysis for physical experiments. Mech Syst Signal Process. **84**, 54–64 (2017)
3. Bureau, E., Schilder, F., Santos, I.F., Thomsen, J.J., Starke, J.: Experimental bifurcation analysis of an impact oscillator—Tuning a non-invasive control scheme. J. Sound Vib. **332**, 5883–5897 (2013)
4. Claeys, M., Sinou, J.-J., Lambelin, J.-P., Alcoverro, B.: Multi-harmonic measurements and numerical simulations of nonlinear vibrations of a beam with non-ideal boundary conditions. Commun Nonlinear Sci Numer Simulat. **19**, 4196–4212 (2014)
5. Ewins, D.J.: Modal testing: theory, practice and application. Research Studies Press Ltd., Baldock, Hertfordshire, England (2000). 0-86380-218-4
6. Ferreira, J.V., Serpa, A.L.: Application of the arc-length method in nonlinear frequency response. J. Sound Vib. **284**, 133–149 (2005)
7. Gnanasambandham, C., Fleissner, F., Eberhard, P.: Enhancing the dissipative properties of particle dampers using rigid obstacle-grids. J. Sound Vib. **484**, 115522 (2020)
8. Hagedorn, P.: Nichtlineare Schwingungen. Akad. Verl. Ges, Wiesbaden (1978). 3-400-00351-4
9. Jahn, M., Stender, M., Tatzko, S., Hoffmann, N., Grolet, A., Wallaschek, J.: The extended periodic motion concept for fast limit cycle detection of self-excited systems. Comput. & Struct. **227**, 106139 (2020)
10. Jahn, M., Tatzko, S., Panning-von Scheidt, L., Wallaschek, J.: Comparison of different harmonic balance based methodologies for computation of nonlinear modes of non-conservative mechanical systems. Mech. Syst. Signal Process. **127**, 159–171 (2019)

11. Kerschen, G., Worden, K., Vakakis, A.F., Golinval, J.C.: Past, present and future of nonlinear system identification in structural dynamics. Mech. Syst. Signal Process. **20**, 505–592 (2006)
12. Kleyman, G., Paehr, M., Tatzko, S.: Application of Control-Based-Continuation for characterization of dynamic systems with stiffness and friction nonlinearities. Mech. Res. Commun. **106**, 103520 (2020)
13. Krack, M.: Nonlinear modal analysis of nonconservative systems: extension of the periodic motion concept. Comput. Struct. **154**, 59–71 (2015)
14. Krack, M., Gross, J.: Harmonic Balance for Nonlinear Vibration Problems. Springer Nature Switzerland AG, Cham, Switzerland (2019). 978-3-030-14022-9
15. Meyer, N., Seifried, R.: Damping prediction of particle dampers for structures under forced vibration using effective fields. Granul. Matter. 23(64), (2021)
16. Nayfeh, A.H., Mook, D.T.: Nonlinear Oscillations. Wiley-VCH Verlag GmbH & Co KGaA, Weinheim (2004)
17. Peeters, M., Viguié, G., Sérandour, G., Kerschen, G., Golinval, J.-C.: Nonlinear normal modes, Part II: Toward a practical computation using numerical continuation techniques. Mech. Syst. Signal Process. **23**, 195–216 (2009)
18. Peter, S.: Nonlinear experimental modal analysis and its application to the identification of nonlinear structures. Mech. Syst. Signal Process. **96**, 139–159 (2017)
19. Peter, S., Leine, R.I.: Excitation power quantities in phase resonance testing of nonlinear systems with phase-locked-loop excitation. Institut für Nichtlineare Mechanik der Universität Stuttgart (2018)
20. Peter, S., Riethmüller, R., Leine, R.I.: Tracking of backbone curves of nonlinear systems using phase-locked-loops. In: Kerschen, G. (ed.) Nonlinear Dynamics, Volume 1. Conference Proceedings of the Society for Experimental Mechanics Series. Springer, Cham (2016)
21. Renson, L., Shaw, A.D., Barton, D.A.W., Neild, S.A.: Application of control-based continuation to a nonlinear structure with harmonically coupled modes. Mech. Syst. Signal Process. **120**, 449–464 (2019)
22. Scheel, M., Kleyman, G., Tatar, A., Brake, M.R.W., Peter, S., Noel, J.-P., Allan, M.S., Krack, M.: A phase resonance approach for modal testing of structures with nonlinear dissipation. J. Sound Vib. **435**, 56–73 (2018)
23. Schilder, F., Bureau, E., Santos, I.F., Thomsen, J.J.: Experimental bifurcation analysis—continuation for noise-contaminated zero problems. J. Sound Vib. **358**, 251–266 (2015)
24. Seydel, R.: Practical Bifurcation and Stability Analysis. Springer, New York Dordrecht Heidelberg London (2010)
25. Sieber, J., Krauskopf, B.: Control-based continuation of periodic orbits with a time-delayed difference scheme. Int. J. Bifurc. Chaos. **17**(8), 2579–2593 (2007)
26. Sieber, J., Krauskopf, B.: Control based bifurcation analysis for experiments. Nonlinear Dyn. **51**, 365–377 (2008)
27. Thomsen, J.J.: Vibrations and stability—advanced theory, analysis, and tools. Springer-Verlag, Berlin Heidelberg GmbH (2003). 978-3-642-07272-7
28. Tomlinson, G.R.: Force distortion in resonance testing of structures with electro-dynamic vibration exciters. J. Sound Vib. **63**(3), 337–350 (1978)
29. Touzé, C., Vizzaccaro, A., Thomas, O.: Model order reduction methods for geometrically nonlinear structures: a review of nonlinear techniques. Nonlinear Dyn. **105**(2), 1141–1190 (2021)
30. Worden, K., Tomlinson, G.R.: Nonlinearity in structural dynamics—detection, identification and modelling. IOP Publishing Ltd, Bristol and Philadelphia (2001). 0-7503-0356-5

Field-Responsive Fluid Based Multi-degree-of-Freedom Dampers for Independently Adjustable Dissipation

Aditya Suryadi Tan and Thomas Sattel

1 Introduction

Damping is a decisive factor that influences the dissipation of energy in a vibrating system. On one hand, lack of damping in a vibrating system could lead to excessive resonance or transients with a low decay rate. In such cases, adaptive damping can potentially *calm* vibrations in the fastest time possible or allow a *smooth* transition to other states. Such options are wanted e.g. to suppress earthquake excitation effects on buildings or to improve ride comfort in passenger cars. Since the vibration excitation could vary in form and magnitude, it is preferable to have a *smart* damper, whereby the damping can be adjusted based on the requirements in different conditions.

Since the 19th century, field responsive fluids, namely the electrorheological (ER) and magnetorheological (MR) fluids, attract the researchers' interest due to their unique property that serves the aforementioned requirement [11, 18]. By utilizing these fluids in damper technology, the damper becomes a *smart* system, whereby the resulted damping can be adjusted in real-time by controlling the strength of the applied field. Up to this point, more and more ER and MR-based damper concepts with various structure designs and mechanical implementations have been investigated [19]. They find successful study applications in different vibration control systems, such as vehicle suspension systems, landing gear systems [2], seismic protection for buildings [7], cable-stayed bridges [17] and advanced prosthetic systems [3]. As commercial products, MR fluid dampers have been implemented in various cars. These pure damper elements do operate only in one direction of motion and require mechanical isolation from torques and transverse forces.

A. S. Tan (✉) · T. Sattel
Technische Universität Ilmenau, Max-Planck-Ring 12, Ilmenau 98693, Germany
e-mail: aditya-suryadi.tan@tu-ilmenau.de

T. Sattel
e-mail: thomas.sattel@tu-ilmenau.de

P. Eberhard (ed.), *Calm, Smooth and Smart*, Lecture Notes in Applied and Computational Mechanics 102, https://doi.org/10.1007/978-3-031-36143-2_16

In most of the aforementioned applications, there exist a movement in more than one spatial Degree-of-Freedom (DOF). In some cases, the spatial movement and therefore the vibration could happen even in all of the six existing spatial DOFs of a rigid body, such as at the driver seat of a truck/bus [8]. For such cases, the damping is required not only in one direction but also in all of the movement directions, yet adjustable. Most commonly, several one-directional MR/ER dampers will be implemented in the vibratory system. In the worst case, at least one element in each operating DOF's axis is required [10]. By doing so, the damping can be provided and controlled in each DOF independently. Another way is to integrate several one-directional MR/ER damper elements in one damper system, so that the one damper system can operate in several spatial DOFs. Such a damper system have been applied as haptic systems, where dissipation in a form of feedback force for the user is required in more than one direction. The number of DOFs could vary from two-DOF [1], three-DOFs [9] or even four-DOFs [12], depending on the requirements. However, using this method, the higher the number of the DOF's operating axes, the bigger the total construction volume of the damper system.

In this work, ideas for new possibilities in expanding the functionality of ER/MR dampers by going beyond the conventional design are explored. The goal of this work is to find a new design of ER/MR damper to have a compact design yet higher DOFs. Appropriate design concepts are constructed and their damping performance is investigated. So far, three general possibilities are explored and investigated, namely the extension (1) by integrating several damper elements, (2) by combining known operating modes, and (3) by adding extra control elements. The damper systems including the investigation results are presented. The advantages and disadvantages for each extension method are compared, discussed, and presented as the main contribution of this work.

2 Electrorheological and Magnetorheological Fluids

ER fluids are composed of smaller micrometer-sized electrically polarizable particles suspended in an electrically insulating liquid, whereas MR fluids consist of larger micrometer-sized suspended multi-domain, magnetically soft particles [6]. Figure 1a—OFF state shows how the particles are spread in the fluid.

These particles will form a chain-like formation when a field (either in a form of electric field strength E or magnetic field strength H) is applied, in parallel to the direction of the applied field. This is illustrated in Fig. 1a—ON state, for an applied field (symbolized by the red arrows) in the vertical direction. Due to this chain-like formation, these fluids possess yield stress $\tau_0(E/H)$, which manifests itself in the shear stress τ-shear rate $\dot{\gamma}$ characteristic shown in Fig. 1b. As it can be seen in the figure, the yield stress value varies in dependence of the applied field (E/H). The

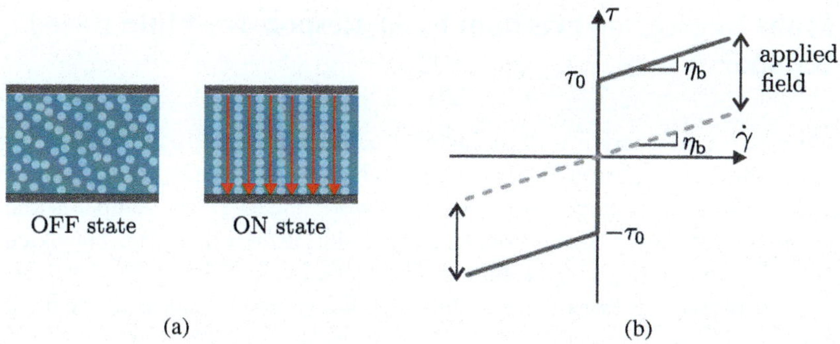

Fig. 1 **a** Principle work of field-responsive fluid and **b** its shear stress-shear rate characteristic line

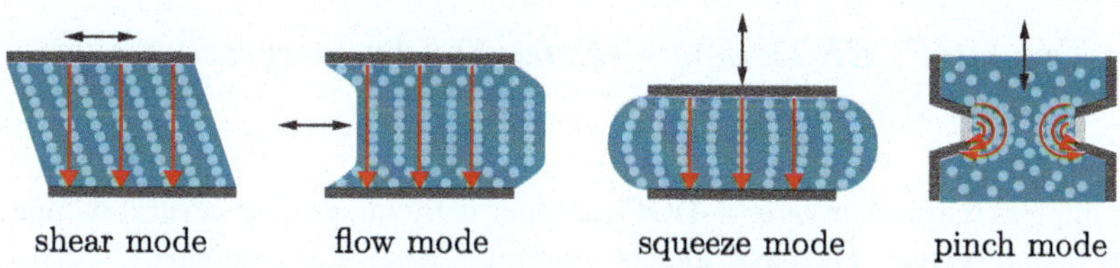

Fig. 2 Configuration of known operating modes of ER- and MR fluids

dynamic viscosity of the fluid η_b defines the gradient of the characteristic line. Due to the nature of these effects, the field-responsive material properties are restricted to the first and third quadrant of the shear stress-shear rate characteristic. The *Bingham model*

$$\tau = \begin{cases} \eta_b \dot{\gamma} + \tau_0(E/H)\,\mathrm{sgn}(\dot{\gamma}) & \text{if} \quad |\tau| \geq \tau_0 \\ G\gamma & \text{if} \quad |\tau| < \tau_0 \end{cases} \tag{1}$$

is sufficient as the first approach to describe the qualitative behavior of the field-responsive fluid. However, there is a wide variety of existing models which allow a more precise description of these fluids, based on its operating mode [16].

The control energy which is required to operate such fluids is solely used to change their material behavior but never transferred into a transducer system. For that reason, the classification as semi-active materials is made.

Irrespective of which fluid type is employed, one of the four different working modes from Fig. 2 can be utilized. The operating and applied field direction is symbolized by the black and red arrows respectively. Depending on the operating mode, the damper will have a different design and behavior.

3 Multi Degrees-of-Freedom Field-Responsive-Fluid Based Dampers

The objective of this work is to explore the possibility of utilizing the ER/MR fluid to have a new design of a Multi DOFs (M-DOF) damper. A M-DOF damper comprises *M* spatial DOF in one damper element. In this work, the M-DOFs damper is realized by expanding the known conventional design of the ER/MR dampers. The extension was done by integrating several damper elements, by combining known operating modes, and by adding extra control elements, whose results are elaborated in this section.

3.1 Two DOFs ER Damper—Extension by Integrating Two Damper Elements

The simplest realization of a M-DOFs damper is by integrating several damper elements in one system. The basic idea is similar to the aforementioned systems, where several stand-alone dampers are integrated and connected into one moving body.

3.1.1 Damper Concept

For this concept, two damper elements utilizing shear operating mode are stacked together in one damper construction [13]. Figure 3a shows the operating concept of the damper.

As can be seen in this figure, the damper is a stack of two pairs of electrodes that are mounted between three parallel plates. Between each pair of the electrode plate, electrorheological fluid is contained. The bottom plate is rigidly fixed to the base and the top plate is the mounting point for a vibrating system. The electrodes are serially

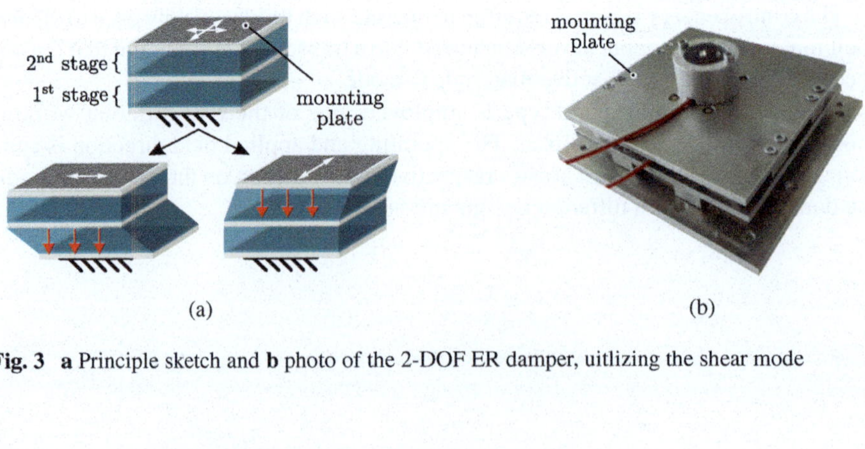

Fig. 3 **a** Principle sketch and **b** photo of the 2-DOF ER damper, uitlizing the shear mode

interconnected through two pairs of linear guides. These pairs of guides are mounted perpendicularly to each other and each pair corresponds to one pair of electrodes. Through such a mechanism, the relative motion will only occur between the two plates in which the movement direction is not constrained, which are illustrated in Fig. 3a. This concept results in a planar damper with two DOFs, whose construction is shown in Fig. 3b.

3.1.2 Operating Mode

The relative motion between two electrodes will shear the ER fluid that is placed between them, resulting in shear stress between these two plates. When this happens, the damper operates in shear mode, where the total output force of the damper can be derived from the Bingham material model in Eq. (1) together with the kinetic and kinematic relations for shear stress τ and shear rate $\dot{\gamma}$

$$\tau = \frac{F}{A_\mathrm{p}} \quad \text{and} \quad \dot{\gamma} = \frac{\dot{x}_i}{h} \,. \tag{2}$$

Here, A_p is the area of the shearing plate, $\dot{x}_i$ is the relative velocity of two moving plates in x- or y-axis and h is the gap between two plates. Inserting those relations in Eq. (1) will result in the ER-damper element model in shearing mode:

$$F_{\mathrm{d},i} = d_i\, \dot{x}_i + F_{0,i}(E_i)\,\mathrm{sgn}(\dot{x}_i) \,, \qquad i = 1, 2 \,, \tag{3}$$

$$\text{with} \quad d_i = \eta_\mathrm{b} \frac{A_{\mathrm{p},i}}{h_i} \,, \qquad F_{0,i} = \tau_{0,i}(E_i)\, A_{\mathrm{p},i} \,. \tag{4}$$

where F_d is the total damping force, d is the damping coefficients and $F_0(E)$ is the blocking forces, that are independently adjustable in each directions by variations in the electric field strengths in each stage.

In most cases, there exist a parasitic friction force that comes from both the mechanical guide and the sealing of the fluid chamber. It could be also a combination of both. These two forces will be generalized as the system friction force F_R, which can be added directly to the Eq. 3, which results in:

$$F_{\mathrm{d},i} = d_i\, \dot{x}_i + \left(F_{0,i}(E_i) + F_{\mathrm{R},i}\right)\, \mathrm{sgn}(\dot{x}_i) \quad \text{for} \quad |F_\mathrm{d}, i| \geq F_{0,i} \tag{5}$$

As it can be seen in the equation, the parasitic friction force is not dependent on the applied field but possesses the same nature as the yield force of the damper. For the case, where $|F_\mathrm{d}, i| < F_{0,i}$, there exist no velocity (the damper is assumed to have no movement).

3.1.3 Experimental Setup and Measurement Results

In this work, the ER fluid RheOil 4.0 from Fludicon is used as the damper medium. The tested damper is shown in Fig. 3b. It has a total dimension of 200 mm × 200 mm × 60 mm with an operating electrode area of 42 mm × 60 mm in each direction and a total in-plane operating range of 30 mm × 30 mm. The lower electrode pair (defined as stage 1) allows a movement along the x-axis, meanwhile the upper electrode pair (defined as stage 2) allows a movement along the y-axis. The first stage and the second stage have an electrode gap of 0.5 mm and 0.35 mm respectively. A voltage will be given at each electrode pair to vary the generated damping force.

Figure 4 shows the plots of the damping force F_d over the damper's velocity for each direction of the damper axis.

The straight line represents the force from the mathematical model in Eq. (5). The markers are the measured force obtained in the experiments. The same color is used in each plot to represent the same electric field strength. It can be seen in the plots, that the measurement results are in agreement with the mathematical model in Eq. (5). The plots show that the Bingham behavior occurs in each stage, where there exists an increase of force generated by the damper as the applied electrical field is increased. The force that exists when there is no applied electrical field (E = 0 kV/mm), is the parasitic friction force F_R that is comprised of the friction of the fluid sealing and the linear guides. It can also be seen in the plots that this friction F_R is higher for the upper stage when compared to the lower stage, which is caused by mechanical parts of the damper itself. The applied field can be increased further up to its dielectric breakdown E_{db}, that lies at E_{db} = 10 kV/mm for this fluid.

Fig. 5 displays the measurement results when the coupling of the forces between the two stages is investigated. In this experiment, the electrical field E_1 was given to the lower stage while the upper stage was moved in a prescribed manner $y(t)$. With the force F_1 in the lower stage acting orthogonally, there is no change in the upper

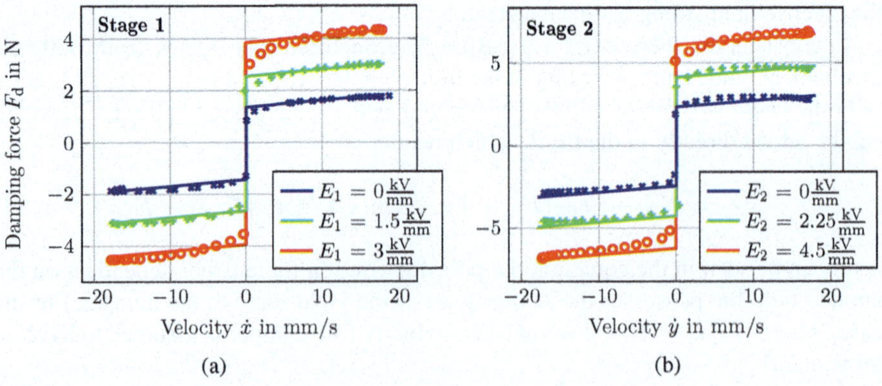

Fig. 4 Comparison between the measured (markers) and calculated (continuous lines) shear force from **a** stage 1 and **b** stage 2 for the different applied current

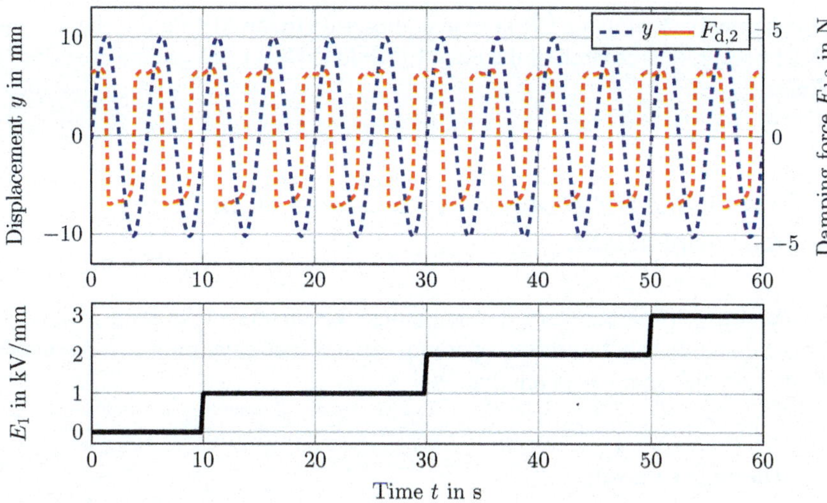

Fig. 5 Measurement result to investigate the force coupling between the stages

stage's force behavior F_2 over time. This experiment was conducted for the other stage and a similar result was obtained. The result of these experiments proved that the damping forces are independently adjustable in each orthogonal damping axis (x and y).

In conclusion, in this work, a new concept of planar electrorheological damper is introduced. This new concept of a planar electrorheological damper offers the possibility to provide two independently adjustable damping forces in two directions of movement through one integrated damping element. Moreover, the performance of such a damper was investigated and verified through experiments. The performance of the damper is shown through simulation in [13] to be promising. The difference in forces between the first and the second stage occurs due to machining tolerance. The mechanical design can be further optimized to have a larger operating range and its dimensions can be up- or downscaled easily, depending on the required damping force in the system. The main difference to the existing M-DOFs field responsive fluid-based damper is that the integration of the damper elements is done structurally, where the damper can be connected to the vibratory system at one point using one connection. Meanwhile, for the existing one, each damper element is connected to the mass individually.

The shear operating mode field responsive fluid-based damper has been proven to be useful as a tuned mass damper [14]. In comparison to a pure frictional damper, ER damper has the flexibility in adapting its yield force, resulting in a better performance. Additionally to that, due to its simple construction, where only a pair of the electrode is required to be the field source in its operation, the field can be segmented to provide another force-velocity characteristic of the damper, as proven experimentally in [15].

The main drawback of using ER fluid in comparison to the MR fluid is the low yield stress. However, the research of the giant ER fluid [5] has a prospect in covering this weakness.

3.2 Two DOFs MR Damper—Extension by Combining Operating Modes

The second extension category is combining two or more operating modes. The idea for this method is to use the operating modes that have the same mechanical configuration, yet operates in another DOF's axis.

3.2.1 Damper Concept

Possible operating modes for this method are the shear and squeeze mode of the field-responsive fluid. As it can be seen in Fig. 2, the shear and the squeeze mode have the same mechanical configuration. In both operating modes, it requires the fluid to be contained between two surfaces that have a relative motion to each other. In both modes, the field is also applied in normal direction of both surfaces. The only difference is in the axis of the relative motion. In shear mode, the relative motion is in the parallel direction to the surface, meanwhile, a relative motion in the normal direction to the surface is required in squeeze mode. Based on these points, both operating modes can be utilized by allowing the damper plate to have movement in both directions.

In order to allow the shear mode and the squeeze mode to be utilized simultaneously, the damper need to be constructed differently than common uniaxial dampers. The concept of the setup is depicted in Fig. 6a.

The idea is to have a construction, where a circular plate can be rotated in θ-direction to shear the MR-Fluid within the chamber. Thus, to allow the squeeze mode to be utilized, this rotating plate should be able to be translated in z-direction as well. This is realized by mounting the shafts using a ball bushing on each side. The circular plate is then inserted in the fluid chamber and held using two shafts, one on each side. Both the fluid chamber and the shaft are made of a non-magnetic material, presented using the yellow color in the Fig. 6. Using this configuration allows the fluid in the chamber to be sheared when the plate is rotated and to be squeezed when the plate is translated. The magnetic field is generated using the electromagnet, surrounding the fluid chamber (as shown in Fig. 6a). The surface of the iron core has the same size as the circular plate. This arrangement allows the magnetic field to be generated perpendicularly over the whole surface of the circular plate. This concept results in a two-DOFs MR damper with one fluid chamber, whose construction is shown in Fig. 6b.

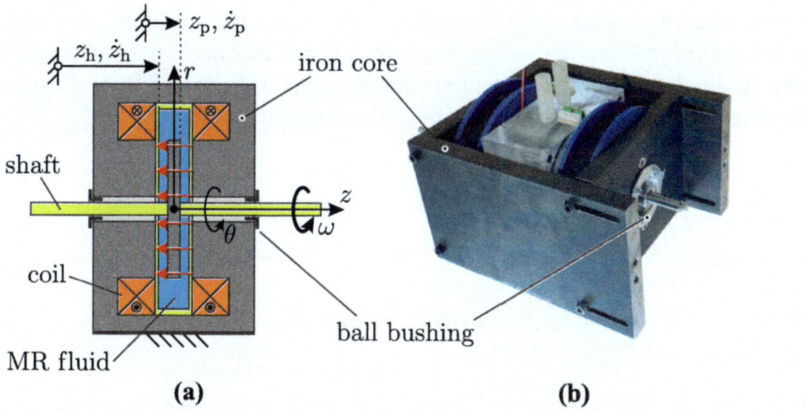

Fig. 6 **a** Principle sketch and **b** photo of the constructed 2-DOF MR damper, utilizing the shear and the squeeze mode

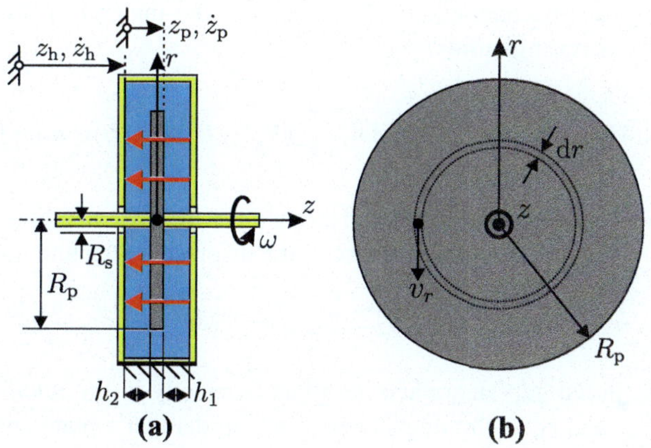

Fig. 7 **a** Cut section and **b** the surface of the 2-DOF MR damper, utilizing the shear and the squeeze mode

3.2.2 Operating Mode

The two operating modes that are utilized in this MR damper, namely the shear and the squeeze operating mode, will be derived separately in this section. Figure 7 shows the details of the damper construction.

It can be seen from this figure that the MR fluid is present on both sides of the damper's plate. Based on this reason, the shear torque and the squeeze force need to be considered from both of the damper's sides.

Shear Mode

Different from the shear mode used in the previous two-DOFs damper, in this damper, the shear mode is caused by a rotational movement. In this case, the tangential velocity varies along the radial axis r of the plate. Therefore, the torque M_d will be the total of all infinitesimal torques $\mathrm{d}M_\mathrm{d}$ from shear stress that acts on a small area with an infinitesimal width of $\mathrm{d}r$. The infinitesimal torque $\mathrm{d}M_\mathrm{d}$ can be derived using Eq. (1) together with the kinematic relations of the tangential velocity

$$v_r = r\,\omega \quad \text{and} \quad \dot{\gamma} = \frac{v_r}{h}, \tag{6}$$

that results in:

$$\begin{aligned} \mathrm{d}M_\mathrm{d} &= r\,\mathrm{d}F(r) = r\,\tau(r)\,\mathrm{d}A(r) \\ &\overset{(1)}{=} \eta\frac{\omega r}{h} + \tau_0(H)\mathrm{sgn}(\omega))(2\pi r\mathrm{d}r)r \\ &= 2\pi\left(\frac{\eta r^3}{h}\omega + \tau_0(H)r^2\mathrm{sgn}(\omega)\right)\mathrm{d}r \end{aligned} \tag{7}$$

The shear torque on one side of the damper is the integral of the infinitesimal torque over the operating radius:

$$M_{\mathrm{d},i} = \int_{R_\mathrm{s}}^{R_\mathrm{p}} \mathrm{d}M_\mathrm{d} = \frac{\pi\eta}{2h_i}(R_\mathrm{p}^4 - R_\mathrm{s}^4)\omega + \frac{2\pi}{3}(R_\mathrm{p}^3 - R_\mathrm{s}^3)\tau_{0_i}(H_i)\mathrm{sgn}(\omega) \text{ for } i \in \{1,2\}. \tag{8}$$

Since the damper plate can be translated along the z-axis, the height on both sides is not necessarily equal. The relation between the displacement z and the gap's height is:

$$h_1 = h_0 - z \quad \text{and} \quad h_2 = h_0 + z, \quad \text{with} \quad z = z_\mathrm{p} - z_\mathrm{h} \tag{9}$$

with h_0 as the initial gap size when the plate is exactly in the middle of the fluid chamber and z_p and z_h as the displacement of the damper's plate and the housing respectively. The total shearing torque

$$M_\mathrm{d} = \left(\frac{1}{h_1} + \frac{1}{h_2}\right)\pi\eta(R_\mathrm{p}^4 - R_\mathrm{s}^4)\omega + \left(\frac{4\pi}{3}(R_\mathrm{p}^3 - R_\mathrm{s}^3)\tau_0(H) + M_\mathrm{R}\right)\mathrm{sgn}(\omega). \tag{10}$$

is the sum of the torque from both sides of the plate added by the parasitic torque M_R due to friction from sealing and guides. It is to be noted that even though the plate is moving along the z-axis, the total magnetic resistance of the magnetic circuit and therefore the magnetic field strength stay unchanged ($H = H_1 = H_2$).

Squeeze Mode

The derivation of the squeeze mode follows the derivation done in [4], where the squeeze force consists of the force due to the viscosity and the rheological effect. As done in the shear mode, the squeeze force needs to be obtained by integrating the infinitesimal force over the operating radius, which comes from multiplying the pressure differences

$$p_\eta(r) = \frac{3\eta r^2}{h^3}\dot{z}, \tag{11}$$

$$p_{\tau_0}(r) = \frac{2\tau_0 r}{h}\operatorname{sgn}(\dot{z}), \tag{12}$$

where $p_\eta(r)$ and $p_{\tau_0}(r)$ are the pressure differences due to the viscous and rheological effect respectively, with the infinitesimal area $\mathrm{d}A = 2\pi r\mathrm{d}r$. Therefore, the infinitesimal forces from each part are:

$$\mathrm{d}F_\eta(r) = \frac{6\pi\eta r^3}{h^3}\dot{z}\mathrm{d}r, \tag{13}$$

$$\mathrm{d}F_{\tau_0}(r) = \frac{4\pi\tau_0 r^2}{h}\operatorname{sgn}(\dot{z})\mathrm{d}r. \tag{14}$$

The total squeeze force on one side of the damper, $F_{\mathrm{d},i} = F_{\eta,i} + F_{\tau_0,i}$,

$$F_{\mathrm{d},i} = \frac{3}{2}\frac{\pi\eta(R_p^4 - R_s^4)}{h_i^3}\dot{z} + \frac{4\pi(R_p^3 - R_s^3)\tau_0(H)}{3h_i}\operatorname{sgn}(\dot{z}) \quad \text{for } i \in \{1, 2\} \tag{15}$$

is the total of the integrated infinitesimal forces in Eqs. (13) and (14) over the operating radius. Considering the forces on both sides of the damper's plate, the total squeeze force

$$\begin{aligned} F_\mathrm{d} = \frac{3}{2}\pi\eta(R_p^4 - R_s^4)\left(\frac{1}{h_1^3} + \frac{1}{h_2^3}\right)\dot{z} \quad \ldots \\ \ldots + \left(\frac{4}{3}\pi(R_p^3 - R_s^3)\tau_0(H)\left(\frac{1}{h_1} + \frac{1}{h_2}\right) + F_\mathrm{R}\right)\operatorname{sgn}(\dot{z}) \end{aligned} \tag{16}$$

is the addition of the squeeze force on both sides of the damper added with the parasitic friction force F_R from the sealing and guides.

3.2.3 Experimental Setup and Measurement Results

The damper plate and the shaft have a radius of 45 mm and 6 mm respectively. The fluid chamber has a height of 10 mm and the damper plate has a thickness of 5 mm respectively. This way, the damper plate will have a distance to the chamber wall of 2.5 mm when it stays in the middle of the chamber. The MR fluid utilized in this damper is the AMT-DAMPRO+ from Arus MR Tech.

For the measurement of the squeeze force F_d, the rotation of the damper is blocked. This is done by fixing the shaft and moving the housing instead. The whole damper is actuated by a linear stepper motor and guided by linear guides. This way, the movement is guaranteed to be perpendicular to the damper's plate. During the movement of the damper housing, the magnetic field is varied and the force is measured.

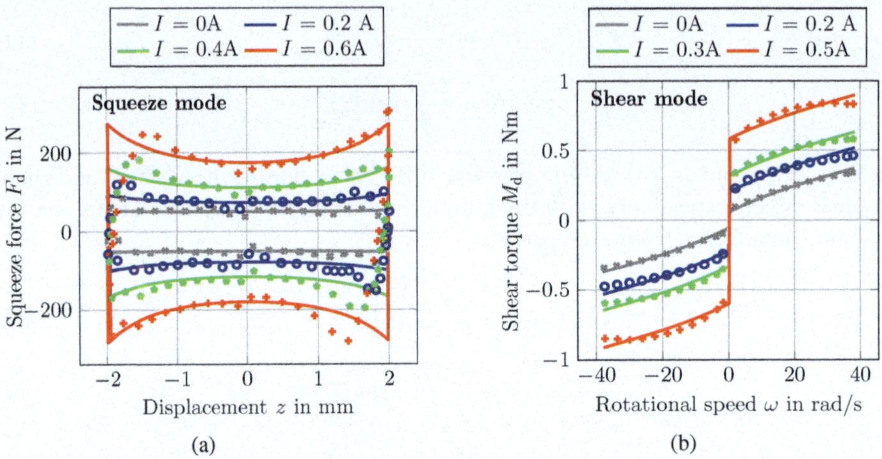

Fig. 8 Comparison between the measurement (markers) and mathematical model (continuous lines) for a **a** squeeze and **b** shear mode of the 2-DOFs MR damper

Figure 8a shows the comparison between the measured squeeze force F_d and the calculated squeeze force from Eq. (16), plotted over the displacement of the damper on the z-axis.

For the measurement of the shear torque M_d, the housing is fixed, and the shaft is rotated by a DC motor. The damper plate is initialized to be in the center of the fluid chamber so that the distance to both walls is equal ($h_1 = h_2$). Both speed rotation and the magnetic field are varied resulting in the plot in Fig. 8b, where the measured torque is compared to the one from the mathematical model in Eq. (10). For both operating modes, the measured force and torque agree with the ones from the mathematical model. In the squeeze mode, the force is increasing as the magnetic field is increased and as the damper gets close to one of the fluid's chamber walls. Thus, the plastic Bingham model in Eq. (1) only able to qualitatively model the squeeze behavior and can not perfectly represent the forces due to the dynamic effect that exists in the second and fourth quadrant of the plot. In the shear mode, the torque is increased as the magnetic field and the rotation speed is increased. In both operating modes, either from the measurement or from the mathematical model, the behavior of the Bingham fluid has been shown.

In comparison to the two-DOFs damper in Sect. 3.1, both operating modes are not completely decoupled from one another. This is the consequence of using the same fluid chamber for multiple operating modes. Therefore, the behavior of the MR damper when the damper is operated using both modes simultaneously is investigated. The investigation is done by rotating the shaft and moving the damper housing along the z-axis together. The rotational speed, the translational speed, and the magnetic field are varied during the whole process. The investigation results are presented in Fig. 9

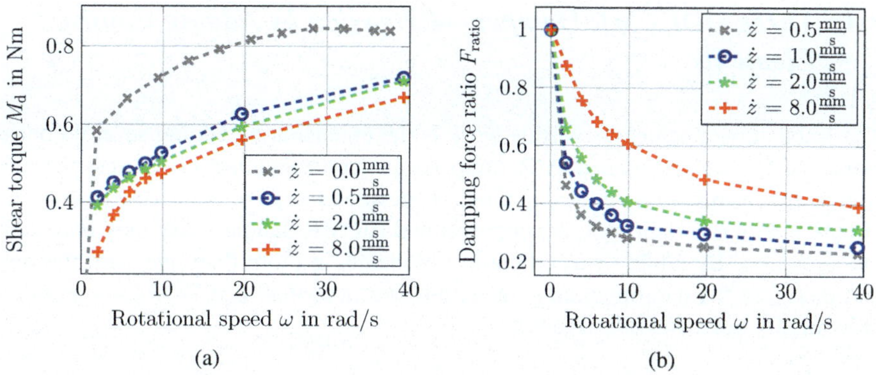

Fig. 9 Measurement results on the coupling between the operating modes

Figure 9a shows the investigation results on how the squeeze force is changed when the damper plate rotates. For that, a reference force is taken for the squeeze force when the damper's plate is at the middle position ($z = 0$). The force ratio

$$F_{\text{ratio}} = \frac{F_{\text{d},z=0}(\omega)}{F_{\text{d},z=0}(\omega = 0)} \tag{17}$$

is defined to be the ratio between the squeeze force when the damper plate is rotated and when the damper's plate is not rotated at position $z = 0$. As it can be seen in the plot, the squeeze force is decreased when a rotation exists. The faster the rotational speed the bigger is the squeeze force decreased. However, the decrement is also dependent on the squeezing speed, where the decrement is larger when the squeezing movement is slower. The same happens to the shearing torque in Fig. 9b. The shearing torque M_d is smaller when the damper plate is translated along the z-axis at the same time. For the shearing torque, the faster the squeezing speed, the bigger is the reduction of the torque.

It can be concluded, that the two-DOFs damper using this extension approach could result in a compact setup, where the same damper arrangement can be used for more DOFs. The only thing required for this improvement is the ball bushing, a mounting that allows the movement of the damper's plate exist in two directions. The damping can be adjusted in both DOFs axes. However, there exists a coupling between both operating modes, and the application of a magnetic field will affect both operating modes at the same time.

3.3 Three DOFs MR Damper—Extension by Adding Control Elements

The last extension category of this work is the extension by adding several control elements. In conventional ER/MR dampers, one control element is used as the field source. Since this single control element is designed to generate a field in only one direction at one certain spot, the conventional ER/MR dampers will only operate in one direction. By adding several control elements in more than one spot, the field and therefore the damping can be generated in more than one DOF axes, not only individually but also collectively.

3.3.1 Damper Concept

The suggested damper concept is with MR-Fluids, with the electromagnet as its control element. Figures 10 and 11 depict the principle sketches and the construction proposed for a three-DOF MR damper.

As it can be seen in this figure, the damper has a shape of a box. At the damper wall, four electromagnets are installed at the four sides of the damper, one electromagnet on each side, except the top and the bottom sides, see Fig. 11. A symbol of an orange triangle with the rotation degree within is given to indicate that the electromagnet's display is actually rotated in the principle sketch by 90° in comparison to the constructed MR damper for understanding purpose, where only the orientation of the electromagnet mounting is changed. The bottom side is used as the mount of the damper and the top side is used as an access for the damper to be connected with a moving system. Each electromagnet is driven by an independent power supply so that the electromagnets can be activated and therefore the magnetic field can be applied individually on the four sides of the box.

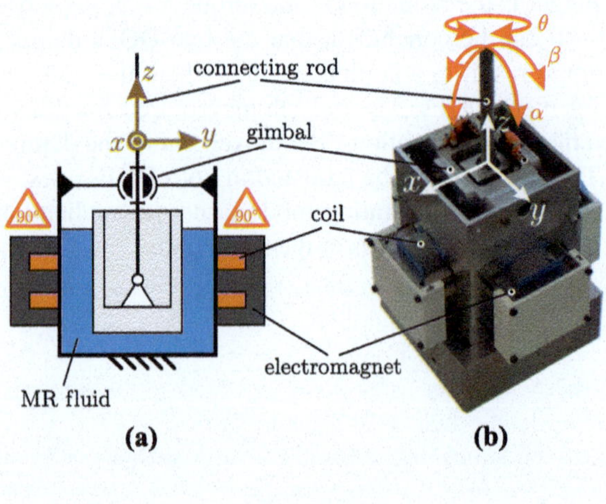

Fig. 10 **a** Principle sketch and **b** photo of the compact 3-DOF MR damper

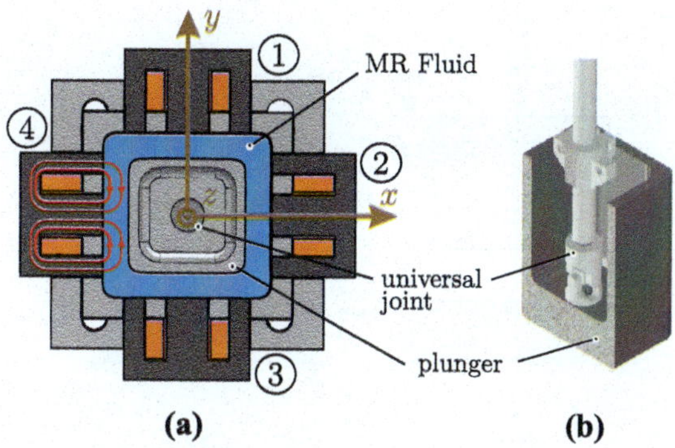

Fig. 11 **a** 2D- and **b** 3D cut section of the damper to show the joint connection

The box itself is the fluid chamber, where the MR fluid is contained. Figure 11 shows the cut section of the box in the xy-plane.

The damper plunger in Fig. 11 can be moved in any direction on the xy-plane, causing the MR fluid to be squeezed in different directions. The state of the squeezed MR fluid in a specific region can be altered by activating the corresponding electromagnet, that is attached to each side of the fluid chamber. As an example, the electromagnet 4 in Fig. 11 is activated, generating a magnetic field only on the left side of the plunger. This solidifies the MR Fluid locally in the region where the magnetic field exists and therefore increases both the flow resistance of the MR fluid in this area and the movement resistance of the plunger in this direction. Additional to that, the plunger is 3D-printed and made of a non-magnetic material. This eliminates the attraction force from the magnet to the plunger. With this configuration, the generated force is a pure damping force from the squeezed MR fluid in the respective movement direction. In Fig. 11b, there is another cut section that shows that the plunger is connected to a rod via a universal joint. This joint allows the plunger surface to be parallel to the wall of the fluid chamber when the plunger is moved by the connecting rod.

3.3.2 Operating Mode

In the previous section, it is mentioned that the MR fluid will be squeezed by the plunger. Therefore, the most suitable operating mode for this damper will be the squeeze operating mode. However, the one used in this damper is not the conventional squeeze operating mode. Figure 12 depicts the configuration comparison between the conventional squeeze mode and the one used in this work.

For this comparison, the fluid is squeezed in the vertical direction (z-axis). As it is shown in Fig. 12a, conventional squeeze mode has a magnetic field applied perpendicular to the flow direction of the fluid and to the squeezing surface. In this work, the field is not always perpendicular to the flow direction of the fluid and to the squeezing surface. As it is shown in Fig. 12b and has been proven by FEM

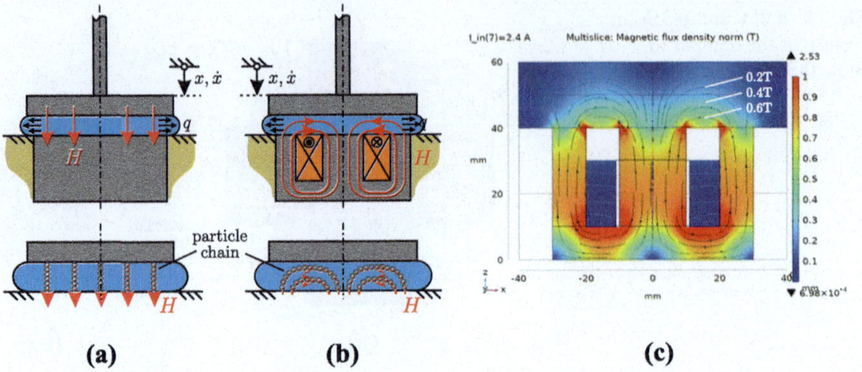

Fig. 12 Configuration comparison between **a** the conventional squeeze mode and **b** the one used for the proposed damper in this work, including **c** the FEM-analysis using COMSOL Multiphysics to investigate the magnetic field range

analysis in Fig. 12c, the magnetic field has a form of an arc. Therefore, for a different location in the fluid chamber, the magnetic field has a different vector. The difference in the magnetic field lines of both configurations made the chain-like structure of the MR fluid to have a different orientation. As it is depicted in the Fig. 12a and b, the conventional one has a chain-like structure in the form of a pillar. Meanwhile, the one with an arc magnetic field will have a chain-like structure in the form of a bridge. Based on that reason, the squeeze mode equation in the literature can not be directly implemented for this setup. This made mathematical modeling difficult and a numerical model is recommended.

However, to give an insight on how is the squeeze force in this configuration, the squeeze force will be compared to the conventional squeeze operating mode. The comparison was done by assuming that the squeezing surface in the conventional one has a radius of

$$l_{\mathrm{m}} = l_{\mathrm{s}} \frac{(1+\sqrt{2})}{4}. \tag{18}$$

where l_{s} is the side length of the quadratic surface of the plate and l_{m} is the average length between the shortest and the longest distance from the middle point to the surface edges, which are $l_{\mathrm{s,short}} = \frac{l_{\mathrm{s}}}{2}$ and $l_{\mathrm{s,long}} = \frac{l_{\mathrm{s}}}{2}\sqrt{2}$ respectively. Using the same derivation steps, as done in Sect. 3.2.2, the squeeze force

$$F_{\mathrm{d}} = \int_0^{l_{\mathrm{m}}} \left(\frac{6\pi \eta r^3}{h^3} \dot{x} + \frac{4\pi \tau_0 r^2}{h} \mathrm{sgn}(\dot{x}) \right) \mathrm{d}r \tag{19}$$

$$F_{\mathrm{d}} = \frac{3}{2} \frac{\pi \eta l_{\mathrm{m}}^4}{h^3} \dot{x} + \frac{4\pi l_{\mathrm{m}}^3 \tau_0}{3h} \mathrm{sgn}(\dot{x}) \tag{20}$$

is the integration of the infinitesimal force $\mathrm{d}F_{\mathrm{d}}$, with l_{m} as the boundary. The calculated conventional squeeze force is then compared with the measured squeeze force

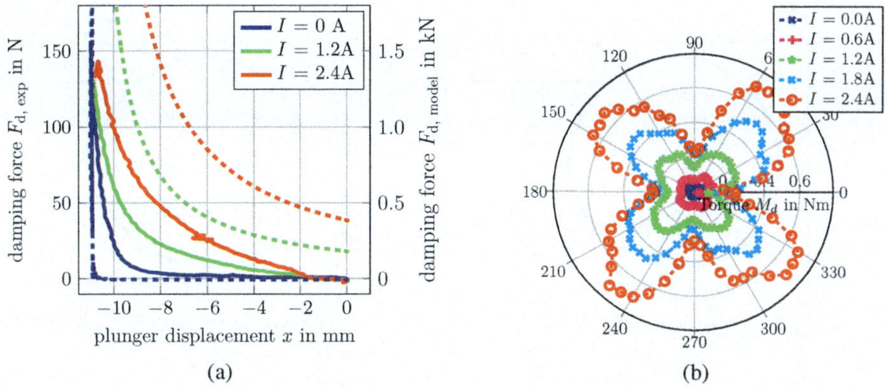

Fig. 13 **a** comparison between the calculated force for conventional squeeze mode (dashed lines) and measured force for the new squeeze mode (straight lines). **b** the measured torque for various applied field

from the new configuration (Fig. 12b), whose comparison will be presented in the next sub-section.

3.3.3 Experimental Setup and Measurement Results

The fluid chamber has a quadratic surface in this plane with a dimension of 78 mm × 78 mm. In the middle of the fluid chamber, a damper plunger with a 56 mm × 56 mm quadratic surface in xy-plane is inserted. This means that each side of the plunger has an equal distance to the wall of 11 mm. The height of the fluid chamber and the plunger can be any length, as long as it is enough to contain the fluid in the chamber during its operation. The fluid chamber is filled with magnetorheological fluid AMT-SMARTEC$^+$ from Arus MR Tech.

For the first investigation, the damper behavior will be analyzed in each DOF. Figure 13 presents the results of the experiments.

Figure 13a presents the measured squeeze force when the plunger is pushed along the x or y-axis. For this investigation, only one control element is activated; the one on the side of the squeezed fluid. For the same displacement, the applied field is varied. As it can be seen in the result, the force is increased as the magnetic field is increased and as the distance between the plunger and the wall is decreased. This reflects the behavior of the squeeze mode. As a comparison, the squeeze force from Eq. (20) is presented in the same plot using dashed lines. Even though they have similar behavior, the new configuration of the squeeze mode in this works results in a smaller squeezing force. This is due to the different orientation of the field and therefore the chain-like structure of the fluid particles that causes less flow resistance in the squeezed spot. In Fig. 13b, the measured torque about the z axis is plotted over its rotational angle. For varied applied fields, the bigger the magnetic field,

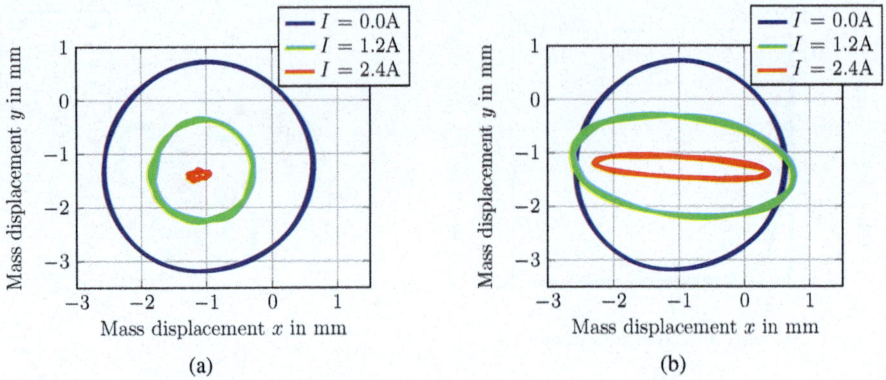

Fig. 14 Measured vibration of the damper's plunger plotted on the xy-plane

the bigger is the measured torque. Using this box configuration, the torque is not evenly distributed about the angle of rotation θ. As it can be seen in the results, the torque reach its maximum when the plunger is rotated by 45° and multiples ($\theta = \frac{\pi}{4} + n\frac{\pi}{2}$ for $\theta \in [0, 2\pi]$ with $n = 0, \ldots, 3$). The minimum torque is reached when the plunger is rotated by 90° and multiples. This is the position where the plunger's wall is parallel to the chamber's wall. These investigations have proven, that the damping can be generated and adjusted in all three DOFs of the MR damper.

As discussed in Sect. 3.2, the usage of one fluid chamber to generate damping in more than one direction has coupled the damping from one DOF to the other DOF. However, in comparison to the previous setup, this MR damper has more than one control element that is installed in different spots of the damper. It opens the possibility to adjust the damping in each DOF separately. Figure 14 is a measurement result for investigating this aspect.

A motor with an imbalance mass is attached to the damper's connecting rod, Fig. 10, resulting in vibrations in the xy-plane. This vibrations are represented by the circular movement of the plunger in this plane in both plots. Figure 14a shows the changes of the plunger movement amplitude, when all four electromagnets are activated. As the field is increased, the damping is increased in all sides of the damper and therefore the vibration amplitude is decreased evenly in all directions. That is the reason why the radius of the circle is smaller for a higher applied current. In Fig. 14b, only one pair of the electromagnet is activated, namely the electromagnets that are installed in the y-axis. This caused the reduction of the vibration amplitude only in y-axis, where the circle becomes an oval. As the field is increased, the oval is smaller. It is also investigated that the reduction of the vibration amplitude exists in the x-axis, yet much less in comparison to the reduction on the y-axis. This shows that using this extension approaches, the damping can be adjusted to be dominant only in one DOF.

4 Conclusion and Outlook

In summary, three extension approaches for MR/ER dampers have been proposed in this work. They were done by integrating several damper elements, combining existing operating modes, and adding control elements in one damper system. The proposed dampers were built and their performance is tested via experiments. The known operating modes are verified with the known mathematical model.

It can be concluded from the results, that M-DOF MR/ER damper can be realized using these extension approaches. Using the proposed methods, there is a prospect of making the damper system to be more compact. By doing so, the damper requires only one fixing point to the vibrating mass, which results in a simpler overall construction. The proposed integration of several elements allows an independent adjustment, whereas the proposed combination of operating modes and the addition of the control elements cause a coupling exists between the damping in different DOFs. Therefore, as the drawbacks of these extensions, the system becomes more complicated as the operating modes are coupled and harder to be modeled mathematically. This might demand a more complex control to regulate the damping in each DOF. Nevertheless, the work has shown that there are still a lot of possibilities for exploring the MR/ER damper design. In the future, a better way to model the damping behavior of new operating modes and the coupling of the damping in the system is required.

Acknowledgements The authors would like to express their gratitude towards the German Research Foundation, *Deutsche Forschungsgemeinschaft* (DFG) for the financial support within the priority program SPP 1897, "Calm, Smooth, and Smart".

References

1. Ahmadkhanlou, F., Washington, G.N., Bechtel, S.E.: Modeling and control of single and two degree of freedom magnetorheological fluid-based haptic systems for telerobotic surgery. J. Intell. Mater. Syst. Struct. **20**(10), 1171–1186 (2009)
2. Choi, Y.-T., Wereley, N.M.: Vibration control of a landing gear system featuring electrorheological/magnetorheological fluids. J. Aircraft **40**(3) (2003)
3. Ekkachai, K., Tantaworrasilp, A., Nithi-Uthai, S., Tungpimolrut, K., Nilkhamhang, I.: Variable walking speed controller of MR damper prosthetic knee using neural network predictive control. In: SICE Annual Conference (SICE) (2014)
4. Farjoud, A., Cavey, R., Ahmadian, M., Craft, M.: Magneto-rheological fluid behavior in squeeze mode. Smart Mater. Struct. **18** (2009)
5. Huang, X., Wen, W., Yang, S., Sheng, P.: Mechanisms of the giant electrorheological effect. Solid State Commun. **139**(11), 581–588 (2006)
6. Janocha, H.: Adaptronics and Smart Structures. Springer, Berlin (2007). https://doi.org/10.1007/978-3-540-71967-0.
7. Lee, S.-H., Youn, K.-J., Min, K.-W.: A decentralized response-dependent MR damper for controlling building structures excited by seismic load. J. Intell. Mater. Syst. Struct. **22**(11), 1913–1927 (2011)
8. Ning, D., Du, H., Sun, S., Li, W., Zhang, B.: An innovative two-layer multiple-DOF seat suspension for vehicle whole body vibration control. IEEE/ASME Trans. Mechatron. **23**(4), 1787–1799 (2018)

9. Nguyen, H.Q., Le, T., Nguyen, D.N., Le, T.D., Lang, T.V., Ngo, T.V.: Development of 3-DOF force feedback system using spherical arm mechanism and MR brakes In: (Chunhui) Yang, R., (ed.), International Journal of Mechanical Engineering and Robotics Research, vol. 9, No. 2 (2020)
10. Oh, J.-S., Sohn, J.W., Choi, S.-B.: Applications of magnetorheological fluid actuator to Multi-DOF systems: state-of-the-art from 2015 to 2021. Actuators **11**(2), 2022 (2015)
11. Rabinow, J.: Magnetic fluid torque and force transmitting device. U.S. Patent 2575360, 1947
12. Song, B.-K., Oh, J.-S., Choi, S.B.:Design of a new 4-DOF haptic master featuring magnetorheological fluid. Adv. Mech. Eng. **6** (2014)
13. Tan, A.S., Ross, K.H., Sattel, T.: A new concept of an electrorheological planar damper for independent damping adjustment in planar movement. In: Borgmann, H., (ed.), ACTUATOR 2018: 16th International Conference on New Actuators, pp. 350–353. VDE Verlag GmbH (2018)
14. Tan, A.S., Aramendiz, J., Ross, K.H., Sattel, T., Fidlin, A.: Comparative study between dry friction and electrorheological fluid switches for tuned vibration absorbers. J. Sound Vibrat. **460**(114874) (2019)
15. Tan, A.S., Belkner, J., Stroschke, A., Sattel, T.: Damping adjustment utilizing digital electrorheological valves with parallely segmented electrodes. Smart Mater. Struct. **28**(075013) (2019)
16. Wang, D.H., Liao, W.H.: Magnetorheological fluid dampers: a review of parametric modelling. Smart Mater. Struct. **20**(2) (2011)
17. Weber, F., Distl, H.: Amplitude and frequency independent cable damping of Sutong Bridge and Russky Bridge by magnetorheological dampers. Struct. Control. Health Monit. **22**(2), 237–254 (2014)
18. Winslow, W.M.: Method and means for translating electrical impulses into mechanical force. U. S. Patent 2417850, 1947
19. Zhu, X., Jing, X., Cheng, L.: Magnetorheological fluid dampers: a review on structure design and analysis. J. Intell. Mater. Syst. Struct. **23**(8), 839–873 (2012)

Granular Mixtures with Tailored Effective Properties

Kianoosh Taghizadeh, Stefan Luding, and Holger Steeb

1 Introduction

In our daily life we are surrounded by granular materials like soil, sand, coffee, nuts, food- or detergent-powders, pharmaceutical products like tablets, and many others. Granular materials constitute over 75% of raw materials feedstock to industry, including pharmaceutical, mining, agriculture, chemical, biotechnological, textile, etc. In spite of their ubiquity and apparent simplicity, their behavior is far from being fully understood. This leads to the loss of about 10% of the world's energy consumption in processing, storage and transport of granular materials, and to various risks related, e.g. to avalanches or landslides due to instability of slopes or foundations.

Granular materials behave differently from usual solids or fluids and show peculiar mechanical properties like dilatancy, history dependence, ratcheting and anisotropy. The behavior of granular materials is highly non-linear and involves irreversibility (plasticity), possibly already at very small strains, due to rearrangements of the elementary particles [1–3]. Furthermore, complex soil behavior also originates from the multi-phase nature of these materials that exhibits both elastic and plastic non-linearities due to solid-fluid coupling.

Mixtures of different type/size particles have become a key research topic in recent years due to their wide range of applications in engineering [4–9], e.g. sand-rubber mixtures in asphalt. The reuse of waste rubber tires creates a win-win situation whereby non-biodegradable tires are given a new lease of life. Understanding gran-

K. Taghizadeh (✉) · H. Steeb
Institute of Applied Mechanics (CE), University of Stuttgart, Stuttgart, Germany
e-mail: kianoosh.taghizadeh-bajgirani@mib.uni-stuttgart.de

H. Steeb
e-mail: holger.steeb@mechbau.uni-stuttgart.de

S. Luding
Multiscale Mechanics (MSM), University of Twente, Enschede, The Netherlands
e-mail: s.luding@utwente.nl

P. Eberhard (ed.), *Calm, Smooth and Smart*, Lecture Notes in Applied
and Computational Mechanics 102, https://doi.org/10.1007/978-3-031-36143-2_17

ular mixtures poses still formidable challenges, in particular for complex mixtures with more than one particle species and phase. However, such granular mixtures can exhibit tailored mechanical properties—better than each of the ingredients—due to the interplay of small-big, soft-stiff, and smooth-angular particles.

In analogy to "classical" solids, the behavior of granular materials depends on the amount of deformation the sample is subjected to. Roughly speaking, we can distinguish (i) an elastic regime at very small strain, (ii) a non-linear elasto-plastic regime that holds from small to intermediate strain, and (iii) a fully visco-elaso-plastic regime at large strain and strain rates, where the material flows (solid to fluid transition) at constant stress and volume—if sheared long and fast enough. A special note on the elastic regime should be given: soil behavior is considered to be truly elastic only in the range of extremely small strains ($\varepsilon \leq 10^{-4}$). For slightly higher strain, soil may exhibit a non-linear stress-strain relationship, stiffness being almost fully recoverable when unloading. For larger strain, the material deforms irreversibly and deformations are permanent (plastic). In all cases, the elastic material stiffness is defined only for small strain and can be found from measurements of the (shear) wave velocity in situ and laboratory tests by probing, i.e. incremental stress-strain tests.

Understanding the effective mechanical properties of closely packed, dense granular systems is of interest in many fields, such as soil mechanics, materials science and physics. The main difficulty arises due to discreteness and disorder in granular materials starting at the particle scale, which requires a multi-scale approach. The concept of an initial purely elastic regime at small strains for granular assemblies is an issue still under debate in the soil mechanics community. Approaches that neglect the effect of elastic stored energy are as questionable as approaches that ignore the dynamics.

For many geotechnical structures under working loads, the inherent deformations are small. The regime of deformation where the behaviour can be considered linear elastic is infinitesimal, with nonlinear and irreversible effects present already at small strains. Nevertheless, characterisation of the stiffness of soils is important, as it provides an anchor on which to attach the subsequent stress-strain response [10, 11]. Features visible in experiments, like the propagation of acoustic waves, can hardly be described without considering an elastic regime. Mechanical waves are perturbations moving through space and time in a medium where the small deformations leads to elastic restoring forces. This causes a transfer of momentum or energy through particle contacts without or with little mass transport. The propagation of the mechanical wave through the medium can give valuable information on the state, the structure and the mechanical properties. Through wave propagation with small amplitudes, one can examine a packing without sample disturbance or destruction which makes it possible to probe different situations.

In recent decades, the Discrete Element Method (DEM) has become increasingly popular as a computational tool to model granular systems in both academia and industry [12–14]. To date, due to an increase of computer power, considerable scientific advances have been made in the development of particle simulation methods resulting in an increasing use of DEM. It is a powerful tool to inspect the influence of the microscopic contact properties of the individual constituents on the effective (bulk) behavior of granular assemblies.

The purpose of this contribution is to improve the understanding of basic mechanisms in particle systems and to guide further developments for new macroscopic constitutive models. It is organised as follows: in Sect. 2, the classical relations between wave velocity and elastic moduli in a solid are reviewed; Sect. 3 introduces the wave propagation technique including the interpretation of experimental results; Sect. 4 investigates the effects of inter-particle contact properties on the elastic bulk and shear modulus by applying isotropic and deviatoric perturbations. Further numerical results are compared with experiments. Finally in Sect. 5, we summarise by mentioning the key points of the research and close with an outlook.

2 Waves and Elasticity

A wave is an elastic perturbation that propagates between two material points through a material body (bulk waves) or on the surface (surface waves) without plastic deformation [15]. In the case of bulk waves the acoustic-elastic effect is related to the change in the wave velocity of small amplitude waves due to the stress state of the body.

In the present section, we will review the relations between the elastic characteristics and the velocities of acoustic waves, in the longitudinal and transversal directions [16]. Let us consider a three-dimensional body with density ρ. It is assumed that the material body is homogenous, isotropic and linear-elastic. A volume element subjected to a force F_i in direction i will undertake a displacement ∂u_i in the same direction. The stress change due to the propagation of the wave in the body is given by the local form of the balance of momentum [17]:

$$\frac{\partial \sigma_{ij}}{\partial x_j} = \rho \ddot{u}_i, \tag{1}$$

with the Cauchy stress σ_{ij} and $\ddot{u}_i$ displacement (second time derivative) of the volume element in directions $i, j = 1, 2, 3$. On the other hand, the constitutive relation for the elastic body holds, that relates the stress tensor to the strain ϵ_{ij} via the 4th order elasticity tensor C_{ijkl}

$$\sigma_{ij} = C_{ijkl}\epsilon_{kl}. \tag{2}$$

In the isotropic case, Eq. (2) becomes Hook's law of linear elasticity

$$\sigma_{ij} = \lambda \epsilon_{kk}\delta_{ij} + 2G\epsilon_{ij}, \tag{3}$$

where summation convention is implied $\epsilon_{kk} = \sum_{i=1}^{3} \epsilon_{ii}$, with G and λ the shear modulus and Lamé coefficients respectively, and the linear strain tensor is given by

$$\epsilon_{ij} = \frac{1}{2}\left(\frac{\partial u_i}{\partial x_j} + \frac{\partial u_j}{\partial x_i}\right). \tag{4}$$

The bulk modulus is related to the previous quantities as $K = \lambda + (2/3)G$.

Using Eqs. (3)–(4) in Eq. (1), the equation of motion becomes

$$\rho \frac{\partial^2 u_i}{\partial t^2} = \frac{\partial}{\partial x_j}\left(\lambda \frac{\partial u_i}{\partial x_i}\right) + G \frac{\partial^2 u_i}{\partial x_j^2} + G \frac{\partial}{\partial x_i}\left(\frac{\partial u_i}{\partial x_j}\right). \tag{5}$$

From Helmholtz decomposition, the displacement vector $\boldsymbol{u}$ can be written in terms of a scalar potential ϕ and a vector potential $\boldsymbol{\psi}$: $\boldsymbol{u} = \nabla\phi + \nabla \times \boldsymbol{\psi}$, where the tensorial notation has been used for the sake of brevity. Thus Eq. (5) transforms to

$$\nabla \left[\rho \frac{\partial^2 \phi}{\partial t^2} - \left(\lambda + \frac{4}{3} G\right) \nabla^2 \phi\right] + \nabla \times \left[\rho \frac{\partial^2 \boldsymbol{\psi}}{\partial t^2} - G \nabla^2 \boldsymbol{\psi}\right] = \mathbf{0}. \tag{6}$$

Equation (6) is known as the wave equation and splits longitudinal and transversal modes of propagation. The first term in Eq. (6) depends only on ϕ, related to propagation of waves in the longitudinal direction, while the second term depends on the vector potential $\boldsymbol{\psi}$, associated with transversal waves. As both terms must be separately zero to satisfy Eq. (6), we obtain the two wave modes for isotropic, homogeneous linear-elastic media. The longitudinal is always faster than the transversal wave mode. Thus, it is also denoted as the P-wave (primary wave). The (slower) transversal wave is known as the secondary wave or S-wave (shear wave).

Finally, if we introduce the longitudinal and shear components of the displacement related to ϕ and $\boldsymbol{\psi}$ as $\boldsymbol{u}_P = \nabla\phi$ and $\boldsymbol{u}_S = \nabla \times \boldsymbol{\psi}$. From Eq. (6) we can derive the velocities of the longitudinal and transversal waves for the isotropic elastic body:

$$V_P = \sqrt{\frac{(\lambda + 4/3G)}{\rho}} \quad \& \quad V_S = \sqrt{\frac{G}{\rho}} \tag{7}$$

Note that local rotations do not occur in longitudinal waves at the material point, while volume changes are not present in transversal waves.

The P-wave can move through solids and fluids (as well as gases), like water or the liquid layers of the earth. P-waves are also known as compressional waves, because of the pushing and pulling responses they generate. Subjected to a P-wave, particles move in the same direction as that of the wave moving (shaking the ground in the direction they are propagating). The second type of body wave is the S-wave or secondary wave or shear wave, which is the second wave one can feel in an earthquake (shaking the ground perpendicular to the direction in which they are propagating).

S-waves are slower than P-waves and can only move through solids, not through liquids medium, i.e. shear waves require a medium with shear stiffness to propagate, $G > 0$; thus, they are filtered in fluids. It is this property of S-waves that led seismologists to conclude that the Earth's outer core is a liquid. When observing Eq. (7), some aspects appear: (i) the propagation velocity increases with the stiffness of the material and decreases with its mass density (inertia) these characteristics being constants in a given solid body; (ii) the velocity of transversal waves is smaller than the velocity of longitudinal waves, given the relative values of the moduli; iii) wave velocities of isotropic, homogeneous, linear-elastic media are not a function of frequency, i.e. the medium is non-dispersive.

3 Experimental Study of Small Strain Stiffness

Characterization of mechanical (elasto-plastic) properties has traditionally been made in a triaxial apparatus using precise displacement transducers or in resonant column devices. In recent years, several methods became commercially available to determine the elasto-plastic response of geomaterials both in the laboratory and in the field [18, 19].

The laboratory tests are classified as dynamic or static, as described in Table 1. Dynamic testing is performed at a strain rate high enough to initialize an inertial effect within the specimen, whereas static testing is conducted at a much lower repetition rate at which inertial effects are negligible.

3.1 Wave Propagation Technique

Among experimental characterization techniques, ultrasound wave propagation is widely accepted for their rapid, non-destructive, and low-cost evaluation. Through transmission techniques for determining one-way travelling times, which includes BE (Bender Elements) and UT (Ultrasonic Transducers) technology, has been gaining popularity as an experimental method due to its relative ease of obtaining the modulus of a material. The instrumentation consists of a pair of piezoelectric transducers,

Table 1 List of experiments classified as static or dynamic and their strain levels [20, 21]

	Type of test	Strain (%)
Static	Triaxial (TX)	$>10^{-4}$
Dynamic	Resonant column (RC)	10^{-5}–10^{-2}
	Bender element (BE)	$<10^{-3}$
	Ultrasonic transducer (UT)	$<10^{-4}$

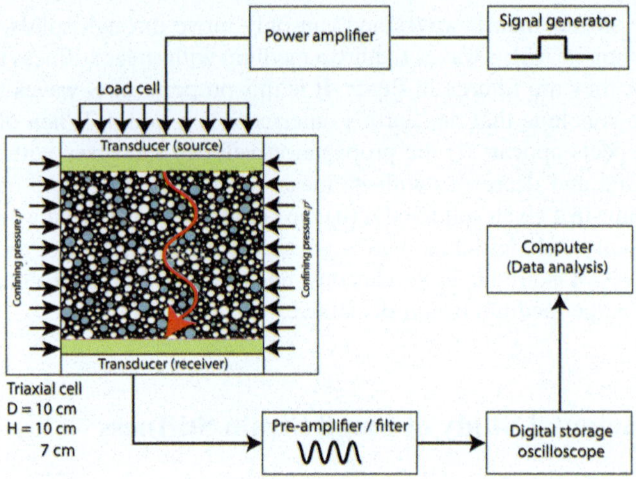

Fig. 1 Schematic drawing of an experimental setup of wave propagation testing [23, 24]

function generator, signal amplifier, voltage divider for the input signals and digital oscilloscope, and signal amplifier/filter. Usually transducers are installed along the axis of the cylindrical sample, i.e. at the top and bottom of a triaxial or oedometric cell, to probe the stiffness of the soil along a given stress path [22]. Figure 1 shows a schematic drawing of the setup and the peripheral electronics. The transmitting transducer transfers the high-voltage electrical signal to a mechanical excitation which is finally propagating through the medium. The receiving piezoelectric transducer receives the propagating mechanical excitation and transforms it back to low-voltage electric signal which will be amplified and acquired with the storage oscilloscope. From the speed-of-sound of the first arrivals and the geometry of the sample, the resulting elastic wave velocities could be obtained. Finally by knowing the elastic wave velocities as measured with the wave-based techniques and total mass density of the media, the stiffness of the materials can be determined.

The signal-to-noise ratio is improved by repetitive averaging of a sufficient number of detected signals using the digital oscilloscope and a computer for further processing. Concerning the travel time (t) and distance (L), necessary to calculate the wave velocity ($V = L/t$), the determination of travel distance (between transducers) is generally considered the less problematic of the two [25, 26].

The longitudinal velocity V_p and the transversal velocity V_s can be measured by using P-wave or S-wave transducers respectively. The longitudinal, P-wave modulus M is related to the P-wave velocity V_p in the medium by

$$M = \rho V_p^2, \tag{8}$$

where ρ is the bulk density of the sample, which is related to porosity ϕ and particle density ρ_p as: $\rho = (1 - \phi)\rho_p$. Note that p stands for rubber or glass, i.e. $p = \{r, g\}$.

3.2 Procedure and Results

Uniform glass and rubber particles with similar size (diameter: $d_r = d_g = 4$ mm) are used in this study to prepare cylindrical specimen with different volume fractions of glass and rubber beads.

Glass-rubber samples were prepared with variable rubber content, $\nu = 0, 0.1, 0.2, ..., 0.9, 1.0$, where $\nu = 0$ is composed of glass particles only and $\nu = 1.0$ of rubber particles only [27, 28]. All specimen are tested in our custom-made triaxial cell with sample diameter equal to $D = 10$ cm and two sample heights, $H = 10$ cm and 7 cm. Granular samples are compressed in the axial direction via the top piston in subsequent stress increments. At each step the radial stress is corrected to match the axial stress. Water is used as confining fluid for the samples enclosed by a rubber membrane. At each pressure level (from 50 to 350 kPa), a high voltage burst signal is excited at top to measure the time of flight.

We limit our studies to the compressional P-waves, since the transducers are mounted on the longitudinal direction. The mass density of samples at different rubber content is $\rho = (1 - \nu)\rho_g + \nu\rho_r$, where ρ_g and ρ_r are true mass densities of glass and rubber beads, respectively.

By means of Eq. (8), we can then calculate the P-wave modulus of the granular mixtures tested in the triaxial cell. Figure 2a shows the evolution of the bulk stiffness with the rubber content for all mixtures at different pressure levels. The figure shows that the compressional modulus remains fairly constant by increasing the volume of the rubber content to $\nu = 0.3$. Since the majority of particles are glass, the wave velocity is controlled by glass medium. In the case of high pressure, adding a small amount of soft particles surprisingly enhances the effective P-wave mod-

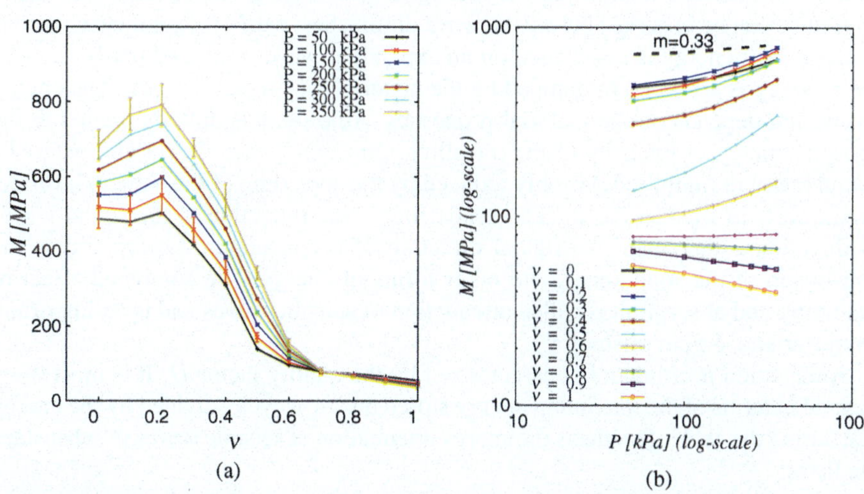

Fig. 2 **a** P-wave modulus against rubber fraction. **b** P-wave modulus against confining stress in log-log, see Ref. [33]

ulus measured from longitudinal ultrasound investigations of the medium and the highest modulus is observed at $\nu \approx 0.2$. Thus, granular mixtures can be manipulated to obtain aggregates with even higher stiffness, but lighter and more dissipative, thanks to rubber, when appropriate external conditions are matched (in this case the pressure) [29]. Between $\nu = 0.3$ and 0.6, there is a considerable drop in the wave velocity which is the transition from glass- to rubber-controlled media. Increasing the amount of rubber particles (i.e. the rubber content ν) reduces the effective stiffness. The modulus is again relatively stable between $\nu = 0.6$ to 1 which is linked to the dominance of the rubber media.

In Fig. 2b, M-modulus is plotted against confining pressure p^c (boundary value problem in Fig. 1). The slopes of the M-lines with P are almost constant for $\nu \leq 0.3$ and it follows the expected scaling $M \propto P^{\frac{1}{3}}$ [2, 30, 31]. While the behavior suddenly tends to change by increasing the rubber content (for $\nu = 0.4$ and 0.5). For these samples with intermediate rubber content, the scaling law of the M-modulus with pressure P is $M \propto P^{\frac{1}{2}}$ with a power higher than the typical scaling ($P^{\frac{1}{3}}$). Such a change in the slope is due to the creation of new particle media by incorporating rubber particles [32]. Further on, the modulus becomes almost independent of pressure for higher rubber content. As already observed in Fig. 2a, samples with $\nu = 0.6$ to 1 do not show significant dependence on pressure. The reason for this is related to the softness of rubber particles which deform so high such that they lose the nature of point-to-point contact, i.e. contacts between particles become surface contact.

3.3 Attenuation

When a mechanical wave propagates through a medium, a gradual decay of the wave amplitude can be observed before the wave diminishes, partly for geometric reasons because their energy is distributed on an expanding wave front, and partly because their energy is absorbed or damped by the material they travel through. The energy absorption depends on the material properties. Amplitude is directly related to the acoustic energy or intensity of the travelling ultrasound wave. In certain materials, sound pressure (amplitude) is only reduced by the spreading of the wave. The effect produced is to weaken the sound. 'Scattering' is the reflection of the sound waves in directions other than its original direction of propagation. 'Absorption' is the conversion of the sound energy to other forms of energy. The combined effect of scattering and absorption is called attenuation of seismic waves and is an important characteristic, e.g. in seismology.

Attenuation is commonly characterized by the quality factor Q. It is most often defined in terms of the maximum energy stored during a cycle, divided by the energy lost during the cycle. Q reflects the energy attenuation of seismic waves in subsurface media as:

$$Q = \frac{\textit{Energy of seisemic wave}}{\textit{Energy dissipated per cycle of wave}} = \frac{2\pi|A(f)|^2}{\Delta|A(f)|^2}, \tag{9}$$

where $|A(f)|^2$ is the energy of the wave, $\Delta|A(f)|^2$ is the change in energy per cycle.

Among various methods [34–40] of measuring attenuation from ultrasound data, the spectral ratio method [41] is most common perhaps because it is easy to use and most stable. Computation of the spectra of the wave and evaluation of the logarithmic ratios for two receivers at depth L_1 and L_2 yields:

$$\ln\left|\frac{A_1(f)}{A_2(f)}\right| = -\frac{\pi(t_2 - t_1)}{Q} f + cte, \tag{10}$$

where $A_1(f)$ and $A_2(f)$ are the amplitude spectra at different lengths, f is the frequency, t_1 and t_2 are the travel time from source to receiver at length L_1 and L_2, and cte is a fitting constant.

To employ the spectral-ratio method for our mixture samples, we perform tests on samples with two different lengths, 7 and 10 cm. As we have found an interesting mechanical response from the stiffness analysis (Fig. 2), we dedicate our attention to samples with low rubber content.

Values of damping (loss factor) Q^{-1} are plotted against the rubber content in Fig. 3a (up to $\nu = 0.3$) for different pressure levels. It is observed that when the amount of rubber increases, the quality factor parameter Q^{-1} decreases with confinig pressure P and increases in a linear fashion, irrespective to the pressure level, i.e. the system is more dissipative by increasing the amount of soft inclusions. Figure 3b demonstrates the systems damping in another fashion where Q^{-1} is plotted versus the confining pressure. As expected, there is a systematical increase of damping by adding soft particle [42, 43]. Combining the observations in Figs. 2 and 3, we summarize that adding roughly about 20% of soft inclusions strongly improves the damping of the system (about 30%), and increases its stiffness (upto 15%) and yet results in a lighter sample (about 15%). Such an effective acoustic behavior of binary mixtures can obviously not be predicted by the application of simple mixture interpolation rules [23].

4 Numerical Modelling—Discrete Element Method

Modelling granular materials can help us to understand their behaviour on the microscopic scale, and to obtain macroscopic continuum relations by a micro- macro transition approach. The Discrete Element Method (DEM) allows to inspect the influence of microscopic contact properties of its individual constituents on the macro bulk behavior of granular assemblies. In this study we use DEM to explore the elastic response of dry frictional granular materials.

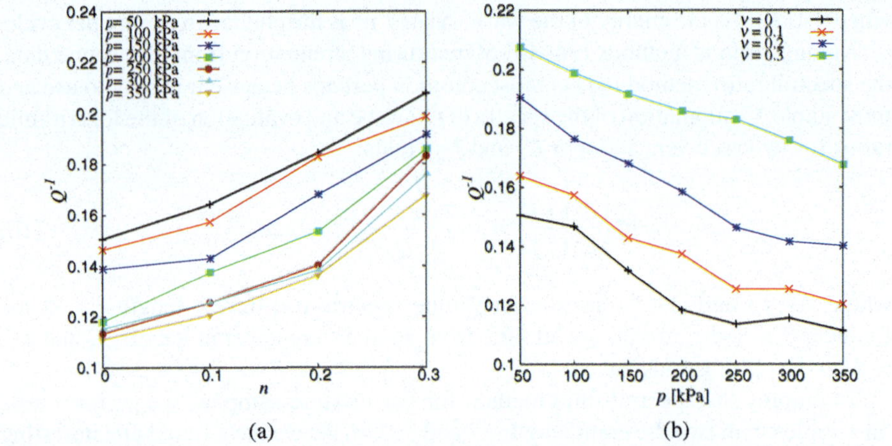

Fig. 3 **a** Attenuation factor Q^{-1} versus rubber content at different pressure levels and **b** attenuation factor Q^{-1} versus confining pressure for different samples with central frequency of measurement $f \approx 10$ kHz

4.1 Discrete Element Method

If the total force f_i acting on particle i, either due to other particles and boundaries or from external forces, is known, then the problem is reduced to the integration of Newton's equations of motion for the translational and rotational degrees of freedom,

$$m_i \frac{\partial^2}{\partial t^2} \mathbf{r}_i = f_i + m_i g, \quad \& \quad I_i \frac{\mathrm{d}}{\mathrm{d}t} \omega_i = t_i, \tag{11}$$

with m_i the mass of particle i, r_i its position, $f_i = \sum_c f_i^c$ the total force acting on it due to contacts with other particles or with the walls, g the acceleration due to volume forces like gravity, I_i the spherical particle's moment of inertia, ω_i its angular velocity and $t_i = \sum_c \left(l_i^c \times f_i^c + q_i^c \right)$ the total torque, where q_i^c are torques/couples at contacts other than the torques due to the tangential force, e.g., due to rolling and torsion, and l_i^c the vector from the particle's centre of mass to the contact point [14].

The equations of motion are thus a system of $\mathcal{D} + \mathcal{D}(\mathcal{D} - 1)/2$ coupled ordinary differential equations to be solved in $\mathcal{D}$ dimensions. The solution of such equations is straightforward, using numerical integration tools such as the ones nicely described in textbooks [44, 45]. The typically short-ranged interactions in granular media allow for further optimizations by using linked-cell spatial structures or alternative methods [44–46] in order to make the search for colliding particles more efficient. In the case of long-range interactions, (e.g. charged particles with Coulomb interaction, or objects in space with self-gravity) this is not possible anymore, so that more advanced methods for optimization have to be applied. Here we restrict ourselves to short-range interactions [47].

Specifically, two spherical particles i and j, with radii a_i and a_j, respectively, interact only if they are in contact, that is, their overlap

$$\delta = (a_i + a_j) - (r_i - r_j) \cdot n \tag{12}$$

is positive, $\delta > 0$, with the unit vector $n = n_{ij} = (r_i - r_j)/|r_i - r_j|$ pointing from j to i. The force on particle i, from particle j, at contact c, can be decomposed into a normal and a tangential part as $f^c := f^c_{ij} = f^n n + f^t t$. In the following, we specify f^c_{ij} for some different models that take into account increasingly complicated grain interactions.

4.2 *Force Laws*

4.2.1 Linear Normal Contact Model

Linear normal forces are the most simple contact laws between two particles [48, 49]. They are based upon a damped harmonic oscillator between two particles and involve a linear repulsive and a linear dissipative force in normal direction:

$$f^n = k\delta + \gamma_0 v_{rel} \tag{13}$$

with a spring stiffness k, a viscous damping γ_0, and the relative velocity v_{rel} in the normal direction $v_{rel} = -(\mathbf{v}_i - \mathbf{v}_j) \cdot \mathbf{n} = \dot{\delta}$. An advantage of the linear contact force law is that the half-period of a oscillation around an equilibrium position resembles the contact duration, $t_c = \pi/\omega$, and can be calculated analytically [50], via:

$$\omega = \sqrt{\frac{k}{m_{ij}} - \left(\frac{\gamma_0}{2m_{ij}}\right)^2} \tag{14}$$

where $m_{ij} = \frac{m_i m_j}{m_i + m_j}$ is the reduced mass and ω is the eigenfrequency of the contact. From Eq. (14) it is possible to obtain the coefficient of restitution which quantifies the ratio of relative velocities after (primed) and before (unprimed) the collision:

$$r_c = \frac{-v'_{rel}}{v_{rel}} = e^{-\frac{\gamma_0 t_c}{2m_{ij}}} \tag{15}$$

For a deeper discussion of the coefficient of restitution and other, more realistic, non-linear contact models, see e.g. [50–52].

The time-step size used in the simulations should be smaller than the contact duration. Here, $\Delta t = \frac{t_c}{50}$ is used for the simulations. Furthermore, notice that in the extreme case of an overdamped spring, t_c can become very large, and therefore the use of neither too weak nor too strong dissipation is recommended.

4.2.2 Hertzian (Non-linear) Contact Force

Instead of linear contact force law, Eq. (13), a non-linear contact force law can be considered, which is based on the elastic contact theory:

$$f^n = E^* \sqrt{d_{eff}} \delta^{\frac{3}{2}}, \tag{16}$$

with $d_{eff} = \frac{4R_i R_j}{R_i + R_j}$ as the effective diameter and the effective Young's modulus for two different material particles $E^* = \frac{2E_i E_j}{E_i + E_j}$.

The linear dissipation used in the linear contact force law can also be applied in the Hertzian contact:

$$f^n = E^* \sqrt{d_{eff}} \delta^{\frac{3}{2}} + \gamma \dot{\delta} \tag{17}$$

Instead of the linear dissipation non-linear options are also available [50]. The Hertzian contact can for example be extended by assuming the material to be non-linierly viscoelastic:

$$f^n = E^* \sqrt{d_{eff}} \delta^{\frac{3}{2}} + \eta \sqrt{\delta} \dot{\delta}, \tag{18}$$

where η is the Hertzian viscosity parameter. This latter force contact law is used in the simulation for the mixtures, as it is known that the linear model represents the experimental dissipation qualitatively wrong as shown in Ref. [50].

4.2.3 Tangential Force Law

The sliding/sticking friction model is based upon Coulomb's law [53]. This law couples the tangential and normal force: For sticking one has static friction $f_t \leq \mu_s f_n$ and for sliding one has dynamic friction, $f_t = \mu_d f_n$. In the static situation a restoring force account for the non-zero tangential force from the surface asperities. To determine if the particle experiences static or dynamic friction a tangential test-force is calculated in the same way as the normal force, which for the linear contact force law gives:

$$f_t = -k_t s - \gamma_t v_{t,rel}, \tag{19}$$

and for the Hertzian normal contact the Mindlin tangential contact force is used:

$$f_t = k_t s \sqrt{d_{eff} \delta} - \gamma_t v_{t,rel}, \tag{20}$$

where k_t is the tangential spring stiffness and s is the relative shear displacement between the two particle centres.

When test-force fulfils $|f_t| \leq f_s^C$ one has static friction. However, when $|f_t| > f_s^C$ is fulfilled, sliding friction becomes active. In Eq. (19) the tangential relative velocity is needed, defined by $v_{t,rel} = v_{ij} - \mathbf{n}(\mathbf{n} \cdot v_{ij})$, where the total relative velocity of the particle surfaces at the contact is given by:

$$v_{ij} = v_i - v_j + a_i'\mathbf{n} \times \omega_i + a_j'\mathbf{n} \times \omega_j \tag{21}$$

with $a_i' = a_i - \frac{\delta}{2}$ is the corrected radius.

4.3 Macroscopic Parameters

4.3.1 Strain

For any deformation, the isotropic part of the infinitesimal strain tensor ϵ_v (in contrast to the true strain ε_v) is defined as:

$$d\epsilon_v = \dot{\epsilon}_v \text{dt} = \frac{\epsilon_{xx} + \epsilon_{yy} + \epsilon_{zz}}{3}, \tag{22}$$

where $\epsilon_{\alpha\alpha} = \dot{\epsilon}_{\alpha\alpha}\text{dt}$ with $\alpha\alpha = xx,\ yy$ and zz as the diagonal elements of the strain tensor ε_{ij} in the Cartesian $x,\ y,\ z$ reference system. The integral of $3d\epsilon_v$ denoted by $\varepsilon_\text{v} = 3\int_{V_0}^{V} d\epsilon_\text{v}$, is the true or logarithmic strain, i.e. the volume change of the system, relative to the initial reference volume, V_0 [54].

4.3.2 Stress

From the simulations, one can determine the stress tensor components (compressive stress is positive as convention):

$$\sigma_{ij} = \frac{1}{V}\left(\sum_{p \in V} m^p v_i^p v_j^p - \sum_{c \in V} f_i^c l_j^c\right), \tag{23}$$

with particle p, mass m^p, velocity v^p, neighbouring particles in contact c, force f^c and branch vector l^c, while Greek letters represent components x, y, and z [55, 56]. The first sum is the kinetic stress (energy density) tensor while the second involves the contact-force dyadic product with the branch vector. Averaging, smoothing or coarse graining [57, 58] in the vicinity of the averaging volume, V, weighted according to the vicinity is not applied in this study, since averages are taken over the total volume.

4.3.3 Structural (Fabric) Anisotropy

Besides the stress of a static packing grains, an important microscopic quantity of interest is the fabric/structure tensor. For disordered media, the concept of a fabric tensor naturally occurs when the system consists of an elastic network or a packing of discrete particles. A possible expression for the components of the fabric tensor is provided in [55, 59]:

$$F_{\alpha\beta} = \langle F^p \rangle = \frac{1}{V} \sum_{p \in V} V^p \sum_{c=1}^{N} n_\alpha^c n_\beta^c, \tag{24}$$

where V^p is the particle volume of particle p which lies inside the averaging volume V, and n^c is the normal vector pointing from the center of particle p to contact c. $F_{\alpha\beta}$ are thus the components of a symmetric rank two 3×3 tensor.

4.4 DEM Simulation of Small-Strain Stiffness

In this section, we show how to use DEM simulations to study the elastic behavior of granular materials. We use Discrete Element simulations to reproduce static experiments for the characterization of the elastic stiffness of the granular packings.

Table 2 Summary and numerical values of particle parameters used in the DEM simulations

Property	Symbol	Value	Units
Time unit	t_u	1	10^{-6} s
Length unit	x_u	1	10^{-3} m
Mass unit	m_u	1	10^{-9} kg
Particle diameter	$\langle d \rangle$	2	x_u
Polydispersity	d_{max}/d_{min}	3	
Number of particles	N	4096	
Particle density	ρ	2000	m_u/x_u^3
Simulation time step	Δt_{MD}	0.0037	t_u
Normal stiffness	k_n	10^5	m_u/t_u^2
Tangential stiffness	k_t/k_2	0.2	
Restitution coefficient	e	0.804	
Coefficient of friction	μ	[0–10]	
Normal viscosity	$\gamma = \gamma_n$	1000	m_u/t_u
Tangential viscosity	γ_t/γ	0.2	
Background viscosity	γ_b/γ	0.1	

We start our simulations from a set of non-overlapping particles randomly generated in a periodical cubic box at an initial volume fraction $\nu = 0.3$. The initial configuration is compressed isotropically by constant compression strain-rate, $\dot{\varepsilon} = 10^{-4}\ \text{s}^{-1}$, until a given volume fraction $\nu = 0.5$. The system is then allowed to relax at constant volume fraction until it reaches a stable state, which means that the particles dissipate kinetic energy and achieve a zero-pressure. This is followed by an isotropic (slow) compression until the desired maximum volume fraction, $\nu = 0.82$ [60]. The same protocol is used for all samples. For the sake of simplicity, the linear contact model was employed as force law between grains. Parameters used in simulations are summarized in Table 2. In order to investigate the elastic response, we perform so-called strain probing tests in several points along the isotropic preparation (pre-strain) compression path above jamming point (a volume fraction where sample is no longer in fluid-like behavior) [61–63].

Here we consider isotropic prepared samples for which only two independent moduli are needed to characterize the bulk material. The elastic constants K (bulk modulus) and G (shear modulus) are chosen here.

After applying the sufficient relaxation, incremental pure volumetric or pure deviatoric strain are applied to the samples, in order to obtain the bulk and shear moduli, respectively. Isotropic compression of samples is such that samples are homogeneously compressed along all directions; during pure shear, samples are compressed along the x-direction and decompressed along the y-direction, while the z-direction is kept stationary. Shear deformations are applied in the form of a pure shear, i.e., by having a displacement in the y-direction imposed on all particles that cross the x-boundary and a displacement in the x-direction on all particles that cross the y-boundary. One can obtain the samples bulk (K) and shear (G) moduli, by measuring the incremental response to isotropic and shear deformation [60, 61, 64].

Since we apply infinitesimal strain perturbations, we do not expect slippage to occur. However, the friction coefficient is set to infinity ($\mu = \infty$) to prevent any sliding at contacts during probing. For each calculation, we verify that the applied strain is small enough to be in the linear response regime, i.e., the coordination number does not vary during the applied incremental deformation and stress increments are linear proportional to strain increments.

After probing the configurations, the effective elastic moduli of the granular assembly are obtained as the ratio between the measured increment in stress and the applied strain:

$$K^* = \delta P^*/3\delta\varepsilon_{vol} \qquad \& \qquad G^* = \delta\left(\sigma^*_{xx} - \sigma^*_{yy}\right)/\delta\left(\varepsilon_{xx} - \varepsilon_{yy}\right) \tag{25}$$

where $P^* = Pd/k_n$ and $\sigma^* = \sigma d/k_n$ are the non-dimensional pressure and stress quantities respectively. Hence, K^* and G^* are dimensionless elastic moduli. The advantage of dimensionless moduli is that one can compare samples with different size and type of particles. Results reported later are dimensionless, and for the sake of simplicity, star superscript symbol * is neglected below.

We scan a wide range of inter-particle friction coefficients and volume fractions, and apply small to very large deformations, in order to understand how the interplay of

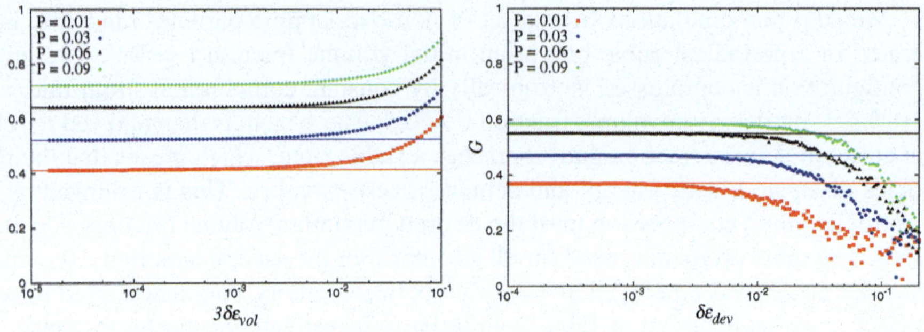

(a) (b)

Fig. 4 Evolution of dimensionless **a** bulk modulus K and **b** shear modulus G with the respective applied isotropic $3\delta\varepsilon_{vol}$ and shear $\delta\varepsilon_{xy}$ strain amplitudes at different pressure $P = 0.01, 0.03, 0.06$ and 0.9, for samples with friction coefficient $\mu = 0.001$. Corresponding dashed lines represent the small strain elastic limit values of K and G

contact and system properties affects the microstructure and thus the elastic moduli. The same procedure is applied for samples created with different values of friction, $\mu = 0 - 10$. As a result, packings at the same density achieve different pressure and microstructure.

As an example, the variation of the bulk and shear moduli with applied strain amplitude is shown in Fig. 4 for $\mu = 0.001$. The elastic moduli stay practically constant for small amplitudes ($3\delta\varepsilon_{vol}$ and $\delta\varepsilon_{xy} < 10^{-4}$ with very slow particles movement rate $\dot{\varepsilon}_{ii} \approx 10^{-5}$) and this can be considered the elastic regime. By increasing the amplitudes of the perturbation, K and G start to increase and decrease non-linearly respectively, i.e. packings are no longer in the elastic regime. The elastic regime becomes larger for higher volume fraction and higher friction (data not shown here) [65].

In Fig. 5, the variations of the bulk K and shear G modulus are plotted with volume fraction for packings with different coefficients of friction μ. As expected, the elastic moduli always increases with increasing density. However, the increase of the moduli is slower for packings with high friction. We can relate this behavior to a lower average number of contacts (i.e. lower volumetric fabric F_v) for samples prepared with higher friction at the same volume fraction. The value of the initial fabric is proportional to the number of contacts, and influences the subsequent evolution of the moduli.

When the elastic moduli are plotted against the isotropic fabric F_v in Fig. 6, the data for the bulk modulus approximately collapse in a unique linear scaling law, implying a general relation between bulk stiffness and isotropic micro-structure: $K = K_0(F_v - F_{v_0})$ with $F_{v_0} \approx 2.2$ and $K_0 \approx 0.1$ for $\mu = 0$ [60] for $\mu > 0$ see [66]. On the other hand, a numerical scaling was not found for the shear modulus G, even if they follow a similar trend with ν. Further investigations are needed. It is worth mentioning that the coefficient of friction has no direct influence on the elastic moduli as sliding is not activated during probing,$\mu_{probing} = \infty$, but rather it effects K and G indirectly through the preparation that leads to a different state variable F_v.

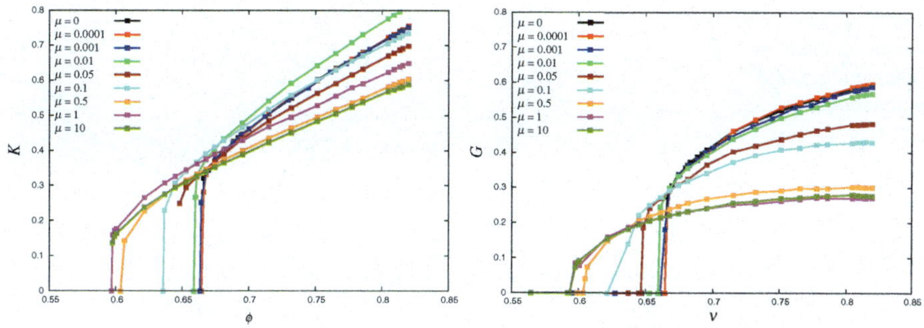

(a) (b)

Fig. 5 Evolution of the normalized **a** bulk modulus K and **b** shear modulus G with volume fraction ν for different coefficients of friction, μ, as shown in the legend

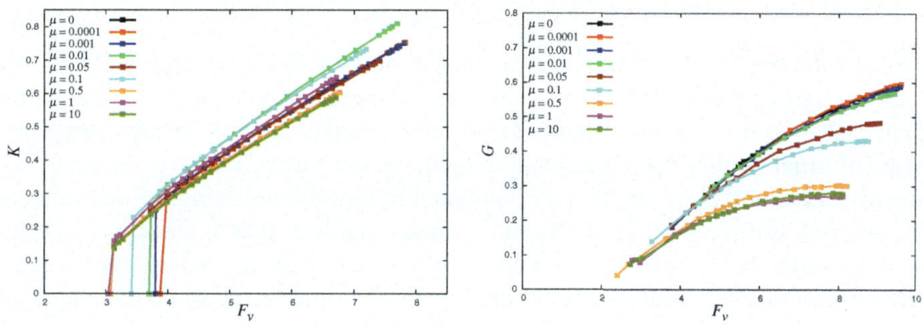

(a) (b)

Fig. 6 Evolution of the normalized **a** bulk modulus K and **b** shear modulus G with volume fraction ν for different coefficients of friction, μ, as shown in the legend

4.5 Comparison Between DEM and Wave Experiments

In this section we show the results of the numerical moduli for different rubber content ν and pressure states. Note that the applied pressure range in simulations is wider than in experiments to gain more insight into the mechanical response of mixtures. Unlike the results shown in the previous section (Sect. 4.4), here the frictional Hertzian contact model was employed for normal contact interactions between two particles; since it captures experimental results better than the linear model. Material properties of glass and rubber are given in Table 3.

In Fig. 7, the P-wave modulus, M, versus pressure is plotted. It is found that the modulus decreases monotonically with the rubber content from $\nu = 0.0$ to $\nu = 1.0$. In this respect, simulations are not able to reproduce the macroscopic behavior observed in Fig. 2, from the experiments with maximum M at $\nu \neq 0$. We associate the mismatch to the adopted contact model not appropriate to properly describe

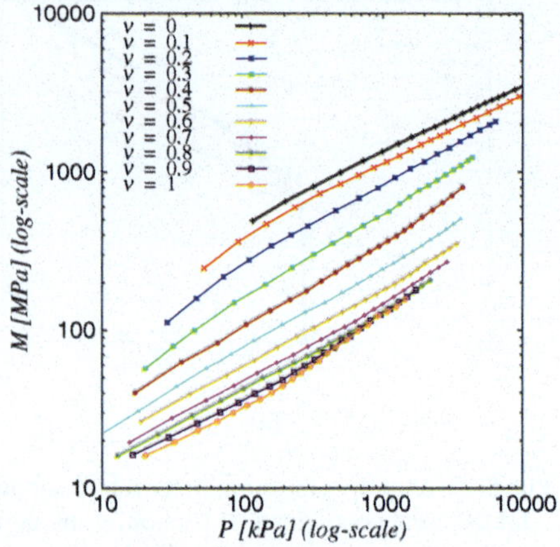

Fig. 7 P-wave modulus, M, versus applied stress P obtained by DEM simulations

rubber-rubber and rubber-glass interactions. Finding a better contact model is subject to ongoing study. The simulations are not expected to exactly reproduce the experimental observations. Rather, the simulations are valuable because they provide information that is not available in the laboratory tests. However, when experiments and simulations are directly compared in Fig. 8, interesting information can be inferred. For the sake of clarity, only three cases are shown, namely, $\nu = 0.05$, $\nu = 0.5$ and $\nu = 1.0$. Simulations with $\nu = 0.05$ capture the experimental data quantitatively, noticeably without any calibration. On the other hand, when looking at the the packing $\nu = 0.5$, the qualitative trend is well captured by simulations even if actual experimental values are higher than the simulated ones.

Finally, for the case $\nu = 1.0$, where the experimental moduli are pressure independent, simulations are far from experiments in qualitative trend.

By summarizing the previous observations, a three regime scenario shows up. In the glass-dominated regime (G), waves do transmit via a glass beads network, where simulations based on Hertzian interactions are able to reproduce the macroscopic behavior. In the intermediate regime (I) waves still have a preferential path via glass

Table 3 Properties of glass and rubber particles. Further simulation information can be found in Refs. [23, 33, 67]

Used material properties	Glass	Rubber
Diameter (mm)	4	4
Mass density (kg/m^3)	1540	860
Young's modulus (MPa)	65000	1.85
Poisson's ratio	0.24	0.46
Friction coefficient	0.2	0.5

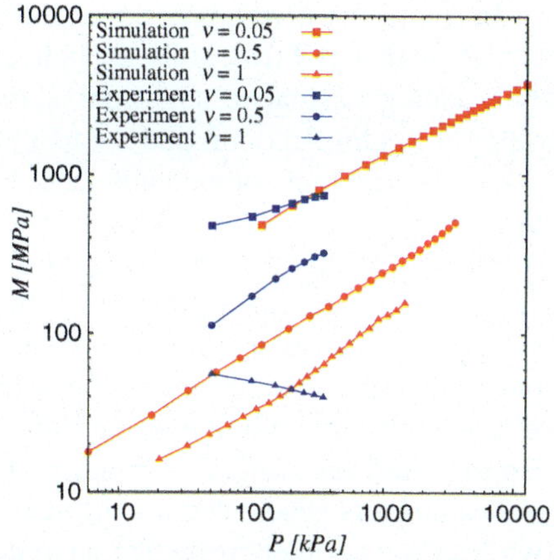

Fig. 8 Experimental and numerical P-wave modulus plotted against pressure; comparison of DEM and experimental glass-rubber mixtures for $\nu = 0.05, 0.5$ and 1.0

bead chains. Here two mechanisms concur to shape the bulk behavior: (i) the density of glass beads in the sample reduces with respect to case G and the actual values of the moduli get lower; (ii) the number of contacts increases with pressure faster than in the G-regime due to easy rearrangement of the rubber particles, that is the slope $M(P)$ gets higher. Finally, in the third regime (R), the behavior of the mixture is dominated by the rubber beads, and the present simple DEM pair contact model cannot offer an accurate representation of the system, because additional high deformation effects need to be considered [67].

5 Summary and Outlook

These days, it has been well understood that grain-scale properties control the bulk-scale behavior of the granular materials. The aim of this contribution was to study and determine the micromechanical mechanisms that govern various phenomena in the world of particles, leading to a better micro-based understanding of the macro-scale response/mechanics of particulate systems.

At first, the elastic response of disordered granular soft-stiff mixtures was investigated experimentally by means of wave-propagation. We found that the behavior is highly non-linear and also non-monotonic with increasing the fraction of soft particles. The uni-axial P-wave modulus is deduced from the linear, uni-axial wave speed and the bulk density of the sample where velocity is the ratio of (variable) sample length and travel time, as measured by the transducers. While it is expected that the stiffness (P-wave modulus) decreases when soft inclusions are replacing the hard particles, interestingly, we observe a nonlinear behavior with a maximum in stiffness at around 20% of soft content, before the stiffness drops and achieves the same rubber-controlled stiffness above 60% soft content.

In contrast to elastic properties, the enhanced dissipative and lightweight properties of the material (like soils, asphalt, etc.) behaves more as expected, when deliberately adding dissipative, soft, light inclusions. This allows for a novel design methodology for calm, smooth, and smart materials that can be better in various aspects than their separate components, e.g., higher stiffness *and* lightweight+damping, at the same time, by adding softer, lighter components.

Numerical simulations using Discrete Element Method (DEM) on the particle scale have revealed the role of the microstructure in characterising the elastic behaviour of granular materials. After preparation by isotropic compression of samples with different inter-particle contact friction, at various volume fraction, the effective elastic moduli are determined from the incremental response by probing isotropic and deviatoric deformations. DEM allows to understand the effects of material parameters, where friction between particles enhances the macroscopic stiffness, while affecting other properties at the same time, such as contact network and coordination number. Thus, selecting/calibrating parameters is an essential task to achieve reliable quantitative agreement with experiments and to allow for predictions for improved materials/designs.

Finally, the P-wave modulus obtained experimentally was compared with the numerical results. While the simulations quantitatively capture the experimental observations for samples at low rubber fraction, the present DEM data show a systematic decrease of the modulus with increasing the rubber fraction unlike the experiments. Future work will focus on improving the contact models towards multi-contact interactions and including the effects of the large deformations of the rubber particles to understand if this improves agreement with experiments; another direction is the extension of our small perturbation approach to elasto-plasticity and then predicting the plastic, irreversible response of a granular assemblies under larger deformation amplitudes.

Acknowledgements Authors acknowledge funding by the German Science Foundation (DFG) through the project STE-969/16-1 within the SPP 1897 "Calm, Smooth and Smart".

References

1. Bardet, J.P.: Numerical simulations of the incremental responses of idealized granular materials. Int. J. Plast **10**(8), 879–908 (1994)
2. Goddard, J.: Nonlinear elasticity and pressure-dependent wave speeds in granular media. Proc. R. Soc. Lond. A **430**(1878), 105–131 (1990)
3. Sibille, L., Nicot, F., Donzé, F.V., Darve, F.: Analysis of failure occurrence from direct simulations. Eur. J. Environ. Civ. Eng. **13**(2), 187–201 (2009)
4. Heimdahl, T.C., Drescher, A.: Elastic anisotropy of tire shreds. JGGE **125**(5), 383–389 (1999)
5. Hınıslıoğlu, S., Ağar, E.: Use of waste high density polyethylene as bitumen modifier in asphalt concrete mix. Mater. Lett. **58**(3–4), 267–271 (2004)
6. Siddique, R., Naik, T.R.: Properties of concrete containing scrap-tire rubber-an overview. Waste Manage. **24**(6), 563–569 (2004)

7. Xiao, F., Amirkhanian, S., Juang, C.H.: Rutting resistance of rubberized asphalt concrete pavements containing reclaimed asphalt pavement mixtures. J. Mater. Civ. Eng. **19**(6), 475–483 (2007)
8. You, Z., Buttlar, W.: Discrete element modeling to predict the modulus of asphalt concrete mixtures. J. Mater. Civ. Eng. **16**(2), 140–146 (2004)
9. Zornberg, J.G., Cabral, A.R., Viratjandr, C.: Behaviour of tire shred sand mixtures. Can. Geotech. J. **41**(2), 227–241 (2004)
10. Kuwano, R., Jardine, R.: On the applicability of cross-anisotropic elasticity to granular materials at very small strains. Géotechnique **52**(10), 727–749 (2002)
11. La Ragione, L., Jenkins, J.T.: The initial response of an idealized granular material **463**(2079), 735–758 (2007)
12. Taghizadeh, K., Luding, S., Magnanimo, V.: DEM applied to soil mechanics. In: ALERT Doctoral School 2017 Discrete Element Modeling, p. 129 (2017)
13. Cundall, P., Strack, O.: A discrete numerical model for granular assemblies. Géotechnique **29**(1), 47–65 (1979)
14. Luding, S.: Introduction to discrete element methods: basic of contact force models and how to perform the micro-macro transition to continuum theory. Eur. J. Environ. Civ. Eng. **12**(7–8), 785–826 (2008)
15. Aki, K., Richards, P.: Quantitative Seismology. Freeman, New York (1980)
16. Magnanimo, V.: In Wave Propagation and Elasticity in Granular Soils: A Numerical Approach for a Micromechanical Perspective: Views on Microstructures in Granular Materials, pp. 107–135. Springer (2020)
17. David, J., Cheeke, N.: Fundamentals and applications of ultrasonic waves (2017)
18. Clayton, C.: Stiffness at small strain: research and practice. Géotechnique **61**(1), 5–37 (2011)
19. Burland, J.: Ninth Laurits Bjerrum memorial lecture: "small is beautiful"-the stiffness of soils at small strains. Can. Geotech. J. **26**(4), 499–516 (1989)
20. Atkinson, J.: Non-linear soil stiffness in routine design. Géotechnique **50**(5), 487–508 (2000)
21. Lee, C., Shin, H., Lee, J.S.: Behavior of sand-rubber particle mixtures: experimental observations and numerical simulations. Int. J. Numer. Anal. Meth. Geomech. **38**(16), 1651–1663 (2014)
22. Sawangsuriya, A.: In Wave Propagation Methods for Determining Stiffness of Geomaterials: Wave Processes in Classical and New Solids. IntechOpen (2012)
23. Taghizadeh, K., Steeb, H., Luding, S., Magnanimo, V.: Elastic waves in particulate glass-rubber mixtures. Proc. R. Soc. A **477**(2249), 20200834 (2021)
24. Ruf, M., Taghizadeh, K., Steeb, H.: Visualizing particle networks in granular media by in situ X-ray computed tomography (2021). arXiv preprint arXiv:2112.12423
25. Viggiani, G., Atkinson, J.: Interpretation of bender element tests. Geotechnique **45**(1), 149–154 (1995)
26. Lee, J.S., Santamarina, J.C.: Bender elements: performance and signal interpretation. J. Geotech. Geoenviron. Eng. **131**(9), 1063–1070 (2005)
27. Platzer, A., Rouhanifar, S., Richard, P., Cazacliu, B., Ibraim, E.: Sand-rubber mixtures undergoing isotropic loading: derivation and experimental probing of a physical model. Granular Matter **20**(4), 81 (2018)
28. Valdes, J.R., Evans, T.M.: Sand-rubber mixtures: experiments and numerical simulations. Can. Geotech. J. **45**(4), 588–595 (2008)
29. Kim, H.K., Santamarina, J.: Sand-rubber mixtures (large rubber chips). Can. Geotech. J. **45**(10), 1457–1466 (2008)
30. Makse, H.A., Gland, N., Johnson, D.L., Schwartz, L.: Granular packings: nonlinear elasticity, sound propagation, and collective relaxation dynamics. Phys. Rev. E **70**(6), 061302 (2004)
31. Somfai, E., Roux, J.N., Snoeijer, J.H., Van Hecke, M., Van Saarloos, W.: Elastic wave propagation in confined granular systems. Phys. Rev. E **72**(2), 021301 (2005)
32. Bassett, D.S., Owens, E.T., Daniels, K.E., Porter, M.A.: Influence of network topology on sound propagation in granular materials. Phys. Rev. E **86**(4), 041306 (2012)

33. Taghizadeh, K., Steeb, H., Magnanimo, V., Luding, S.: Elastic waves in particulate glass-rubber mixture: experimental and numerical investigations/studies. In: EPJ Web of Conferences, vol. 140, p. 12019. EDP Sciences (2017)
34. Jongmans, D., Campillo, M.: The determination of soil attenuation by geophysical prospecting and the validity of measured Q values for numerical simulations. Soil Dyn. Earthq. Eng. **12**(3), 149–157 (1993)
35. Tonn, R.: Comparison of seven methods for the computation of Q. Phys. Earth Planet. Inter. **55**(3–4), 259–268 (1989)
36. Wang, Z.J., Cao, S.Y., Zhang, H.R., Qu, Y.M., Yuan, D., Yang, J.H., Shao, G.M.: Estimation of quality factors by energy ratio method. Appl. Geophys. **12**(1), 86–92 (2015)
37. Cheng, P., Margrave, G.F.: Estimation of Q: a comparison of different computational methods. CSPG/CSEG/CWLS GeoConvention **12**, 2–5 (2013)
38. Kim, E., Martínez, A.J., Phenisee, S.E., Kevrekidis, P., Porter, M.A., Yang, J.: Direct measurement of superdiffusive energy transport in disordered granular chains. Nat. Commun. **9**(1), 640 (2018)
39. Toksöz, M., Johnston, D.H., Timur, A.: Attenuation of seismic waves in dry and saturated rocks: I. Laboratory measurements. Geophysics **44**(4), 681–690 (1979)
40. Johnston, D.H., Toksöz, M., Timur, A.: Attenuation of seismic waves in dry and saturated rocks: II. Mechanisms. Geophysics **44**(4), 691–711 (1979)
41. Sears, F.M., Bonner, B.P.: Ultrasonic attenuation measurement by spectral ratios utilizing signal processing techniques. IEEE Trans. Geosci. Remote Sens. **2**, 95–99 (1981)
42. Hovem, J.M., Ingram, G.D.: Viscous attenuation of sound in saturated sand. J. Acoust. Soc. Am. **66**(6), 1807–1812 (1979)
43. Cha, M., Cho, G.C., Santamarina, J.C.: Long-wavelength P-wave and S-wave propagation in jointed rock masses. Geophysics **74**(5), E205–E214 (2009)
44. Allen, M.P., Tildesley, D.J.: Computer Simulation of Liquids. Oxford University Press, Oxford (1987)
45. Rapaport, D.C.: The Art of Molecular Dynamics Simulation. Cambridge University Press, Cambridge (1995)
46. Krijgsman, D., Ogarko, V., Luding, S.: Optimal parameters for a hierarchical grid data structure for contact detection in arbitrarily polydisperse particle systems. Comput. Particle Mech. **1**(3), 357–372 (2014)
47. Luding, S., Rivas, N., Weinhart, T.: From soft and hard particle simulations to continuum theory for granular flows. ALERT Geomaterials Doctoral School (2017)
48. Luding, S., Taghizadeh, K., Cheng, C., Kondic, L.: Understanding slow compression and decompression of frictionless soft granular matter by network analysis. Soft Matter **18**(9), 1868–1884 (2022)
49. Taghizadeh, K., Combe, G., Luding, S.: ALERT Doctoral School 2017 Discrete Element Modeling (2017)
50. Luding, S.: In Collisions and Contacts Between Two Particles: Physics of Dry Granular Media, pp. 285–304. Springer (1998)
51. Singh, A., Magnanimo, V., Luding, S.: Mesoscale contact models for sticky particles. Submitted to Powder Technology (2013)
52. Thornton, C., Cummins, S.J., Cleary, P.W.: On elastic-plastic normal contact force models, with and without adhesion. Powder Technol. **315**, 339–346 (2017)
53. Luding, S.: Cohesive, frictional powders: contact models for tension. Granular Matter **10**(4), 235–246 (2008)
54. Göncü, F., Durán, O., Luding, S.: Constitutive relations for the isotropic deformation of frictionless packings of polydisperse spheres. Comptes Rendus Mécanique **338**(10–11), 570–586 (2010)
55. Luding, S.: Anisotropy in cohesive, frictional granular media. J. Phys.: Condens. Matter **17**(24), S2623 (2005)
56. Imole, O.I., Kumar, N., Magnanimo, V., Luding, S.: Hydrostatic and shear behavior of frictionless granular assemblies under different deformation conditions. Kona Powder Part. J. **30**, 84–108 (2013)

57. Weinhart, T., Thornton, A.R., Luding, S., Bokhove, O.: From discrete particles to continuum fields near a boundary. Granular Matter **14**(2), 289–294 (2012)
58. Weinhart, T., Hartkamp, R., Thornton, A.R., Luding, S.: Coarse-grained local and objective continuum description of three-dimensional granular flows down an inclined surface. Phys. Fluids **25**(7), 070605 (2013)
59. Madadi, M., Tsoungui, O., Lätzel, M., Luding, S.: On the fabric tensor of polydisperse granular materials in 2D. Int. J. Solids Struct. **41**(9), 2563–2580 (2004)
60. Kumar, N., Luding, S., Magnanimo, V.: Macroscopic model with anisotropy based on micro-macro information. Acta Mech. **225**(8), 2319–2343 (2014)
61. Magnanimo, V., et al.: Characterizing the shear and bulk moduli of an idealized granular material. EPL (Europhys. Lett.) **81**(3), 34006 (2008)
62. Kumar, N., Luding, S., Magnanimo, V.: Macroscopic model with anisotropy based on micro-macro informatics (2015). arXiv preprint arXiv:1506.02992
63. Calvetti, F., Viggiani, G., Tamagnini, C.: A numerical investigation of the incremental behavior of granular soils. Rivista italiana di geotecnica **37**(3), 11–29 (2003)
64. Taghizadeh, K., Kumar, N., Magnanimo, V., Luding, S.: Understanding the effects of inter-particle contact friction on the elastic moduli of granular materials. In: IOP Conference Series: Earth and Environmental Science, vol. 26, p. 012008. IOP Publishing (2015)
65. Taghizadeh Bajgirani, K.: Elasticity and wave propagation in granular materials. Ph.D. Thesis. University of Twente (2019)
66. Göncü, F., Luding, S.: Effect of particle friction and polydispersity on the macroscopic stress-strain relations of granular materials. Acta Geotech. **8**(6), 629–643 (2013)
67. Giannis, K., Schilde, C., Finke, J., Kwade, A., Celigueta, M., Taghizadeh, K., Luding, S.: Stress based multi-contact model for discrete-element simulations. Granular Matter **23**(2), 1–14 (2021)

The manufacturer's authorised representative in the EU is Springer Nature Customer Service Centre GmbH, Europaplatz 3, 69115 Heidelberg, Germany. If you have any concerns regarding our products, please contact ProductSafety@springernature.com

Printed and bound by CPI Group (UK) Ltd, Croydon, CR0 4YY
28/09/2026
02224085-0010